ASTRONOMICAL AND ASTROPHYSICAL OBJECTIVES OF SUB-MILLIARCSECOND OPTICAL ASTROMETRY

INTERNATIONAL ASTRONOMICAL UNION

UNION ASTRONOMIQUE INTERNATIONALE

ASTRONOMICAL AND ASTROPHYSICAL OBJECTIVES OF SUB-MILLIARCSECOND OPTICAL ASTROMETRY

PROCEEDINGS OF THE 166TH SYMPOSIUM OF THE INTERNATIONAL ASTRONOMICAL UNION, HELD IN THE HAGUE, THE NETHERLANDS, AUGUST 15–19, 1994

EDITED BY

ERIK HØG

Copenhagen University Observatory,
Copenhagen, Denmark

and

P. KENNETH SEIDELMANN

U.S. Naval Observatory,
Washington, DC, U.S.A.

KLUWER ACADEMIC PUBLISHERS

DORDRECHT / BOSTON / LONDON

Library of Congress Cataloging-in-Publication Data

International Astronomical Union. Symposium (166th : 1994 : Hague, Netherlands)
Astronomical and astrophysical objectives of sub-milliarcsecond optical astrometry : proceedings of the 166th Symposium of the International Astronomical Union held in the Hague, Netherlands, August 15-19, 1994 / edited by Erik Høg and P. Kenneth Seidelmann.
p. cm.
Includes indexes.
ISBN 0-7923-3442-6 (alk. paper)
1. Astrometry--Congresses. I. Høg, E. II. Seidelmann, P. Kenneth. III. Title.
QB807.I57 1994
522'.7--dc20 95-11844

ISBN 0-7923-3442-6 (HB)
ISBN 0-7923-3443-4 (PB)

Published on behalf of
the International Astronomical Union
by
Kluwer Academic Publishers, P.O. Box 17, 3300 AA Dordrecht, The Netherlands.

Kluwer Academic Publishers incorporates
the publishing programmes of
D. Reidel, Martinus Nijhoff, Dr W. Junk and MTP Press.

Sold and distributed in the U.S.A. and Canada
by Kluwer Academic Publishers,
101 Philip Drive, Norwell, MA 02061, U.S.A.

In all other countries, sold and distributed
by Kluwer Academic Publishers Group,
P.O. Box 322, 3300 AH Dordrecht, The Netherlands.

Printed on acid-free paper

Printed in the Netherlands

TABLE OF CONTENTS

2.4 REFERENCE FRAMES AND SOLAR SYSTEM

3. EXPECTED DEVELOPMENTS IN HIGH PRECISION ASTROMETRY

POSTERS

1.1 DEVELOPMENTS IN GROUND-BASED ASTROMETRIC TECHNIQUES AND LARGE CATALOGUES

2.1 STELLAR ASTROPHYSICS

PREFACE

Astrometry is on the threshold of great changes due to the fact that this decade, alone, is witnessing an improvement of stellar positions equivalent to the total improvement of the previous two centuries. The Hipparcos Satellite has concluded its observations, and the catalog is in preparation. Preliminary results assure that the Hipparcos catalog will provide positions, parallaxes and annual proper motions for over 100,000 stars with accuracies of 1.5 milliarcseconds. In addition, the Tycho catalog will provide positions of about 30 milliarcseconds accuracy for over 1 million stars, and annual proper motions with 3 milliarcsecond accuracy will subsequently be obtained by means of first epoch positions from the Astrographic Catalog.

Optical interferometers on the ground are beginning operation, and these instruments can provide observational accuracies of approximately one milliarcsecond. Also, the traditional reference frame based on the Fundamental Catalog of bright stars is being replaced by the extragalactic reference frame, based on radio sources with accuracies of one milliarcsecond. Thus, astrometry will change from a fundamental reference frame defined in terms of the dynamical reference frame of the solar system with accuracies of 100 milliarcseconds to a space-fixed, extragalactic reference frame with accuracies of one milliarcsecond. Future astrometric observations should be in the 1 – 100 milliarcsecond accuracy range.

There are a number of concepts for future astrometric instruments in space. Most of these can provide sub-milliarcsecond astrometric accuracies. Many of them are optical interferometers. Thus, it was time to hold a symposium that could summarize the present status of astrometry, consider the scientific benefits from increased accuracy in astrometry, and review the concepts for sub-milliarcsecond astrometry.

This symposium was one of six symposia scheduled at the same location as the IAU General Assembly, and overlapping in time with the General Assembly. It was held from Monday, August 15, through Friday, August 19, 1994, with a break for the opening session of the General Assembly. The symposium was sponsored by Commissions 4, Ephemerides; 8, Positional Astronomy; 9, Instruments and Techniques; 19, Rotation of the Earth; 24, Photographic Astrometry; 26, Double and Multiple Stars; 27, Variable Stars; 35, Stellar Constitution; 42, Close Binary Stars; 44, Astronomy from Space; and 45, Stellar Classification.

The Scientific Organizing Committee was chaired by J. Kovalevsky from France, and included F. Arias, Argentina, A. Baglin, France, J. Einasto, Estonia, M. Grenon, Switzerland, E. Høg, Denmark, Y. Kondo, U.S.A., M. Miyamoto, Japan, L.V. Morrison, U.K., M.A.C. Perryman, U.K., J.H. Schrijver, Netherlands, P.K. Seidelmann, U.S.A., C. de Vegt, Germany, and S.H. Ye, China, Nanjing. The Local Organizing Committee was chaired by J.H. Schrijver. The sessions were chaired by E. Høg, L.V. Morrison, S.H. Ye, M. Miyamoto, J.H. Schrijver, J. Kovalevsky, F. Arias and P.K. Seidelmann.

The Symposium was organized into three broad subjects. The first was Current Advances in Astrometry and included developments in ground-based astrometric techniques in large catalogs, space missions, and extragalactic reference frames. The second general topic was the Current and Future Needs for Very Accurate Astrometry. This included stellar astrophysics, galactic populations, kinematics and dynamics, extragalactic astrometry, and reference frames and solar system. The final area was the Expected Developments in High Precision Astrometry. The highlights of the meeting are summarized in the final paper by the chairman of the Scientific Organizing Committee.

In addition to 65 oral presentations and discussions there were approximately 50 Poster Papers. These proceedings are organized in a similar manner to the meeting. As usual there are some papers that were presented orally, but that have not been successfully put into a written version. The editors wish to thank all the authors for their efforts in preparing timely, written versions of their papers, and in following the composition instructions necessary for the preparation of these proceedings.

The Symposium agreed for the sake of uniformity to adhere to the following abbreviations. Milliarcsecond should be abbreviated to milliarcsec or mas, without any hyphens; but sub-milliarcsecond is written with a hyphen; and microarcsecond becomes μarcsec or μas.

December 1994

Erik Høg *P. Kenneth Seidelmann*

1. CURRENT ADVANCES IN ASTROMETRY

1.1 DEVELOPMENTS IN GROUND-BASED ASTROMETRIC TECHNIQUES AND LARGE CATALOGUES

CCD ASTROMETRY

R.C. STONE AND C.C. DAHN
U.S. Naval Observatory, Flagstaff Station
P.O. Box 1149 Flagstaff, AZ 86002-1149

Abstract.
CCD transit telescopes are now determining star positions at faint magnitudes (i.e., down to V $\sim$ 17 mag) and with accuracies as good as $\sim \pm 150$ mas. CCD parallaxes with errors $\sim \pm 1$ mas are now being determined routinely in time periods less than 3 years.

1. Introduction

The Charge-Coupled Device (CCD) has become the detector of choice in optical astronomy because of its high quantum efficiency, linearity, and direct imaging capabilities. The growth in CCD technology has been dramatic over the past decade, and thinned 1024^2 and 2048^2 devices with excellent cosmetics are now available. The imaging area of a single device is now almost 50 by 50 mm, and work has already begun at a number of observatories on constructing large arrays of CCDs which can image over 1 deg^2 of the sky. The nature of these improvements and the bright future for CCD astrometry are the subject of this paper.

2. Wide-Angle CCD Astrometry

Several observing strategies have been developed for making CCD observations. In stare mode the telescope tracks a given star field, and the length of the exposure is controlled by the shuttering. Unfortunately, only a small amount of sky is subtended by a single typical CCD. In drift scan mode the telescope is kept stationary and the sky image is clocked across the CCD at the diurnal rate. This method enables large regions of the sky to be scanned rapidly, but is subject to several disadvantages. Namely, the exposure is controlled by the physical width of the CCD in right ascension, star

E. Høg and P. K. Seidelmann (eds.),
Astronomical and Astrophysical Objectives of Sub-Milliarcsecond Optical Astrometry, 3–8.

images away from the celestial equator can be distorted by their nonuniform transits across the chip, and scanning at high declinations becomes very inefficient since the scan rate is proportional to cosδ. A third strategy, driving scan mode, overcomes the last disadvantage cited above in that the telescope is driven across the sky at a non-diurnal rate with modifications to the scan rate and direction. This latter technique will be used by the Sloan Digital Sky Survey (SDSS).

A number of telescopes employing CCD drift scanning are now being used for measuring the positions of stars, galaxies, asteroids, and comets (e.g., Gehrels *et al.* 1986, Stone and Monet 1990, and Benedict *et al.* 1991) and other instruments are either being developed or are in the planning stage. The SDSS 2.5 m telescope is nearing completion and will determine accurate star positions for about 10,000 square degree of the sky centered on the North Galactic Pole (Gunn and Knapp 1993) using an array of 30 2048^2 photometric and 17 2048 $\times$ 512 astrometric CCDs. This array will scan the sky in 2.3^o wide bands in different passbands and reach a limiting magnitude of V $\sim$ 23 mag. The targeted accuracy of the survey is $\pm$30 mas in each coordinate.

The Flagstaff Astrometric Scanning Transit Telescope (FASTT) has been described previously (Stone and Monet 1990, Stone 1993) and has been used in making CCD observations of stars and minor planets (Stone 1994, Monet *et al.* 1994). It is a 20-cm (f/10) meridian refractor equipped with a thick front-illuminated CRAF/Cassini 1024 x 1024 (12μ pixels) and observes in drift scan mode. The limiting magnitude of the telescope is V $\sim$ 17.5 mag, and about 9000 star hr^{-1} can be observed while scanning. The best currently available reference objects are in the VLBI catalog (Ma *et al.* 1990), and the FASTT measures star positions relative to them. Since these observations are made over wide arcs in the sky, efforts have been made to reduce refractive and instrumental errors, through the use of a laser metrology system and the application of corrections for room refraction and instrumental motions. Columns 4 and 5 in Table 1 give FASTT positional errors for these observations. The increase in the errors with magnitude can be explained by Poisson statistics affecting weak star images. For well-measured objects, the positional errors are $\sigma \sim \pm$130 mas in both coordinates. The FASTT has also been used to scan repeatedly regions in the sky at the same zenith distance. The internal error in these star positions is dominated by atmospheric errors which agree with the empirical relation given by Høg (1968). These errors are $\sigma \sim \pm$80 mas and can only be improved by lengthening the exposure time. The remaining $\pm$100 mas error in FASTT observations is caused by uncorrected instrumental errors.

Errors in star position can be reduced dramatically if reference objects are observed routinely in the course of scanning. Columns 2 and 3 of Table

TABLE 1. Distribution of FASTT errors.

V-Magnitude (mag)	Relative σ_x (mas)	Relative σ_y (mas)	Wide-Angle σ_x (mas)	Wide-Angle σ_y (mas)
9.0 - 9.5	±38	±35	±131	140
9.5 - 10.0	51	40	131	140
10.0 - 10.5	28	38	130	140
10.5 - 11.0	35	40	128	141
11.0 - 11.5	33	47	129	142
11.5 - 12.0	37	38	130	141
12.0 - 12.5	34	42	137	146
12.5 - 13.0	36	38	143	152
13.0 - 13.5	35	40	142	151
13.5 - 14.0	40	40	144	145
14.0 - 14.5	46	43	143	144
14.5 - 15.0	56	51	142	142
15.0 - 15.5	62	62	154	152
15.5 - 16.0	90	81	161	159
16.0 - 16.5	118	107	200	185
16.5 - 17.0	169	158	229	214
17.0 - 17.5	219	205	270	260

1 give the expected errors for this type of observing based on FASTT differential reductions. The error for well-exposed star images is $\sigma \sim \pm 38$ mas. Since program and reference objects are observed in the same field, both atmospheric and instrumental errors are greatly reduced. Unfortunately, the currently available star catalogs are neither very dense nor accurate. This situation will dramatically improve with the release of the HIPPARCOS/TYCHO catalogs. The SDSS 2.5-m telescope under construction has a 2.3^o by 2.3^o field of view, and accordingly many TYCHO stars will be observed while scanning. Simulations including atmospheric errors (Lindegren 1980, Han 1989) and errors in the TYCHO catalog have been used to compute positional errors for various field sizes and exposures. According to these simulations, a telescope with a 1° by 1° field of view and an exposure time of 100^s could theoretically measure star positions differentially to the TYCHO catalog at the ±11 mas level. Many existing meridian telescopes are being modernized with CCD detectors and will probably measure star positions differentially at the ±50-150 mas level. New technology telescopes, like the SDSS instrument, will do much better.

3. Narrow-Angle CCD Astrometry

The application of CCDs to ground-based, narrow-field, differential astrometric observations includes direct measures of more widely separated binary star components, speckle interferometric measures of close binary pairs, planetary satellite observations, and classical stellar trigonometric parallax determinations, of which only the latter will be discussed here. Those reporting CCD parallax measures to date include Ianna (1993), Ruiz *et al.* (1990), Tinney (1993), and the U.S. Naval Observatory (Monet *et al.* 1992). Since the first three programs employ general-user telescopes, their observational opportunities are restricted significantly. Consequently, together they have reported results for only about 20 stars and the accuracies achieved to date generally have been in the ± 2–5 mas range. The USNO program, on the other hand, has the 61-in Astrometric Reflector dedicated primarily to parallax efforts. The first USNO CCD parallaxes, obtained using a Texas Instrument 800^2 chip, were presented by Monet *et al.* (1992) and included 23 stars with precisions < 1.0 mas.

Starting in the spring of 1992, the USNO program began using a Tektronix 2048^2 CCD for parallax determinations. This device has 24 micron square pixels, providing a sampling of 0.325 arcsec/pixel and a field of view roughly 11 x 11 arcmin. Compared with limited 2.7 x 2.7 arcmin field of view presented by the TI 800^2 chip employed earlier, many more and brighter parallax targets became observable with much better quality reference star frames.

As of July 1994, 12 or more observations (126 maximum) have been obtained on each of 142 stars out of the 155 objects currently targeted. These stars have brightnesses largely in the $13 < V < 18$ range, but include a few as bright as $V \sim 12.5$ and a few as faint as $V \sim 20.5$. Preliminary reductions of this material to relative parallaxes have been carried out neglecting corrections for differential color refraction since the required photometry has not yet been obtained. The epoch ranges of the observations (i.e., time differences between the first and last observations for each star) varied from 1.06 to 2.36 years. The distribution of formal mean errors of the derived relative parallaxes is shown in Fig. 1. Clearly, a significant number of relative parallaxes (69) with precisions in the range from $\pm$ 0.4 mas to $\pm$ 1.0 mas have already been obtained with observational time spans less than 2.5 years.

Several solutions have been examined in detail in an attempt to understand why sub-mas precisions were or were not achieved. First of all, comparison of two subsets of the 142 solutions – those with epoch ranges from 1.06 to 1.42 yr (34 stars) versus those with epoch ranges from 1.85 to 2.36 yr (108 stars) – showed that the number distributions of relative par-

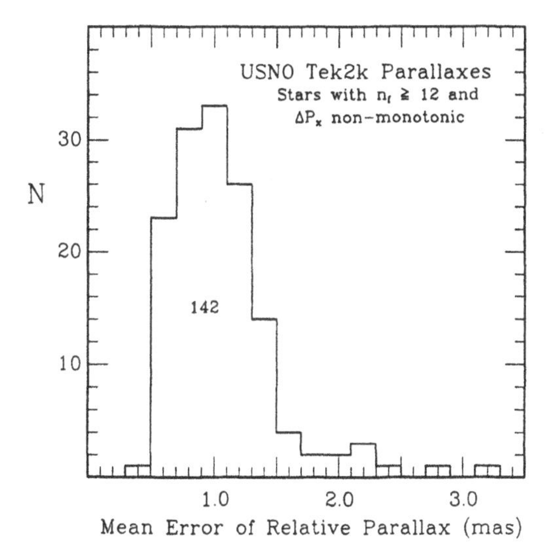

Figure 1. Distribution of formal mean errors for 142 preliminary USNO Tektronix 2048^2 CCD relative parallax solutions

allax errors were virtually the same (i.e., essentially like Fig. 1). Hence, the length of the observational series is not the major factor in producing sub-mas results. Next, the 'evolution' of solutions for ten stars was studied as data were added, one observation (CCD frame) at a time. In all instances, the error in the relative parallax decreases very rapidly for the first 10–15 observations but then decreases only very slowly thereafter. Although this behavior is roughly consistent with $\sqrt{n}$ statistical improvement, the ensemble of solutions suggests that additional factors are at play.

Detailed examination of the reference stars employed for approximately 30 fields suggests that the overall quality of the reference frame – the angular extent on the sky, the configuration with respect to the parallax star, and ability to employ well-exposed stars – is a, if not *the*, major factor in producing sub-mas results. Atmospheric turbulence is recognized as a limiting factor in differential astrometry, producing errors that scale by the angular separation to the ≈ 0.3 power and by the inverse square root of the exposure (integration) time (Lindegren 1980, Han 1989). This dependence is seen qualitatively in the present solutions; that is, fields employing reference frames of larger angular extent and requiring shorter exposure times (e.g., less than about 2 minutes) produce obviously poorer results.

The qualitative impression given by the reductions performed to date suggests that relative parallaxes with formal mean errors in the range $\pm$ 0.5 to $\pm$ 1.0 mas can readily be obtained. Just how much further improvement is possible is yet to be determined. However, the prospects for reaching

$\pm$ 0.3 mas for fields with optimal reference star configurations and for reasonable observational times intervals (e.g., < 5 yr) seems at least possible at this juncture. Astrophysical problems that can only be addressed directly by parallaxes with precisions ≤ 0.5 mas include: 1) the distance to F–G subdwarfs of varying metallicity for calibration of the distances to globular clusters; 2) the delineation of interior composition loci within the degenerate star sequence; and 3) accurate determinations of the luminosities of a sample of dwarf carbon stars to clarify the nature of the components of these supposedly binary systems. CCD parallaxes for the F–G subdwarfs will require a technique for magnitude compensation (i.e., for selectively dimming the bright target star while simultaneously exposing on faint reference stars). A neutral density spot (3 mm in diameter, deposited on an optically flat quartz substrate, and producing 6.0 magnitudes of attenuation) has been ordered and tests using it will start this fall.

References

Benedict, G.F., McGraw, J.T., Hess, T.R., Cawson, M.G.M., and Keane, M.J. 1991, AJ, 101, 279.

Gehrels, T., Marsden, B.G., McMillan, R.S., and Scotti, J.V. 1986, AJ, 91, 1242.

Gunn, J.E., and Knapp, G.R. 1993, in *Sky Surveys: Protostars to Protogalaxies*, ASP Conf. Series, No. 43, ed. B.T. Sofer, 267.

Han, I. 1989, AJ, 97, 607.

Høg, E. 1968, ZfAp, 69, 313.

Ianna, P.A. 1993, in *Developments in Astrometry and Their Impact on Astrophysics and Geodynamics*, IAU Symp. No. 156, ed. I.I. Mueller and B. Kolaczek, 75.

Lindegren, L. 1980, A&A, 89, 41.

Ma, C., Shaffer, D.B., deVegt, C., Johnston, K.J. and Russell, J.R. 1990, AJ, 99, 1284.

Monet, A.K.B., Stone, R.C., Monet, D.G., Dahn, C.C., Harris, H.C., Leggett, S.K., Pier, J.R., Vrba, F.J., and Walker, R.L. 1994, AJ, 107, 2290.

Monet, D.G, Dahn, C.C., Vrba, F.J., Harris, H.C., Pier, J.R., Luginbuhl, C.B., and Ables, H.D. 1992, AJ, 103, 638.

Ruiz, M.T., Anguita, C., and Maza, J. 1990, AJ, 100, 1270.

Stone, R.C. 1993, in *Developments in Astrometry and Their Impact on Astrophysics and Geodynamics*, IAU Symp. No. 156, ed. I.I. Mueller and B. Kolaczek, 65.

Stone, R.C. 1994, AJ, 108, 313.

Stone, R.C., and Monet, D.G. 1990, in *Inertial Coordinate System on the Sky*, IAU Symp. No. 141, ed. J.H. Lieske and V.K. Abalakin, 369.

Tinney, C.G. 1993, AJ, 105, 1169.

RELATIVE ASTROMETRY WITH IMAGING TRANSIT TELESCOPES

G. F. BENEDICT
McDonald Observatory
University of Texas
Austin, TX 78712

AND

J. T. MCGRAW AND T. R. HESS
Institute for Astrophysics
University of New Mexico
Albuquerque, NM 87131

Abstract. A CCD/Transit Instrument (CTI) has produced relative astrometry with standard errors less than 2.6% of a 1.55 arcsecond pixel for stars with $V \leq 17$. Additional astrometric studies with existing data are required to better understand the ultimate contribution these devices can make to our science.

The CTI is presently dismantled, awaiting a move to a new site. We briefly discuss the potential astrometric scientific returns from the exisiting data set, from a refurbished CTI, and from a similar device emplaced on the Moon.

1. Introduction

It may be that an Imaging Transit Telescope that patiently collects data night after night can, through brute force $\sqrt{n}$, approach the lofty goals of our symposium title. While van Altena (1994) discusses possible roles for CCD drift-mode devices dedicated to astrometry, few have been built specifically for astrometry. Generally, the astrometrist must make do with devices built to satisfy other scientific goals (*e.g.*, McGraw and Benedict, 1990). We contend that astrometry can and should come from any such device built for whatever astronomical purpose. We shall explore the astro-

E. Høg and P. K. Seidelmann (eds.),
Astronomical and Astrophysical Objectives of Sub-Milliarcsecond Optical Astrometry, 9–12.

metric scientific returns from existing CTI data, from a refurbished CTI, and from a Lunar-based Imaging Transit Telescope.

2. The Steward Observatory CCD/Transit Instrument (CTI)

The CTI (1.8 m f/2.2) is an example of an Imaging Transit Telescope and its operation. The principal motivation for the CTI (McGraw *et al.*1986) is to monitor a strip of the sky nightly in an imaging survey to detect and characterize objects variable on time scales of days to years. The goal is to measure the light curves of a variety of objects, including variable stars, supernovae, and QSOs, and to provide high S/N time-averaged magnitudes, colors and positions for these objects. Because CTI measures variability, brightness cannot be used as a matching criterion when comparing objects from night to night. Matching must be done by positional coincidence. Sub-arcsecond accuracy astrometry is required for the proper function of the CTI system.

3. Present and Future Astrometric Capabilities of the CTI

Though not built specifically for astrometry, several attributes of the CTI make it astrometrically attractive. These include use of stable, thermally-controlled CCD detectors, observations which are always made within a few degrees of the zenith, measurement to faint limiting magnitudes, and the structural integrity of the telescope/detector combination. Benedict *et al.* (1991) presented a preliminary assessment of the CTI as an astrometric device. Our primary conclusion was that overlapping plate astrometry with data from ten nights of CTI operation yielded relative positions with formal standard errors of less than 40 mas (2.6% of a 1.55 arcsec CTI pixel) for stars $V \leq 17.0$.

The stars at the top and bottom of the CCD drift across at different rates, due to declination differences. Monet (1994) singles this out as a primary concern for CCD-based drift mode devices. CTI smearing amounts to 0.3 arcsec at the top of the field, compared to the center. We saw no patterns in the residuals indicating poorer results for the smeared stars. Evidently, when used for differential astrometry within a fixed declination strip each night, our simple first moment centering algorithm provides repeatable centroids.

Having only explored 10 nights out of hundreds available, additional astrometric studies with existing data are required to better understand the ultimate contribution these devices can make to our science. These include running models with "plates" constructed from over 50 data sets, to determine the level at which we no longer obtain a $\sqrt{n}$ decrease in positional error. This study is currently underway in support of a project to identify (Wetterer *et al.*1994) many new RR Lyrae stars from colors and magnitudes and to confirm those identifications through spectroscopy and proper motions.

The test area represents 0.056% of the total CTI surveyed strip. Since this random piece of the sky netted us three stars out of 61 whose motions were in excess of 2 arcsec/century, continuing the exploration of the entire strip might result in the discovery of more than 3500 stars with similar proper motions in the range $12.5 < V < 19$.

The CTI is presently shut down. Once it is moved to a site in New Mexico and recommissioned, we would like to upgrade the CTI with modern CCD technology, providing smaller pixel size and far lower read-out noise. For example, using CCDs with 15 micron pixels would allow centroiding individual objects to approximately 15 mas, assuming we can continue to achieve 2% pixel precision. The smaller pixels would more adequately sample the seeing disk, and allow experimentation with point spread function fitting algorithms to more accurately estimate centroids. Lowering the read-out noise provides sky noise limited photometry over all bandpasses. This would in turn allow higher precision centroids to be determined, independently of the details of the centering algorithm. Sampling the point spread function more densely might allow us to better centroid the smeared images at the declination extremes of the CCD. Finally, the CTI provides a test-bed for alternative area detectors (Baron and Priedhorsky, 1993) that could electronically compensate for varying drift rates across the field of view.

Finally, a once again operational CTI would, after a year of data collection, have a data base with a time span approaching 9 years. This would allow precise proper motion determinations for stars at V=17 with proper motions $> 6\ mas\ yr^{-1}$. The Tychocatalog will provide a wealth of high-precision local astrometric reference stars against which to measure these motions. The Tycho catalog density on the sky will reduce the along-track size of the sample required to obtain a suitable reference frame.

4. The Lunar Ultraviolet Transit Experiment (LUTE)

The Moon is a gravity-gradient stabilized satellite from which to accomplish a wide variety of astrophysical observations. It presently lacks instrumentation. For many compelling reasons (McGraw, 1994) the first telescope on the Moon could be an Imaging Transit Telescope, not necessarily using a CCD detector. Paramount among them is simplicity. Space is a harsh environment. Simple often survives. No moving parts is the epitome of simplicity. Simplicity means relatively low cost.

The scientific return from a lunar-based UV-sensitive Imaging Transit Instrument can include (McGraw and Benedict, 1990); discovery and statistics of stars with active chromospheres, studies of interstellar dust and reflection nebulae, statistics of flaring on dwarf M and related stars, and stellar necrology (white dwarfs).

Obtaining proper motions and distances from astrometry of any objects found peculiar or otherwise interesting from a LUTE mission with a 2 year or longer duration is an obvious major increase in the scientific return.

The technology to produce a Lunar Transit Telescope has been investigated (McBrayer *et al.*, 1994). Research is currently underway (Baron and Priedhorsky, 1993) to provide area-format detectors capable of meeting the astrophysical constraints of accurate and precise astrometry and photometry, while being robust enough to withstand the high-energy particle flux on the lunar surface. This is the last remaining technological hurdle to implementing the next version of an imaging transit telescope.

In conclusion Imaging Transit Telescopes are relatively cheap to build and operate. Someday there may be many more. Astrometrists should know of their existence and limitations and be prepared to exploit them.

References

Baron, M. and Priedhorsky, W. (1993). "Crossed-delay Line Detector for Ground- and Space-based Applications" in *SPIE Proceedings for the EUV, X-Ray, and Gamma-Ray Instrumentation IV*, 2006, 188

Benedict, G. F. and Shelus, P. J. (1978). "Applications of Automated Inventory Techniques to Astrometry" in *Proc. IAU Colloq. 48 on Modern Astrometry*

Benedict, G. F., McGraw, J. T., Hess, T. R., Cawson, M. G. M., and Keane, M. J. (1991) "Relative Astrometry with the Steward Observatory CCD/Transit Instrument", *Astron. J.*, 101, 279

McBrayer, R. O., *et al.*(1994) "Lunar Ultraviolet Telescope Experiment (LUTE) - Phase A Final Report", *NASA Technical Memorandum 4594*

McGraw, J. T., Cawson, M. G. M., and Keane, M. J. (1986) "CCD Transit Telescopes", *SPIE*, 627, 60.

McGraw, J. T., and Benedict, G. F. (1990) "Scientific Programs of a Lunar Transit Telescope", in AIP Conf. Proc. 207, *Astrophysics from the Moon*, eds. Mumma, M.J. and Smith, H. J., p. 433

McGraw, J. T. (1994) "Optical Astronomy from the Earth and Moon", *ASP Conf. Ser. vol. 55*, eds. Pyper and Angione, p. 283

Monet, D. (1994) "Digitization in Astronomy", in Astronomy from Wide-field Imaging, *Proc. IAU Symp. 161*, Kluwer, Dordrecht, p. 163

van Altena, W. (1994) "Image Detection, Characterization and Classification, and the Role of CCDs in Wide Field Astrometry", in Astronomy from Wide-field Imaging, *Proc. IAU Symp. 161*, Kluwer, Dordrecht, p. 193

Wetterer, C. J., McGraw, J. T., Hess, T. R., and Grashuis, R. (1994) " Variable Stars in the CCD/Transit Instrument (CTI) Survey", *BAAS*, 26, 899

NARROW-ANGLE AND WIDE-ANGLE ASTROMETRY VIA LONG BASELINE OPTICAL/INFRARED INTERFEROMETERS

XIAOPEI PAN, SHRI KULKARNI
California Institute of Technology, Pasadena, CA 91125

AND

MICHAEL SHAO, M. MARK COLAVITA
Jet Propulsion Laboratory, Pasadena, CA 91109

ABSTRACT. Long baseline optical/infrared interferometers, such as the Mark III Stellar Interferometer[1] on Mt. Wilson and the ASEPS-0 Testbed Interferometer[2] on Palomar Mountain, California, have good capabilities for narrow-angle and wide-angle astrometry with very high precision. Using the Mark III Interferometer many spectroscopic binaries became "visual" for the first time. The measurement accuracy of angular separation is 0.2 mas, the smallest separation measured between two components is 2 mas, the maximum magnitude difference is 4 mag, and the smallest semi-major axis is 4 mas. Such high angular resolution and dynamic range have been used to determine stellar masses with precision of 2% and differential stellar luminosities to better than 0.05 mag for separations of less than $0.''2$. For some binary stars, not only have the systems been resolved, but also the diameter of the primary component has been determined, yielding direct measurements of stellar effective temperature with high accuracy. For parallax determination, the precision is 1 mas or better and is unaffected by interstellar extinction. For wide-angle astrometry with the Mark III interferometer, the observation results yielded average formal 1σ errors for FK5 stars of about 10 mas. Presently a new infrared interferometer, the ASEPS - 0 Testbed Interferometer on Palomar Mountain is under construction, and is being optimized to perform high accuracy narrow-angle astrometry using long baseline observations at 2.2 μm, with phase referencing for increased sensitivity. The goal is to demonstrate differential astrometric accuracies of

E. Høg and P. K. Seidelmann (eds.),
Astronomical and Astrophysical Objectives of Sub-Milliarcsecond Optical Astrometry, 13–18.

0.06–0.1 mas[3] in order to allow for detection of extra-solar planets in the near future.

Key words: astrometry - optical/IR interferometer

1. Introduction

Long baseline optical interferometers are the only instruments which are capable of both narrow-angle and wide-angle astrometry with milliarcsec precision. The Mark III Stellar Interferometer is a good example of such an instrument with important features which include active fringe tracking, two color astrometry, and full automation. This instrument has two fixed baselines (12 m N-S and E-S) for wide-angle astrometry, and a variable baseline (3 – 32 m N-S) for narrow-angle astrometry. Typical measurements of fringe visibility are simultaneously made at 800 nm, 550 nm, and 500 nm with a bandwidth of 22 nm, 25 nm and 25 nm respectively. The Mark III Interferometer measures 160 – 220 stars per night and was in daily operation for more than 7 years. Brief descriptions of results for narrow- and wide-angle astrometry will be given in the following sections.

2. Parallax with Precision of $< 5\%$

Parallax is one of the fundamental parameters for a star. Trigonometric parallax achieves a typical precision of 10 mas with the classical photographic technique. New CCD parallax programs have reached a precision at the 1 mas level. However it is limited to very faint stars (15 mag or dimmer) because it is extremely difficult to find reference stars in the small field of view. Another technique, the Multichannel Astrometric Photometer, provides relative parallax to the 1 mas level for many interesting bright stars. However, it is a difficult task to convert the relative parallax to the absolute value. For binary stars, the Mark III has determined the angular semi-major axis to 0.2 mas[4,5,6], and has yielded parallaxes with a precision of 1 mas after incorporating the linear semi-major axis from spectroscopy. This absolute parallax is not affected by interstellar absorption, and can be used as a calibrator and a check with HIPPARCOS's results. The largest distance obtained is 270 pc, corresponding to a parallax of 3.7 mas. Some results of parallax determination are listed in Table 1, and are compared with the corresponding values from the photographic techniques. It is interesting to note that the brightest binary in the Hyades cluster, θ^2 Tau, has its distance determined as 44.1 pc$\pm$2.2 pc. This star is about 1 pc in front of the center of the cluster. The estimated distance of the center of the Hyades is 45.5 pc, which is consistent with the latest results from other techniques.

Table 1. Parallax Determination using the Mark III Interferometer

Star	Semimajor Axis a''(mas)	Parallax (mas) Mark III	Phgr.
HR 15	24.15 ± 0.13	29.6 ± 1.0	24
HR 154	6.82 ± 0.16	3.7 ± 0.1	0
HR 271	10.81 ± 0.20	14.3 ± 0.6	9
HR 553	37.02 ± 0.23	58.1 ± 1.1	63
HR 622	9.02 ± 0.19	30.6 ± 1.3	12
HR 936	94.61 ± 0.22	35.4 ± 1.1	31
HR 1412	18.60 ± 0.18	22.4 ± 1.1	29
HR 5054	10.74 ± 0.13	44.1 ± 0.8	37
HR 5793	7.75 ± 0.13	38.9 ± 0.9	43
HR 6927	122.5 ± 0.20	118.7 ± 1.0	120
HR 7478	23.7 ± 0.40	12.9 ± 0.2	7
HR 8131	11.99 ± 0.08	18.1 ± 0.8	13

3. Stellar Masses with 2% Accuracy

The mass of a star determines its fate and evolutionary course. Stellar masses with an accuracy of 1–2% are very critical to check evolutionary theory and modeling. There are fewer than 150 individual stars in binary systems with masses determined to better than 15%, and only 45 stars have masses with a precision of 2%. However, most of these stars are in the main sequence, and the evolved stars are very rare. The Mark III has determined inclinations[7,8] to better than 1%, and provided stellar masses to about 2%; most of these stars are giants and sub-giants which are lacking among current measurements.

Table 2. Accurate Stellar Masses with the Mark III Interferometer

Star	Spectrum	a''(mas)	i°	K_1(km/s)	K_2(km/s)	$m_1(m_\odot)$	$m_2(m_\odot)$
HR5793	B9.5IV,G5	7.75 ± 0.13	88.30 ± 0.07	35.35 ± 0.50	99.00 ± 0.50	2.58 ± 0.04	0.92 ± 0.0
HR6927	F7V,K3V	122.5 ± 0.20	74.76 ± 0.05	17.87 ± 0.10	24.20 ± 0.20	1.08 ± 0.03	0.78 ± 0.0
HR7478	G8III,G8III	23.7 ± 0.40	78.37 ± 0.40	26.79 ± 0.05	27.88 ± 0.05	2.26 ± 0.05	2.17 ± 0.0

4. Precise Photometry within a Separation of 0.″1

It has often been noted in the literature that there is a lack of precise measurements of stellar luminosities and colors along with mass determination. Interferometer is the only technique, besides Lunar Occultation, which can provide direct photometric measurements within small angular separations. In the periodic changes of fringe visibility of a binary star as measured with an interferometer, the peak-to-peak change is proportional to the intensity ratio of the two components. Since the typical precision of visibilities can reach 2% with the Mark III, the corresponding magnitude differences are readily determined to a precision of better than 0.05 mag. Another important characteristics of the Mark III is simultaneous measurements at multiple wavelengths. It is easy to calculate the color index for both components when combined with classical photometric results. It is worth noting that the traditional spectrophotometric method of analyzing spectral lines has had obvious biases as large as 1 mag. In contrast, lunar occultations do provide accurate magnitude differences for some binary systems, which agree well with the results of the Mark III Interferometer[7]. The comparisons between the Mark III and spectrophotometry are provided in Table 3.

Table 3. Comparison of Magnitude Differences measured with the Mark III Interferometer and Spectrophotometry

Star	Mark III		Spectrophotometry
	at 800 nm	at 550 nm	at 550 nm
HR 15	$1.^m82 \pm 0.^m04$	$1.^m99 \pm 0.^m04$	$1.^m35$
HR 154	$0.^m97 \pm 0.^m10$	$0.^m92 \pm 0.^m10$	$3.^m17$
HR 271	$0.^m94 \pm 0.^m06$	$1.^m01 \pm 0.^m05$	$0.^m29$
HR 553	$2.^m63 \pm 0.^m22$	$3.^m30 \pm 0.^m30$	$2.^m80$
HR 622	$0.^m52 \pm 0.^m04$	$0.^m44 \pm 0.^m04$	$1.^m19$
HR 936	$2.^m63 \pm 0.^m09$	$2.^m92 \pm 0.^m15$	$2.^m60$
HR 6927	$2.^m02 \pm 0.^m06$	$2.^m44 \pm 0.^m17$	$1.^m99$
HR 8131	$1.^m23 \pm 0.^m03$	$0.^m47 \pm 0.^m06$	$0.^m60$

5. Resolution of Radio Stars

Radio stars play important roles in connecting stars with the extragalactic reference frame. VLBI observations have constructed a quasi-inertial reference frame to an accuracy of about $0.''001$. The only way to link the optical reference frame to the radio reference frame is to observe these radio stars at both optical and radio wavelengths. The HIPPARCOS mission emphasized necessity to observe radio stars. However, many radio stars are binaries or in a triple system, and their orbital motion is not negligible if astrometric precision at the milliarcsec level is necessary.

The Mark III Interferometer has successfully resolved two radio stars[9], Algol and λ And. For the well-known triple system Algol, the AB-C system has had its geometric and physical parameters determined with higher resolution and accuracy than obtained with other techniques. Since the presence of the third component seriously complicates the analysis of both the light curves and the spectra of the central eclipsing pair, the results from the Mark III are very useful. The mass ratio ($r = 2.63 \pm 0.20$) indicates that the central pair will have displacements as large as 52 mas, which must be accounted for. For λ And, the angular separations are about 3 mas, and the magnitude difference is about 4.5 mag at 550 nm.

6. Wide-Angle Astrometry

Modern meridian circles are the main astrometric instruments for measuring star positions at present, and have reached precisions of 0.″1 to 0.″2 for a single star transit. The analysis of observations from meridian circles has disclosed systematic errors in R.A. and Decl. in the Fundamental Star Catalogue (FK5) which reach about 0.″1. Astrometric observations with the Mark III interferometer in 1988[10] demonstrated average formal 1σ errors of 6 mas in Decl. and 0.6 ms (= 9 mas) in R.A. for a group of stars covering a range of 90° in R.A. and 45° in Decl. Such high precision comes from the good thermal control of the instrument, monitoring of variations of the delay offset, and the two color data analysis technique. Repeated observation were made in 1989 and had the same average formal 1σ errors as in 1988. Examination of the errors for the difference of the positions for the two years (see Table 4 and the figures below) indicates that the formal errors of each year are underestimated by a factor of 1.5, suggesting average external errors of approximately 9 mas in Decl. and 13 mas in R.A. for each year. As expected, observations with the Mark III Interferometer indicate that most of the FK5 stars in this group have positions within their estimated error ranges (about 50 mas in both coordinates in 1988 and 1989), and some stars do have position errors as large as 100 mas at present.

Table 4. Wide-angle Measurements of FK5 Stars in 1988 & 1989 with the Mark III Interferometer

FK5	Offset in R.A.(ms)				Offset in Decl.(mas)			
	1988	1989	Mean	Δ	1988	1989	Mean	Δ
19	-2.0 ± 0.5	-0.7 ± 0.5	-1.3	-1.3	-2 ± 5	-18 ± 6	-10	16
33	-1.8 ± 0.6	-1.7 ± 0.6	-1.7	-0.1	15 ± 5	13 ± 6	14	2
52	-2.2 ± 0.7	-2.7 ± 0.8	-2.5	0.5	25 ± 6	29 ± 8	27	-4
66	7.0 ± 0.5	6.7 ± 0.6	6.8	0.3	3 ± 6	14 ± 8	9	-11
79	-0.5 ± 0.6	1.7 ± 1.0	0.6	-2.2	-90 ± 6	-124±10	-107	34
835	2.4 ± 0.5	2.1 ± 0.4	2.3	0.3	66 ± 5	80 ± 4	73	-14
848	6.2 ± 0.7	4.3 ± 0.6	5.2	1.9	-46 ± 6	-43 ± 5	-45	-3
862	-5.3 ± 0.5	-4.2 ± 0.4	-4.7	-1.1	-35 ± 5	-32 ± 4	-34	-3
1534	2.5 ± 0.6	0.3 ± 0.3	1.4	2.2	-10 ± 5	-7 ± 4	-9	-3
1568	0.6 ± 0.7	-0.7 ± 0.5	-0.0	1.3	25 ± 5	34 ± 4	30	-9
1619	-5.8 ± 0.7	-6.4 ± 0.7	-6.1	0.6	-7 ± 5	2 ± 7	-3	-9

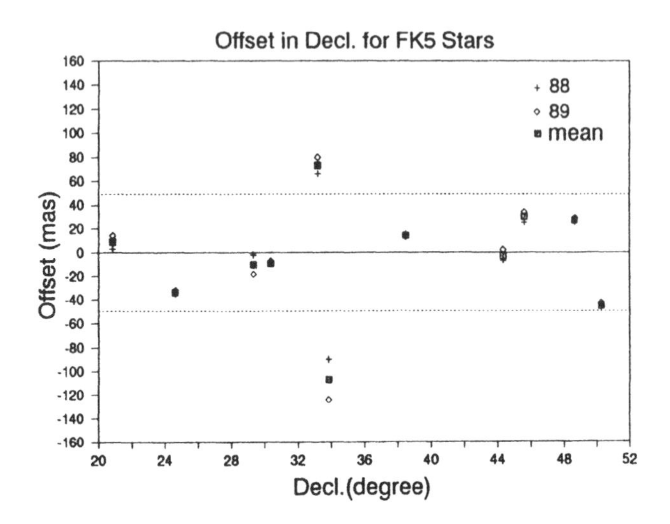

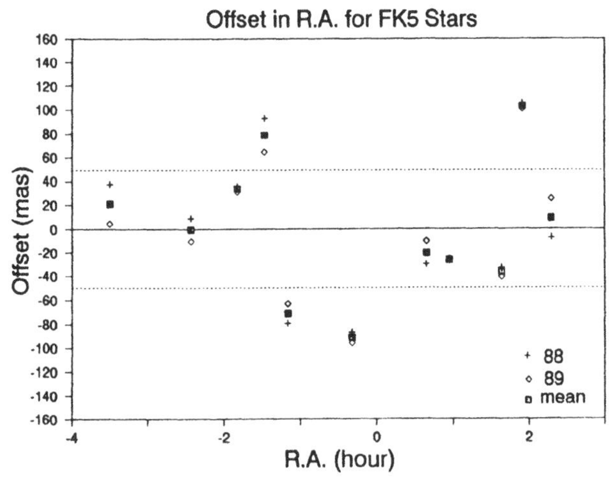

References

1. Shao, M. et al. 1988, A & A, **193**, 357
2. Colavita, M. M. et al. 1994, Proc. SPIE, **2200**, 89
3. Colavita, M. M. 1994, A & A, **283**, 1027
4. Pan, X.P. et al. 1992, ApJ, **384**, 624
5. Pan, X.P. et al. 1990, Proc. SPIE, **1237**, 301
6. Pan, X.P. et al. 1993, IAU Coll. **135**, 502
7. Pan, X.P. et al. 1994, Pro. SPIE, **2200**, 360
8. Armstrong, J.T. et al. 1992, AJ, **104**, 2217
9. Pan, X.P. et al. 1993, ApJ, **413**, L129
10. Shao, M. et al. 1990, AJ, **100**, 1701

SINGLE AND DOUBLE STAR ASTROMETRY WITH THE MARK III INTERFEROMETER

C.A. HUMMEL
Universities Space Research Association
NRL/USNO Optical Interferometer Project
c/o U.S. Naval Observatory - AD 5
3450 Massachusetts Av. NW, Washington, DC 20392, USA

Abstract. Present construction of large multi-element optical interferometers is based on experience with several pioneering instruments, a very successful one being the Mark III interferometer on Mt. Wilson, California. With it, over the last few years, some 26 spectroscopic binaries were resolved at separations as small as 3 milliarcseconds (mas) and orbital elements have been derived. In addition, the installation of vacuum delay-lines and multi-color fringe-detection allowed the derivation of dispersion-corrected geometrical delays of stars, which were used to determine relative stellar positions over wide angles with a precision of about 20 mas. We present and discuss recent results with respect to the determination of stellar positions, distances, masses, and luminosities.

1. Introduction

The Mark III Optical Interferometer was operated by the Remote Sensing Division of the Naval Research Laboratory (with funding from the Office of Naval Research) for six years between 1987 and 1992. During this time, the Mark III contributed direct measurements of stellar diameters and of visual orbits of spectroscopic binaries to astrophysical research. The usefulness of an interferometer for astrometry was demonstrated by measuring star positions with better than FK5 accuracy. See Shao et al. (1988) for a description of the Mark III.

E. Høg and P. K. Seidelmann (eds.),
Astronomical and Astrophysical Objectives of Sub-Milliarcsecond Optical Astrometry, 19–22.

For diameter and binary observations, a variable baseline was used; lengths ranged from 3 m to 31 m, but only one configuration was used at a time. For the astrometric observations, we used the 12 m N-S and E-S baselines realized by three high-precision siderostats mounted on massive concrete piers in climate-controlled protective huts. Fringes could be detected on stars as faint as $m_V = 5$. Typically, we obtained 150 to 200 75-sec measures of the delay and the fringe visibility amplitude in a night on a list of eight to 15 stars, half of which were small-angular-diameter stars for calibration. Uniform-disk models were fitted to the calibrated visibility data.

2. Wide angle astrometry

The geometrical delay $d_i(t)$ for a star i at time t over a baseline $\mathbf{B}$ is related to its position $\mathbf{S}_i(t)$ through

$$d_i(t) = \mathbf{B} \cdot \mathbf{S}_i(t) + C(t).$$

This delay can be derived to high precision (typically one to three microns) from the readings of the laser-controlled delay lines of the Mark III, based on the cancellation of the effects of refraction in an interferometer design featuring evacuated pipes, three-channel detection of the fringe position (at $\lambda\lambda$ 800 nm, 550 nm, and 500 nm) for the correction of the atmospheric refractive index fluctuations, and an internal white light source for the measurement of path length variations $C(t)$ induced by temperature drifts. Star positions and baseline coordinates are obtained from a least-squares fit to the delay data.

We show here (Fig. 1) the measured corrections to the FK5 catalog positions for a group of twelve stars, as obtained with the Mark III (Hummel *et al.* 1994b) and with ESA's astrometry satellite HIPPARCOS (M. Perryman, priv. comm.). The latter data were taken from a preliminary solution based on 18 months worth of observations. We note that a degeneracy of the Mark III solution due to the small declination range of the observed stars prevented the determination of "absolute" declinations. Furthermore, the right ascension zero point is arbitrary. The RMS difference between the Mark III and HIPPARCOS positions is about 20 mas. It is somewhat larger than expected from the formal errors of the Mark III results. Probable causes have been identified as unmonitored baseline drifts during the night, mechanical imperfections of the siderostats, and water vapor fluctuations not solved for in the dispersion correction. However, the agreement between the results of these fundamentally different observing techniques is a "proof-of-concept" for astrometric measurements with optical interferometers.

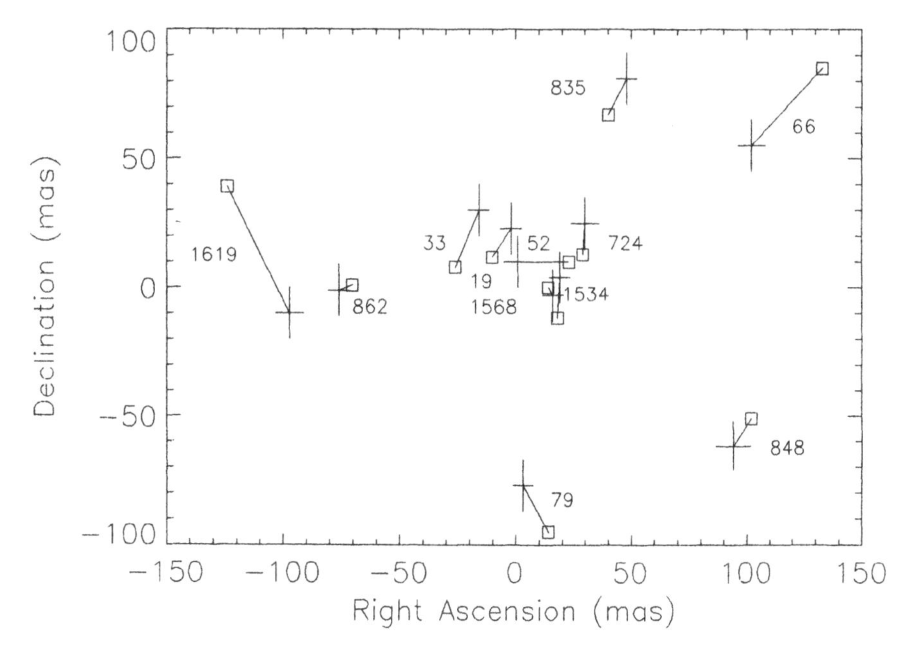

Figure 1. Corrections to the FK5 catalog position for a group of twelve stars. Plus symbols indicate the average results from four years of measurements with the Mark III; square symbols indicate data from the HIPPARCOS satellite.

3. Binary stars

Determinations of stellar masses should be better than about 2% in accuracy and should be accompanied by a determination of the absolute luminosities of the components to help distinguish between stellar evolution models (Andersen 1991). With the Mark III, we were able to resolve orbits of spectroscopic binaries with major axes down to about 3 mas, if the magnitude difference did not exceed about $3^{m}_{.}5$. In addition to the orbital elements, we measured the magnitude difference of the components at $\lambda\lambda$ 800 nm, 550 nm, 500 nm, and 450 nm (with a 25 nm bandwidth of each channel). Of the twenty-six spectroscopic binaries which we have resolved and for which we have determined orbital elements, seven are single-lined (π And, 6 Lac, τ Per, α Dra, η Peg, α And, and 113 Her). In six cases, the eccentricity derived from spectroscopy is inconsistent with the Mark III value (χ Dra, β Tri, ϕ Cyg, θ^2 Tau, ξ Cep A, and 6 Lac). Thus, we were able to determine masses and absolute luminosities for only a half of the binaries, but in reasonable agreement with recent stellar evolution models. The accuracy for the masses ranges from about 1%-4% (six binaries) to about 10%-20% (seven binaries), and is in most cases limited by the accu-

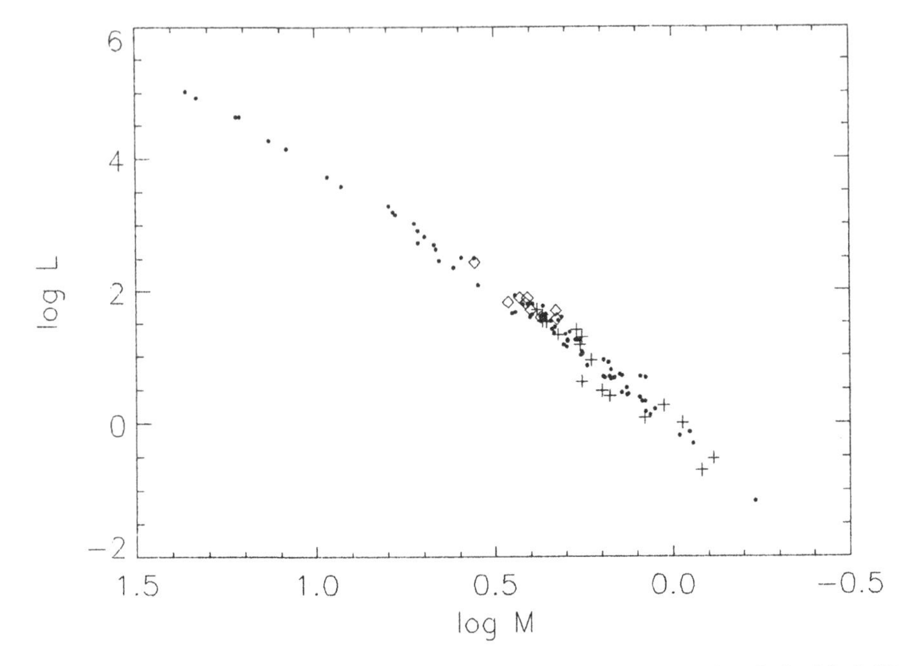

Figure 2. Mass-luminosity diagram for double-lined binaries observed with the Mark III. Plus symbols denote dwarfs or sub-giants; diamonds denote giants. Dots denote eclipsing binary data from Andersen (1991).

racy of the spectroscopy. If the latter were free of errors, mass uncertainties would be accurate to better than 7% in all cases. In Fig. 2, we plot masses and luminosities for 17 dwarf and 9 giant components, together with data from Andersen (1991) for eclipsing main-sequence binaries. We have not included error bars in the plot due to the large fraction of preliminary orbits. Part of the Mark III results has been published: β Ari by Pan *et al.* (1990), α And by Pan *et al.* (1992), α Equ by Armstrong *et al.* (1992a), ϕ Cyg by Armstrong *et al.* (1992b), η And by Hummel *et al.* (1993), β Per by Pan *et al.* (1993), and α Aur by Hummel *et al.* (1994a).

References

Andersen, J. (1991), *Astron. Astrophys. Rev.*, 3, 91
Armstrong, J.T., et al. (1992a), *Astron. J.*, 104, 241
Armstrong, J.T., et al. (1992b), *Astron. J.*, 104, 2217
Hummel, C.A., et al. (1993), *Astron. J.*, 106, 2486
Hummel, C.A., et al. (1994a), *Astron. J.*, 107, 1859
Hummel, C.A., et al. (1994b), *Astron. J.*, 108, 326
Pan, X.P., et al. (1990), *Astrophys. J.*, 356, 641
Pan, X.P., et al. (1992), *Astrophys. J.*, 384, 624
Pan, X.P., et al. (1993), *Astrophys. J.*, 413, L129
Shao, M., et al. (1988), *Astron. Astrophys.*, 193, 357

THE USNO ASTROMETRIC OPTICAL INTERFEROMETER

D.J. HUTTER, K.J. JOHNSTON
United States Naval Observatory
Washington, DC 20392-5420

AND

D. MOZURKEWICH
Remote Sensing Division,
Naval Research Laboratory
Washington, DC 20375

Abstract. The U.S. Naval Observatory Astrometric Optical Interferometer (AOI) began operation on Anderson Mesa, near Flagstaff, Arizona, in the autumn of 1994. The AOI incorporates four siderostats that are located in a Y-shaped configuration, and features a full-array laser metrology system to monitor baseline motion. The AOI incorporates state-of-the-art delay lines and a real-time fringe-tracking system. The AOI will have a limiting visual magnitude of 10, under typical observing conditions, and will produce star positions accurate to a few milliarcseconds (mas). With a planned operational lifetime of several decades, this instrument will be capable of maintaining the optical reference frame by improving the proper motions of thousands of the brighter HIPPARCOS stars through repeated observations.

1. Introduction

The U.S. Naval Observatory Astrometric Optical Interferometer (AOI) is the dedicated astrometric subarray of the new Navy Prototype Optical Interferometer (NPOI) at Lowell Observatory, which is being built in collaboration with the Naval Research Laboratory. The Naval Research Laboratory is constructing the imaging subarray of the NPOI, the "Big Optical Array" (BOA; Armstrong 1994).

The AOI was built using the experience gained from the Mark III Interferometer on Mt. Wilson, CA, which demonstrated "proof of concept" of wide-angle astrometry by a long baseline optical interferometer. The Mark III is a broad-band phase-tracking optical interferometer, which was operational from late 1986 through 1992 (Shao *et al.* 1988). The Mark III is capable of single baseline measurements on either a north-south or east-south baseline, each of about 12 m in length. The Mark III incorporates evacuated delay lines with automated high speed servo systems, allowing the detection and tracking of white-light fringes

E. Høg and P. K. Seidelmann (eds.),
Astronomical and Astrophysical Objectives of Sub-Milliarcsecond Optical Astrometry, 23–29.

in real time. Measurements at two different wavelengths are used to correct the measured delays for refractive index fluctuations in the turbulent atmosphere.

Measurements of stellar positions by the Mark III in a subset of the FK5 catalog, made at several epochs over a four-year period, have uncertainties of 10 to 20 mas (Hummel *et al.* 1994). The accuracy of the measurements is limited by systematic errors largely due to unmonitored changes in the baseline coordinates. These errors are due to mechanical and thermal motions within the instrument, principally in the siderostats. Therefore, in the design of the AOI, major consideration was given to an extensive baseline metrology system. This system will monitor fully the motions of the baselines in three dimensions with respect to a reference system fixed to the Earth's crust. The AOI also includes new multi-way beam combination and fringe detection techniques, along with an increase in the number of siderostats from two to four. These improvements allow simultaneous observations on six baselines (of lengths between 18 m and 35 m, three of which are geometrically independent). Such observations will permit the total separation of the baseline components and star position corrections in the astrometric solutions. The addition of a dispersed fringe detection technique allows unambiguous central fringe identification, and better correction for atmospherically induced delay fluctuations. Additionally, the increased aperture, state-of-the-art delay lines, and greater detector sensitivity will permit the AOI to measure the positions of some of the brighter stellar radio sources, whose positions have been tied to the radio reference frame defined by extragalactic sources.

The combination of these improvements will permit the AOI to produce highly accurate catalogs of star positions on a nearly inertial frame. The initial goal of the AOI is to establish, by mid 1997, a catalog of the order of one thousand stars with statistical and systematic errors in the range of 1 to 3 mas. This catalog will be more than an order of magnitude more precise than the FK5 and of the same order as expected from space-based systems. With an anticipated operational lifetime of decades, the AOI will significantly improve the measured proper motions of these (and additional) stars. Stellar positions measured over a period of ten years will have proper motions known to a few 100 μas/year. Position measurements repeated at regular intervals will also allow unambiguous separation of binary motion from proper motion, an accomplishment that might be difficult to achieve from space-based observations that are likely to be repeated only at intervals of decades.

Thus, the AOI, with a capability of milliarcsecond astrometry and a planned operational lifetime of several decades, will be capable of maintaining the optical reference frame by improving the proper motions of thousands of the brighter HIPPARCOS stars through repeated observations.

2. Description of the Instrument

The light-collecting apertures of the AOI consist of four siderostats located in a Y-shaped array. Starlight collected by the siderostats is directed into the "feed system" vacuum pipes that carry the light to the center of the array. Additional mirrors at the array center redirect the light along the vacuum lines to the optical laboratory. The laboratory contains the delay lines, beam-combining optics, fringe detection system, and the star-tracking sensors. All interferometer systems are remotely controlled from a separate operations facility.

2.1. SIDEROSTATS

The siderostats are housed in roll-off-roof enclosures, which will be climate-controlled to approximate average nighttime temperatures. The siderostats consist of permanently mounted, high-precision, two-axis alt-az mountings which each support a flat mirror of 50-cm clear aperture. The siderostats are driven in each axis by a novel drive system consisting of in-line dual harmonic drive gear reduction units driven by stepper motors. The yoke and mirror cell of each siderostat include provisions for accurately intersecting the azimuth and elevation axes, and for placing the mirror surface on the elevation axis (both to within a few microns).

A wide field-of-view star acquisition capability is also incorporated via a CCD camera located near each siderostat. A two-axis gimbal-mounted flat mirror, with piezo-electric actuators, is located in the stellar light path near each siderostat. These mirrors are steered by the star-tracking sensors to provide the small angle, high speed tip-tilt correction necessary to compensate for random image motion due to atmosphere turbulence.

Provision is made within the siderostat enclosures for the addition of a 35-cm aperture afocal beam compressor next to each siderostat. When the beam compressors are installed in 1995, the limiting sensitivity of the AOI will be increased from V = 8.5 (for the current aperture of 12.5 cm, set by the feed system optics) to V = 10 under typical observing conditions (V = 11 with excellent seeing).

2.2. BASELINE METROLOGY SYSTEM

For the AOI to achieve its design goal of milliarcsecond astrometry, a baseline laser metrology system must be employed that can measure the three-dimensional motions of the baselines to an accuracy of approximately 0.1 μm (Elias 1994). This system consists of a number of laser interferometers tied to four temperature-stabilized super-Invar reference plates, one next to each siderostat. Five laser interferometers measure the position of a "cat's-eye" retroreflector (Danchi *et al.* 1986) at the intersection of the rotation axes of each siderostat, relative to the adjacent reference plate. The rotation and tilt of the reference plates will, in turn, be measured by two other metrology subsystems. In the first case, six laser interferometers will monitor changes in the distances of three points on each plate from retroreflectors embedded in a deep subsurface layer. The second subsystem consists of interferometers along lines of sight between the reference plates that detect motions of the plates in the horizontal plane. Most of the length of each laser interferometer line of site is in vacuum. Corrections are applied to the laser metrology data for changes in the optical path length due to variations of the temperature of the transmissive optics (and the variations in temperature, pressure, and relative humidity of the small remaining air paths).

Together, these various subsystems contain 56 laser interferometers that will allow a continuous measurement of the time evolution of the baselines with respect to an Earth-fixed reference system with submicron error.

Changes in the delay zero-point offset on each baseline will also be continuously monitored by infrared laser metrology beams that traverse the same optical path through the instrument as the stellar beams. The absolute value of the delay-offset will be periodically measured by fringe tracking on an internal white-light source, in a manner similar to that employed on the Mark III (Hummel *et al.* 1994).

2.3. DELAY LINES

The AOI shares with the BOA six delay lines, each providing up to 35 m of optical delay. The prototype of these delay lines was developed under contract at the Jet Propulsion Laboratory. These delay lines are similar to those used with the Mark III interferometer, but have smoother support and drive mechanisms that allow operation at the higher speeds and greater range of motion required by the AOI. As in the Mark III, the delay lines are entirely in vacuum to permit the wide observational bandwidth necessary for astrometric observations.

Within each delay line vacuum tank is a pair of precision rails upon which rolls the retroreflector cart. The retroreflector consists of a parabolic mirror with a piezoelectrically driven flat secondary mirror at the focus. Starlight enters the retroreflector in a beam parallel to, but above, the axis of the paraboloid, and exits in a parallel beam below the axis. The retroreflector is held in a flexible parallelogram frame on the cart; its position on the cart is adjusted by a voice coil. The cart itself is moved along the rails by a stepper motor on an auxiliary cart, which is coupled to the retroreflector cart by a second voice coil. The position of the retroreflector is continuously monitored by a laser metrology system. Corrections to all three positions - the cart position along the rails, the retroreflector position on the cart, and the position of the secondary mirror with respect to the primary - are generated based on the position of the central interference fringe. These delay lines can actively fringe track at typical sidereal fringe motion rates with a measured RMS smoothness of 10 nm.

In addition to introducing the correct delays, the delay lines also introduce modulations in the delays. In the initial, three-delay-line system, one-wavelength amplitude, 500 Hz triangle-wave modulations are introduced in the positions of two of the three piezo-mounted secondaries. In the six-delay-line system, multiwavelength modulations at 5 kHz will be needed to deconvolve the several baselines from each output of the beam combiner.

2.4. BEAM COMBINER

Unlike the Mark III, which was limited to tracking stellar fringes on only a single baseline at a time, the AOI will be capable of simultaneous observations on all its six available baselines. This is made possible by the larger number of delay lines employed, and by significantly improved beam combination and fringe detection systems.

After exiting the delay lines, the starlight beams from the several siderostats enter the beam combiner. The prototype beam combiner for the AOI and BOA produces three pairwise combinations of the input beams from three siderostats. The beam combiner will be upgraded by early 1995 to allow multi-way combination of light from as many as six siderostats for both astrometric and imaging observations (Mozurkewich 1994). With six input beams, each output of this combiner will have light from four siderostats. The fringe patterns on each baseline will be distinguished by modulating each delay line at a unique frequency.

2.5. FRINGE DETECTION

The output beams from the beam combiner are each dispersed in a spectrometer onto a linear array of 32 optical fibers that carry the light to each of 32 avalanche photo-diodes (APDs). The intensities in the spectral channels (spanning a wavelength range of 450 to 900 nm) are detected synchronously with the delay line modulations. The result for each baseline is a two-dimensional array of photon count data, where one dimension is channel number and the other is the delay difference between the two input beams forming that baseline. A two-dimensional Fourier transform of this array yields the delay error.

Although the beam combiner and fringe-sensor systems were originally conceived for imaging work with the BOA, they offer significant advantages for astrometric observations over the techniques employed on the Mark III. In addition to the multi-way beam combiner allowing simultaneous observations on all the available baselines, the dispersed fringe detection technique permits unambiguous central fringe identification, and better correction for atmospherically induced delay fluctuations.

3. Milestones

Construction at the interferometer site on Anderson Mesa began in September 1992 and the first phase (including all the infrastructure for the AOI) is now nearly complete (White *et al.* 1994). Access roads and the interferometer control building were completed in the spring of 1993, the optical laboratory in July, the array piers in December, and the siderostat shelters in January 1994.

The installation of those subsystems of the AOI necessary for first light was completed in August 1994. This work included the installation of the first three siderostats and the assembly and vacuum testing of the feed system. The motor and retroreflector carts were installed in the first three delay lines and tested. The beam-combiner hardware and the fringe detection electronics were also assembled and tested. The hardware and software of the interferometer control system are now essentially complete.

First light on the AOI (tracking stellar fringes on one baseline) occurred on October 28, 1994. Three-element operation is expected early in 1995. The baseline metrology system will be fully operational in mid-1995; routine astrometric observations with four siderostats will then commence.

4. Conclusion

The AOI, with a capability of milliarcsecond astrometry and a planned operational lifetime of several decades, will be capable of maintaining the optical reference frame by improving the proper motions of thousands of the brighter HIPPARCOS stars through repeated observations.

References

Armstrong, J.T. (1994) Progress on the Big Optical Array (BOA), in *Amplitude and Intensity Spatial Interferometry II*, ed. J.B. Breckinridge, *Proc. SPIE*, **2200**, 62.

Danchi, W.C., Arthur, A., Fulton, R., Peck, M., Sadoulet, B., Sutton, E.C., Townes, C.H., and Weitzmann, R.H. (1986) A High Precision Telescope Pointing System, in *Advanced Technology Optical Telescopes III*, ed. L.D. Barr, *Proc. SPIE*, **628**, 422.

Elias, N.M. (1994) Baseline Metrology System of the USNO Astrometric Interferometer, in *Amplitude and Intensity Spatial Interferometry II*, ed. J.B. Breckinridge, *Proc SPIE*, **2200**, 71.

Hummel, C.A., Mozurkewich, D., Elias, N.M., Quirrenbach, A., Buscher, D.F., Armstrong, J.T., Johnston, K.J., Simon, R.S., and Hutter, D.J. (1994) Four Years of Astrometric Measurements with the Mark III Optical Interferometer, *Astron. J.*, **108**, 326.

Mozurkewich, D. (1994) A Hybrid Design for a Six Way Beam Combiner, in *Amplitude and Intensity Spatial Interferometry II*, ed. J.B. Breckinridge, *Proc. SPIE*, **2200**, 76.

Shao, M., Colavita, M.M., Hines, B.E., Staelin, D.H., Hutter, D.J., Johnston, K.J., Mozurkewich, D., Simon, R.S., Hershey, J.L., Hughes, J.A., and Kaplan, G.H. (1988) The Mark III Stellar Interferometer, *Astron. Astrophys.*, **193**, 357.

White, N.M., Millis, R.L., Franz, O.G., Loven, J.M., Hutter, D.J., Johnston, K.J., Armstrong, J.T., and Mozurkewich, D. (1994) Progress Report on the Construction of the Navy Prototype Optical Interferometer at the Lowell Observatory, in *Amplitude and Intensity Spatial Interferometry II*, ed. J.B. Breckinridge, *Proc. SPIE*, **2200**, 242.

NEW ASTROMETRIC INSTRUMENTATION IN JAPAN

M. YOSHIZAWA
National Astronomical Observatory
Mitaka, Tokyo 181, JAPAN

1. Introduction

The meridian circle is one of the most fundamental instrument in the field of astrometry where the astronomical objects are studied observationally for positions and their changes on the celestial sphere. At the Tokyo Astronomical Observatory (= National Astronomical Observatory since July, 1988) the Gautier Meridian Circle of 1903 was used until 1982 for various international meridian circle observations like SRS and NPZT programs, as well as for observations of OB stars, and the Moon and planets.

In the field of plate astrometry the Schmidt telescope at Kiso observatory of the University of Tokyo has been used for wide-field plate astrometry in Japan since 1974. Nakamura and Sekiguchi (1993) summarizes the accuracy of the plate astrometry with the Kiso Schmidt telescope. Recently, a mosaic CCD camera of eight 1024 $\times$1024 CCDs became available for test use of CCDs in wide-field astrometry, in addition to the traditional plate materials (Kiso Observatory 1991).

2. Tokyo PMC

In 1982 a new meridian circle, the Tokyo Photoelectric Meridian Circle (Tokyo PMC) manufactured by Carl Zeiss Oberkochen, was installed at Mitaka campus. Since then the Tokyo PMC has been utilized as the main instrument for global astrometry in Japan. The idea to perform meridian observations by using a photoelectric device was first proposed with the concept of the multi-slit photoelectric micrometer (Høg 1970). The photoelectric micrometer of the Tokyo PMC is similar in the basic idea to the multi-slit micrometers, but has an oscillating V-shaped slit plate instead of a fixed multi-slit plate.

An activity report of the Tokyo PMC for the first ten years from 1982

E. Høg and P. K. Seidelmann (eds.),
Astronomical and Astrophysical Objectives of Sub-Milliarcsecond Optical Astrometry, 31–34.

to 1993 is summarized in Yoshizawa et al. (1994). The limit of magnitude of the faintest stars reached with the Tokyo PMC is $V = 12.2$ mag. The magnitude limit was too bright to perform direct meridian observations of compact extragalactic objects. After running the regular Tokyo PMC program for 5 to 6 years, the urgent task was to develope a new micrometer which can extend the magnitude limit into the fainter range than the Hipparcos/Tycho catalogs'.

3. New CCD Meridian Circle

3.1. DRIFT-SCANNING CCD MICROMETER

The objective of the development of CCD micrometer is to extend the capability of a meridian circle both in limiting magnitude and observational efficiency. The so-called drift-scanning method was adopted so that the photons from stars can be accumulated as many as possible while keeping the positional information precisely. The second model of the CCD micrometer named DISC-II (DIgital Strip scanning Ccd micrometer) consists of a CCD chip of EEV (U.K.) cooled by liquid nitrogen down to around 150 K and driving clock and A/D conversion boards. The size of the CCD is 1242×1152 pixels of 22.5μm square. It is possible for us to observe stars down to 16th mag with the Tokyo PMC equipped with DISC-II. A more detailed description of the CCD micrometer is found in Yoshizawa (1994).

3.2. REDUCTION OF RELATIVE COORDINATES

When we reduce the positions given in CCD coordinates to celestial coordinates a reference catalog is needed to find scales and orientation of the CCD coordinates. Yoshizawa (1994) gave an analysis on the use of the Guide Star Catalog (AURA 1992) for that purpose. It is found in Yoshizawa (1994) that the random errors of the GSC positions within a small region are around 0.2 arcsec, and the GSC positions can be used for determination of the scale and orientation of a CCD frame. However, there is a serious imperfection in the GSC positions because of large local systematic errors. The systematic errors must be removed when the observed positions are presented in the FK5 system.

It is possible to utilize the ACRS (Corbin and Urban 1991) or PPM (Röser and Bastian 1989) catalogs as a fainter-extended realization of the FK5 catalog. In that case caution must be paid again to the local systematic errors of the ACRS and/or PPM catalogs of the order of 0.1 arcsec (cf. Miyamoto 1994). The problem of local systematic errors of reference catalogs will be solved or diminished considerably when the Hipparcos/Tycho catalogs become available.

3.3. PERFORMANCE OF DISC-II

Shown in Tab. 1 are the performance of DISC-II measured in mas expected in single observations in right ascension and declination for stars of various magnitude ranges. The results of Tab. 1 are based on the observations made

TABLE 1. Internal error of a single observation vs. brightness of stars

Star brightness	Standard Deviation		Star brightness	Standard Deviation	
(mag)	R.A.(mas)	Dec.(mas)	(mag)	R.A.(mas)	Dec.(mas)
9.0 - 9.5	38	48	12.5 - 13.0	90	63
9.5 - 10.0	61	70	13.0 - 13.5	87	78
10.0 - 10.5	63	110	13.5 - 14.0	115	92
10.5 - 11.0	58	58	14.0 - 14.5	99	110
11.0 - 11.5	76	74	14.5 - 15.0	119	122
11.5 - 12.0	57	61	15.0 - 15.5	194	188
12.0 - 12.5	68	61			

in winter, spring, and summer seasons and for a wide variety of declination zones from $-10°$ to $+42°$. For a region closer to the zenith the internal errors become smaller in all magnitudes.

The field of view covers the width of 37 arcmin in declination. Thus, within an hour, DISC-II digitizes all the position and brightness information down to 16th mag, contained in a long strip whose area is equivalent in an equator region to 9.3 square degrees. In this size of area about 7500 stars brighter than 16th mag are expected to be found, or roughly 60,000 stars during eight consecutive hours. It is seen from Tab. 1 that single observation errors (including photon noise error plus atmospheric fluctuation effect) are smaller than 100 mas for most of stars brighter than 15th mag. Four fold observations of the same strip thus enable us to determine the positions of tens of thousands of stars with internal precision better than 50 mas.

3.4. SCIENTIFIC TARGETS

With the introduction of CCD micrometers the scientific targets of meridian observations must be examined carefully, and may be different from the traditional ones as suggested by Høg (1994). Specifically programs involving the observations of stars fainter than 12th mag will be one of the most interesting works. Firstly, such programs enable us to engage in a deep survey of proper motions of young stars and old giants for studying the kinematics of the galactic disk and halo. Secondly, such magnitude ranges are complements to the Hipparcos/Tycho catalogs and must be observed from the ground for a while. Furthermore, the accumulation of direct observations of compact extragalactic objects like QSOs yields a new definition of a stellar reference frame based on the geometrical concept of inertial system.

4. A Plan for Japanese Astrometric Satellite

The basic concept of a future astrometric satellite is the combination of CCD technologies and scanning satellite to observe a huge amount (order of 10^8) of objects that are brighter than 18th mag with internal precision better than 0.1 mas. For this end the size of satellite must have a scale larger than, or at the least comparable to, that of Hipparcos (cf. Roemer/GAIA projects; Høg 1994, Lindegren and Perryman 1994).

It is possible to launch this size of satellite by using a new satellite launcher (called M-V) of ISAS, the Institute of Space and Astronautical Science. The new launcher will attain the thrust which is enough to launch a payload of 2000 kg into LEO (Low Earth Orbit), or about 600 kg payload into a geostationary orbit. About one third of the payload can be occupied by scientific instrumentation. ISAS is going to select within the next five years several satellite projects that are going to be launched with M-V for the years 2000 to 2005.

References

AURA (1992) *The Guide Star Catalog Version 1.1.* ST ScI, Baltimore.

Corbin, T.E. and Urban, S.E. (1991) *Astrographic Catalog Reference Stars (ACRS).* U.S. Naval Observatory, Washington D.C.

Høg, E. (1970) A Theory of a Photoelectric Multislit Micrometer, *Astron. Astrophys.*, **Vol. 4**, pp. 89–95

Høg, E. (1994) A New Era of Global Astrometry and Photometry from Space and Ground, *Proceedings of the G. Colombo Mem. Conf. at Padova*, in press

Lindegren, L. and Perryman, M.A.C. (1994) A Small Interferometer in Space for Global Astrometry: the GAIA Concept, this volume.

Kiso Observatory (1991) *Annual Report of Kiso Observatory.* Inst. of Astronomy, University of Tokyo, Mitaka.

Nakamura, T. and Sekiguchi, M. (1993) Error Analysis of Full-Field Astrometry with a Schmidt Telescope, *Publ. Astron. Soc. Japan*, **Vol. 45**, pp. 119–128

Röser, S. and Bastian, U. (1989) *Catalogue of Positions and Proper Motions (PPM).* Astron. Rechen-Inst., Heidelberg.

Miyamoto, M. (1994) Galactic Kinematics on the Basis of Modern Proper Motion Data, this volume.

Yoshizawa, M. (1994) Astrometry Accuracies with Drift Scanning CCD Meridian Circle, *Celestial Mechanics*, in press

Yoshizawa, M., Miyamoto, M., Sôma, M., Kuwabara, T., and Suzuki, S. (1994) Ten years of the Tokyo Photoelectric Meridian Circle (Tokyo PMC): An activity report for the years from 1982 to 1993, *Publ. National Astron. Obs. Japan*, **Vol. 3**, pp. 289–304

LARGE FUNDAMENTAL AND GLOBAL TRANSIT CIRCLE CATALOGS

T. E. CORBIN
U. S. Naval Observatory
Washington D. C. 20392-5420

Abstract. The fundamental system is currently based on the Dynamical Reference Frame as defined by the motions of the Earth and other Solar System objects. The link to this frame has traditionally been made by the observation of these objects and of stars in absolute transit circle programs. The zero points of the link are applied to quasi-absolute catalogs, and these are combined with the absolute catalogs to define the fundamental system. The individual positions and motions of the fundamental stars are then strengthened by incorporating differential catalogs reduced to the fundamental system. The system can be extended to higher densities and fainter magnitudes by further systematic reduction and combination of differential catalogs. The fundamental system itself, however, can only be extended through a planned series of observations resulting in absolute stellar positions over a range of epochs. A new fundamental catalog being compiled at the U.S. Naval Observatory is discussed and compared with the existing standard, the FK5.

1. INTRODUCTION

The two main problems that compilers of star catalogs have always faced are, first, to form an internally consistent set of observed positions referred to standard coordinates at the mean epochs of observation, and, second, to make those positions usable at other epochs without introducing distortions or rotations. These problems have traditionally been solved by referring observations taken at many epochs to a standard frame of reference and then combining the observations to give positions at a central epoch as

E. Høg and P. K. Seidelmann (eds.),
Astronomical and Astrophysical Objectives of Sub-Milliarcsecond Optical Astrometry, 35–42.

well as proper motions. These positions and motions can then be used to represent the standard frame at all epochs.

When the standard frame is the Dynamical System, and the stellar positions and motions are linked to it through absolute and quasi-absolute observational catalogs, the result is known as a fundamental catalog.

2. TYPES OF CATALOGS

2.1. ABSOLUTE CATALOGS

Catalogs that have been observed and reduced without dependence on existing catalogs and that include observations of Solar System objects are referred to as absolute. These catalogs have been derived almost exclusively from transit circle observations.

The process of producing an absolute transit circle catalog involves two main components: observations that can be reduced to give a consistent, instrumental frame derived from independently reduced nights of observation, and observations of Solar System objects that can be used to tie the instrumental system to the dynamical one.

In order to meet the first requirements the observational program must be conducted in a way that permits each night's observations to be independently reduced to a frame that is consistent from night to night. This is achieved by accurate calibration of the instrument's mechanical configuration and orientation, independent determinations of the azimuth of the instrument, corrections to refraction based on environmental quantities at the time of observation, and reference to an initial equinox point, usually based on the current fundamental catalog. Some of the mechanical calibrations are made relatively frequently. Quantities such as the departure of the axis from the horizontal plane (level), rotation of the axis in the horizontal plane (azimuth), orientation of the lens (collimation), and direction of the vertical (nadir), need to be made every two to three hours. The bending of the tube (flexure) is regularly measured during the course of the program as well. Other measurements, such as calibration of the circles, errors in the axis pivots, and changes of scale in the micrometer (mechanical or photoelectric) can generally be performed two or three times during a program.

The mechanical calibrations serve to fix the instrument within its own frame of reference as well as motion within that frame during the course of each night's work. The link to the celestial zero points defined by the Earth's rotation and orbital motion requires additional measurements. The mechanical calibration of the azimuth mentioned above is generally made with marks which are point sources of light about 100 meters from the instrument. These can give the axial motion in azimuth but not an absolute calibration. An absolute azimuth must come from observations of circumpolar stars made above and below the pole on the same night. This process fixes both the azimuth of the instrument and the pole to which the observations in declination are referred, the IERS Celestial Reference Pole.

In right ascension the procedure is generally to reduce each night's work using a standard catalog, such as the Fifth Fundamental Catalog (FK5) Parts I and II (Fricke et al. 1988, 1991), to link the observations made each night. At the end of the program the observations are first used to derive individual and systematic corrections to the stars of the standard catalog. The tie to the Dynamical System comes from an adjustment of the equinox and equator of the catalog based on the observations made of the Solar System objects and a standard ephemeris, currently DE200 (Standish 1990). This process is much more difficult in practice than in principle. The main problem is that the observations of the stars are made at night, while the Solar System objects that carry the highest weight in the solution, the Sun, Mercury, and Venus, can only be observed during the day. The brightest stars are observed to link the day and night observations, but historically this has been the weakest part of forming an absolute catalog. Observations of the minor planets can help the situation considerably since they are observed at night, but until now they have not been included in the standard ephemerides.

Currently the accuracy that can be achieved for the various calibrations and adjustments of a night's observations on the Washington 6-Inch Transit Circle is in the 20 to 80 mas range. The quantities that are the least reliable are the day-night differences, the flexure, and the refraction. Solution for the refraction has been shown (Høg and Fabricius 1988) to be affected by the presence of internal refraction (INR) in many instruments. INR is also being investigated at the USNO as a possible source of systematic error in the observations of the Sun relative to those of the day stars. In addition, individual nights' observations can suffer from offsets due to heat islands and other sources of anomalous refraction (Hughes and Kodres 1991) With the exception of the day-night differences, improved values for these quantities can be determined at the end of the program through combined solutions.

2.2. QUASI-ABSOLUTE CATALOGS

These catalogs result from observing programs that contain the procedures described above for absolute programs except that the zero points are defined relative to some other catalog, and determination of the fundamental azimuth is not always included in the program. These catalogs have internally consistent systems, but require an external calibration source to fix the zero points. Quasi-absolute catalogs are not necessarily derived from transit circle observations, and in fact some of the best, most notably HIPPARCOS and the astrolabe catalogs, are derived from other instrumentation.

2.3. DIFFERENTIAL CATALOGS

Catalogs that have been compiled using the star positions of another catalog as reference stars are termed differential. Each night's observations are generally a mix of program stars and reference stars, and the reductions are made relative to the reference catalog. These comments apply to observations made over both wide angles by transit circles and narrow fields by astrographs.

The three basic kinds of catalogs just described are summarized in Table 1.

TABLE 1. CHARACTERISTICS OF CATALOGS

TYPE OF CATALOG	CHARACTERISTICS	CONTRIBUTION TO THE SYSTEM
ABSOLUTE	Rigid frame	
	Instrumental calibration	Zero points
	Nightly calibration	
	Fundamental azimuth	System definition
	Clock star observations	
	Linkage of nights	Individual positions
	Latitude, flexure and refraction solutions	and proper motions
	Solar System objects	
QUASI-ABSOLUTE	Rigid frame	
	Same calibrations as absolute	System Definition
	Clock star observations	
	Linkage of nights	Individual positions
	Flexure, refraction solutions	and proper motions
DIFFERENTIAL	Basic instrumental calibration	Individual positions
	Reference stars	and proper motions

3. THE WASHINGTON FUNDAMENTAL CATALOG

The FK5 represents a considerable improvement over its predecessor, the FK4, with a corrected zero point in right ascension, incorporation of new astronomical constants, and significant improvements in the individual positions and motions of the stars, especially in the Southern Hemisphere. In compiling the catalog Solar System observations from absolute catalogs were combined with other data to make the correction to the FK4 equinox. The stellar data of the absolute and quasi-absolute catalogs were then used to correct warps in the system of FK4 positions and motions. Differential catalogs were reduced to the new system to improve the positions and motions of the individual stars (Schwan 1988).

The system of the FK5 was developed by using catalogs not available at the time the FK4 was compiled to correct the FK4 system. Thus the desired continuity in the FK series was maintained. The FK5 has now served the astronomical community extremely well for the better part of a decade.

The decision to begin work on a fundamental catalog at the U.S. Naval Observatory in Washington, the WFC (Cole and Yao 1989), was based on two factors. First, it was felt that it would be desirable to compile a fundamental catalog whose zero points are defined only by direct comparison between the absolute catalogs compiled since 1900 and the standard ephemerides for Solar System objects, currently DE200. Second, there are a number of important catalogs that can be used for the compilation of the WFC that were not available when the FK5 was prepared. In particular, the USNO Pole-to-Pole program (absolute) and HIPPARCOS (quasi-absolute) can be incorporated in the WFC and, respectively, will make very strong contributions to the zero points and definition of the system. Finally, recent work at the USNO has shown that FK5 Equinox appears to be in error by about −130 mas at the epoch of the W1J00, the absolute catalog observed on the Washington 6-Inch Transit Circle from 1977 to 1982. The work has been planned to proceed in three steps which will define the zero points, the system and the individual positions and motions of the WFC.

3.1. STEP 1

The zero points of the WFC will be determined through a comparison of Solar System observations and DE200. The observational data are found mainly in the Cape, Greenwich, and Washington series of absolute catalogs observed since 1900, and in all 33 catalogs are currently being analyzed. In each case it is first necessary to ensure that the Solar System observations are on the same instrumental system as that of the stars. Each observation is then differenced from the DE200 ephemeris values at the time of transit. This provides a direct link between the instrumental system and the Dynamical System, the results of which can be applied to the observed positions of the stars.

Each of the catalogs is evaluated separately and the zero point corrections derived for the mean epoch of observation. Minor planets are to be included if observed in the catalog, but many of the earlier catalogs either have too few of these to be useful or do not include them at all. A solution is made for each of the objects separately in each of the catalogs and then combined to give final corrections to each catalog's equinox and equator. This procedure has been developed by Z. Yao at the USNO using the data of the W1J00.

When the catalogs have been adjusted to the zero points of DE200, they are combined to give stellar positions and proper motions. The zero points of these positions and motions should coincide with those of the DE200 over a wide range of epochs. The positions and motions will first be used to make a final adjustment to the equinox and equator of each of the Step 1 catalogs. This prepares each catalog for Step 2 by bringing it into coincidence with the mean zero points of the system.

3.2. STEP 2

The data of step 2 come from a combination of absolute and quasi-absolute catalogs. The same procedure that was applied to the absolute catalogs at the end of Step 1 must also be used for the quasi-absolute catalogs. This again is a rotation into the zero points of the mean system at the observed epoch of each catalog using the positions and proper motions from Step 1.

An important part of the Step 2 data will come from sources other than transit circles. First of all HIPPARCOS is expected to receive extremely high weight in both the Step 2 and 3 solutions. This will undoubtedly mean that the mean epoch of the system will be within a few years of that of HIPPARCOS. Second, there are results from astrolabes that are also expected to make a significant contribution to the quasi-absolute data of Step 2, in particular the series of observations from China and France. The quasi-absolute transit circle data come mainly from the republics of the former Soviet Union, Chile, Japan, South Africa, Australia Argentina, France, Germany, England, Denmark, Spain, and the United States.

It should be noted that, as in Step 1, the adjustments to the catalogs are rotations, and no corrections are made to the catalog systems. The combination of this considerable body of data will give positions and proper motions that will define the system of the WFC.

3.3. STEP 3

At this stage the differential catalogs are reduced using the output of Step 2. In this step the catalogs are reduced in detail to the WFC system, and these catalogs contribute only to the positions and motions of the individual stars. However, there is a large body of differential data. Many of the catalogs are at early epochs and will be important for the quality of the individual proper motions.

Each of the above steps will include faint (to 9th magnitude) as well as bright stars. Thus another important way in which the WFC will differ from FK5 will be in the magnitude range and number of stars that define the system. The FK5 has been issued in two parts. The first part contains the same stars as the FK4, and it is these stars that define the system of the FK5. The second part, The FK5 Extension, has been compiled differentially and reduced to the FK5 system. Thus the system of the FK5 is defined by the bright stars, and this system is extended differentially to the 9th magnitude. The FK5 and the WFC are compared in Table 2.

If the goal of including the fainter stars in the definition of the zero points and system of the WFC is to be realized, then there must be sufficient observational histories for these stars in Steps 1 and 2. Preparation of the database for the WFC by C. Cole at the USNO Black Birch, New Zealand, station indicates that Step 1 will contain about 15,000 stars, most of which are 7th magnitude or fainter. The first version of Step 1 is scheduled for completion in the first half of 1995. Since the final catalog will contain the stars of the AGK3R+SRS (IRS), FK5 Basic, and FK4 SUP, it will contain almost 42,000 stars, and some 85% to 90% of these will make up Step 2, or at least 35,000 stars.

TABLE 2. Comparison of FK5 and WFC

	FK5	WFC
ZERO POINTS	Corrected FK4 equinox	Defined by absolute catalogs and DE200
SYSTEM	Improved FK4	Defined by combined absolute and quasi-absolute catalogs
SYSTEM DEFINITION	Basic - bright ($m < 7$) 1,535 stars	Bright and faint 35,000 stars
SYSTEM EXTENSION	FK5 Extension (to $m = 9$) Differential 3,117 Stars	Additional 6,000 stars from step 3

4. THE FUTURE OF THE WFC

When work on the WFC based on the Dynamical System is completed, it will be necessary to consider a revision based on the extragalactic reference frame. If the HIPPARCOS proper motions can successfully maintain the HIPPARCOS frame up to 80 or 90 years from the mean epoch, then HIPPARCOS rotated into the extragalactic frame can be used to reduce the earlier catalogs to that frame. However, the proper motions should be tested first by new, high-precision results from optical interferometers. Results from such instruments as the USNO Optical Interferometer over the next two or three years should have the necessary accuracy and be far enough from the HIPPARCOS epoch to provide the necessary check. If necessary, the system of HIPPARCOS proper motions could be adjusted by incorporating these new results. At that point the whole of the WFC database could be converted and thus the highly accurate WFC proper motions would be brought into the extragalactic frame.

5. ACKNOWLEDGMENT

The conception and initial direction of the WFC project came from Dr. C. A. Smith. Until the time of his death in May of 1993 he continued to contribute valuable insights and suggestions. Those of us who continue the compilation of the WFC do so with the acknowledgment of his importance to the success of the work.

References

Cole, C. S., and Yao, Z. G. 1989, Bull. Am. Astron. Soc., 21, 1105.

Fricke, W., Schwan, H., and Lederle, T. 1988, Veroff. Astron. Rechen-Inst., No. 32, Fifth Fundamental Catalog Part I, G. Braun, Karlsruhe

Fricke, W., Schwan, H., and Corbin, T. 1991, Veroff. Astron. Rechen-Inst., No. 33, Fifth Fundamental Catalog Part II, G. Braun, Karlsruhe.

Hughes, J. A., and Kodres, C. A. 1991, In IAU Colloquium No. 112, Light Pollution Radio Interference, and Space Debris, ed. D. L. Crawford, ASP, San Francisco, p. 326.

Høg, E. and Fabricius, C. 1988, Astron. & Astrophys., 196, p. 301.

Schwan, H. 1988, In IAU Colloquium 100, Fundamentals of Astrometry, Astron. Rechen-Institut Preprint Series, No. 16.

Standish, E. M. 1990, Astron. & Astrophys., 233, p. 252.

ASTROMETRY OF LARGE SKY SURVEYS

W.F. van ALTENA, T.M. GIRARD, I. PLATAIS

Abstract. One of the major uses of the Large Sky Surveys such as the Palomar and SERC Schmidt Surveys is the derivation of positions for faint objects discovered in the course of other surveys. The HST Guide Star Catalogue has proved to be of great help in deriving positions for such objects, but the well-known, and unavoidable, systematic errors in the GSC positions limit the accuracy of the final positions. In this Review, we will describe techniques that can be used to successfully transfer a reference frame from the bright FK5 stars to the small confines of a CCD chip containing only faint objects with the goal of maintaining an accuracy of approximately 0.1 arcsecond in the final positions.

E. Høg and P. K. Seidelmann (eds.),
Astronomical and Astrophysical Objectives of Sub-Milliarcsecond Optical Astrometry, 43.

THE LICK NORTHERN PROPER MOTION PROGRAM

A.R. KLEMOLA, R.B. HANSON AND B.F. JONES
UCO/Lick Observatory
University of California, Santa Cruz, CA 95064 USA

1. Introduction

The Lick Northern Proper Motion (NPM) Program will provide absolute proper motions (referred to faint galaxies), equatorial coordinates, and two-color photographic photometry for some 300,000 stars with $8 < B < 18$ covering the 70% of the sky north of declination $-23°$. Part 1 of the NPM program (NPM1), recently completed, covers the 72% of the northern sky (899 of 1,246 fields) outside the Milky Way. Two catalogs result from NPM1: The NPM1 Catalog (Klemola *et al.* 1993a, Hanson 1993a) contains 149,000 stars. The NPM1 Reference Galaxy List (Klemola *et al.* 1993b, Hanson 1993b) contains 50,000 faint galaxies. Klemola *et al.* (1987, 1994, 1995) describe the NPM program. Hanson *et al.* (1994) describe the NPM1 Catalogs.

In the present paper we give an overview of the stellar content of the NPM1 Catalog, describe current and future applications of the NPM data, and outline plans for the Part 2 of the NPM program (NPM2) – the Milky Way sky.

2. Content of the NPM1 Catalog

The NPM1 Catalog is arranged in one-degree zones of declination, from $+89°$ to $-23°$. Each star is assigned a running number within its zone, in right ascension order. Multiple measures from overlapping fields have been averaged to give one entry per star. Where available, the AGK3 (north of $-2°$) or SAO (south of $-2°$) number is given.

The NPM1 Catalog contains three general categories of stars (Table 1). Large samples of anonymous stars, in two magnitude ranges ($10 < B < 13$ and $14 < B < 17$) were selected for use in the astrometric reductions and for statistical studies of solar motion and galactic rotation. Samples of stars

E. Høg and P. K. Seidelmann (eds.),
Astronomical and Astrophysical Objectives of Sub-Milliarcsecond Optical Astrometry, 45–48.

TABLE 1. Major Components in NPM1 Catalog

Component	Number*	Comment
Anonymous Stars	93,935	Mag. 12 & 16
Positional Reference	23,242	AGK3 & SAO
Special Stars	34,717	Various classes
Stars (total)	148,940	NPM1 Catalog

*Numbers exceed total due to duplicate entries.

from the AGK3 and SAO catalogs were used for astrometric and photometric reductions. These catalog stars will also be useful for galactic and astrometric studies. Finally, many stars of astrophysical interest, selected by Klemola from the astronomical literature and complied in the Lick *Input Catalog of Special Stars*, were measured for the NPM1 Catalog. Counts of some of the more numerous classes are given in Table 2. Certain stellar classes form the basis of studies in progress, or planned, at Lick and elsewhere. Some such studies are noted in the next section.

To enhance the utility of the NPM1 Catalog, Klemola has assembled a file giving some 42,000 cross-identifications between the NPM1 Catalog and the original source identifications for each of the 35,000 Special Stars. This *NPM1 Cross-Identifications File* will be deposited with the data centers in 1994, allowing users to select all stars of a given type from the NPM1 Catalog, as well as locating individual stars of interest.

TABLE 2. Some Classes of Special Stars in NPM1 Catalog

Class	Number	Class	Number
FBO/FBS	4805	Variables (all)	3020
White Dwarf	1405	RR Lyr	1059
Dearborn (gKM)	1846	Semi-Regular	453
Horizontal-Branch	995	Eclipsing	537
UV-Excess	309	Mira	357
Subdwarf	191	Irregular (L)	297
Metal-Deficient	169	UV Cet	131
Carbon & S	158	Irregular (I)	43

3. Applications of the NPM1 Catalog

Several immediate applications of the NPM1 absolute proper motions are already underway at Lick and in collaborations with other institutions.

Hanson (1987, 1989) used faint anonymous stars from a partial version of the NPM1 catalog, then about 60% complete, to study the solar motion and galactic rotation. These studies now continue with the full NPM1 Catalog, including the brighter anonymous and catalog stars. The catalog stars will also be used to redetermine the corrections to the constants of precession.

The NPM1 Catalog contains over 1000 RR Lyrae stars, whose proper motions will be valuable for statistical studies. Layden *et al.* (1994) have combined the NPM1 data for 200 RR Lyraes with new metal abundances and radial velocities in a major new statistical parallax luminosity calibration.

The NPM1 Catalog contains over 5000 halo population candidate stars of several types. Reduced proper motion diagrams are being used to isolate sub-samples (e.g. ~1000 field horizontal-branch stars) for galactic studies and further observations.

Another important application of the NPM1 results is to help link the Hipparcos proper motions with the extragalactic reference frame. The NPM1 Catalog has nearly 13,000 stars in the Hipparcos Input Catalogue. Intercomparison will give a major check of the Hipparcos proper motion frame.

4. Steps Toward the NPM2 Catalog (Milky Way)

Part 2 of the NPM program (NPM2) will cover the remaining 28% of the sky north of $-23°$, the Milky Way zone (347 NPM fields). Here the lack of galaxies requires an alternative means of defining the proper motion reference frame. We intend to use the Hipparcos results to supply the final correction from relative to absolute proper motions for NPM2.

Plate surveys and measurements for NPM2 will begin in 1995, using the Hipparcos Input Catalogue to select reference stars. In addition, large samples of stars will be taken from the Tycho program. We expect NPM2 will take 3–5 years to complete. As in NPM1, data reductions will proceed in step with the measurements, except that the proper motion zero-point reductions will await the availability of the final Hipparcos results at the end of 1996.

The total number of stars in NPM2 will roughly equal the 150,000 in NPM1. The the number of stars *per field* will be much higher. The Lick *Input Catalog of Special Stars* will be updated and expanded to include more fully the many classes of stars which are concentrated towards the

Galactic plane, or toward the Galactic center. Hence NPM2 will contain a much higher fraction of Special Stars than NPM1.

5. Acknowledgements

We thank the National Science Foundation for its continued support of the Lick program. Current work is supported by grant AST 92-18084.

References

Hanson, R.B. 1987, AJ, 94, 409
Hanson, R.B. 1989, BAAS, 21, 1107
Hanson, R.B. 1993a, Lick Northern Proper Motion Program: NPM1 Catalog – Documentation for the Computer-Readable Version, National Space Science Data Center Document No. NSSDC/WDC-A-R&S 93-41
Hanson, R.B. 1993b, Lick Northern Proper Motion Program: NPM1 Reference Galaxies – Documentation for the Computer-Readable Version, National Space Science Data Center Document No. NSSDC/WDC-A-R&S 93-35
Hanson, R.B., Klemola, A.R., & Jones, B.F. 1994, BAAS, 26, 898
Klemola, A.R., Hanson, R.B., & Jones, B.F. 1993a, Lick Northern Proper Motion Program: NPM1 Catalog. National Space Science Data Center/Astronomical Data Center Catalog No. A1199
Klemola, A.R., Hanson, R.B., & Jones, B.F. 1993b, Lick Northern Proper Motion Program: NPM1 Reference Galaxies. National Space Science Data Center/Astronomical Data Center Catalog No. A1200
Klemola, A.R., Hanson, R.B., & Jones, B.F. 1994, in Galactic and Solar System Optical Astrometry, eds. L.V. Morrison & G.F. Gilmore, (Cambridge University Press, Cambridge), p. 20
Klemola, A.R., Hanson, R.B., & Jones, B.F. 1995, in preparation
Klemola, A.R., Jones, B.F., & Hanson, R.B. 1987, AJ, 94, 501
Layden, A.C., Hanson, R.B. & Hawley, S.L. 1994, BAAS, 26 911

LARGE PHOTOGRAPHIC CATALOGUES

S. RÖSER
Astronomisches Rechen-Institut
Mönchhofstr. 12-14, 69120 Heidelberg, Germany

AND

V.V. NESTEROV
Sternberg Astronomical Institute, Moscow University
Universitetsky 13, 119899 Moscow V-234, Russia

Abstract. Large photographic catalogues still have their importance in modern astrometry. Old catalogues are used to derive good-quality proper motions. Present-day catalogues may be used to extend the Hipparcos and Tycho systems to fainter magnitudes.

1. Introduction

For more than a century, photographic astrometry has been *the* method for aquiring large numbers of star positions. Soon after the introduction of dry emulsion plates, photographic techniques were introduced into astronomy. At this early stage, the largest project of international cooperation in astronomy was started, the *Astrographic Catalogue* (AC), (Congrès Astrophotographique International 1887). The aim was a complete survey of the sky down to 13th or 14th magnitude with 0."3 rms error per star position. Later on somewhat smaller surveys were carried out: AGK2 (Schorr and Kohlschütter 1951) and AGK3 (Heckmann et al. 1975) on the northern hemisphere, CPC (Jackson and Stoy 1954), CPC2 (Zacharias et al. 1992) and FOKAT-S (Bystrov et al. 1989) on the southern hemisphere. A series of zonal catalogues has been produced at the Yale Observatory. In this paper we shall demonstrate that these photographic catalogues make an important contribution for the determination of precise proper motions.

E. Høg and P. K. Seidelmann (eds.),
Astronomical and Astrophysical Objectives of Sub-Milliarcsecond Optical Astrometry, 49–52.

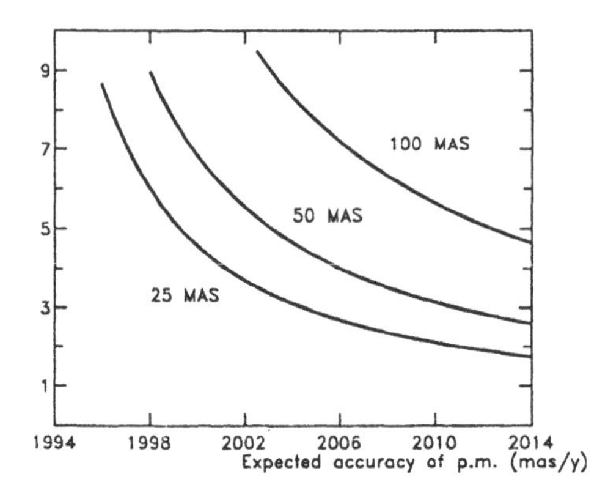

Figure 1. Expected accuracy of proper motions (mas/year), which can be achieved by combining Tycho measurements with new measurements of a given accuracy, plotted as a function of epoch. The individual curves refer to the assumed rms-errors per coordinate of these new measurements.

2. Status of astrometry today

All coordinates of celestial objects must refer to an appropriate coordinate system. This has been established in the past by the Fundamental Catalogues, the last of this series being FK5 (Fricke et al. 1988). From 1997 onwards, the Hipparcos catalogue, tied to an extragalactic reference frame, will be the reference for optical astrometry. Its coordinate system will be defined by the positions and proper motions of the 120 000 Hipparcos stars. The individual rms-errors of the Hipparcos stars will be 2 mas per coordinate in position and 2 mas/year per coordinate in proper motion. There is indication that the Hipparcos catalogue will be even more accurate.

The Tycho experiment on Hipparcos will provide us with the positions of one million stars with an accuracy of 30 mas per coordinate, but, unfortunately, the accuracy of the Tycho proper motions, 30 mas/year, will be rather poor (Høg et al. 1992).

The use of CCDs in astrometry is going to be a new tool for the acquisition of high quality positions for large numbers of stars. Van Altena (1994) describes two methods of CCD astrometry: CCDs operated in drift-scan mode are expected to obtain individual star positions with 100 mas rms-error per coordinate. If operated in the classical stare mode, 50 mas may be achieved under favourable conditions (van Altena 1994). Let us assume that the measurements can be tied to a reference system without loss of accuracy. If sky surveys of this kind should be carried out in the near future, we can ask about their impact on proper motions if the Tycho measurements are used as a first epoch. Figure 1 shows the rms-error of a proper motion per coordinate, expected from a combination of Tycho with future observations of the 100, resp. 50 mas accuracy level. If e.g. a survey with 50 mas would be performed in 2002 the resulting proper motions would only be good to 5.5 mas/year. Were it possible to increase the accuracy to 25 mas in the future, observations from 2002 onwards would make sense. However, 100 mas and even 50 mas observations before 2010 cannot compete with old photographic catalogues for the determination of proper motions, as we will show below.

3. The role of old photographic catalogues

Old photographic catalogues usually do not reach an accuracy comparable to modern ones or CCD techniques. However, the latter still have to prove their potential, when actual data will be published. Old photographic catalogues profit from large epoch differences when combined with Tycho. In Table 1 we present the performances of selected photographic catalogues. AGK2 in the north, and CPC2 in the south yield fairly good proper motions when combined with Tycho. However, they only supply positions for roughly half of the Tycho stars. Better results are expected when combining Tycho with AC. Essentially all Tycho stars are contained in AC, and proper motions with rms-errors of 2 - 3 mas/year are expected, with the Paris and Sydney zones at the extremes (see Table 1).

TABLE 1. Accuracy of proper motions to be achieved by combining the catalogue in column 1 with the Tycho observations. The columns have the following meaning: catalogue, number of stars, decl. zone, epoch, rms-error of a catalogue position in mas (in the case of AC two measurements enter a catalogue position), expected rms-error of a proper motion in mas/year.

Catalogue	Number of stars	Declination Zone	Epoch	ε_p	ε_μ
AGK2	180 000	+90 -02	1930	180	3.0
CPC2	250 000	+00 -90	1967	100	4.3
AC-Paris	340 000	+24 +17	1897	160	1.7
AC-Sydney	380 000	-52 -65	1907	300	3.6

More than 100 years after the start of the AC project, it can now be fully exploited because all the published measures have been put into machine-readable form (Nesterov et al. 1990). The project of combining AC with Tycho is described in Röser and Høg (1993) and is called *Tycho Reference Catalogue* (TRC).

Table 2 describes the situation regarding proper motions to be expected within the next 3 years. Progress will be in number and accuracy. Large photographic catalogues contribute to the last two lines of this table. The catalogue (GSC-AC) is described by Röser et al. (1994). Proper motions for all AC stars are determined using a systematically improved version of the *Guide Star Catalogue* (GSC) as second epoch. These catalogues of proper motions will have an impact on stellar kinematics up to a distance of one kpc, although the $sub-milliarcsecond$ regime is not yet reached.

TABLE 2. Accuracy of proper motions to be expected within a few years. ε_μ is the rms-error of proper motion. Unit: mas/year.

Catalogue	Number of stars	ε_μ
Hipparcos	120 000	1.5
TRC	1 million	2-3
GSC-AC	4 million	4-6

So far we have not touched the problem of the systematic accuracy of the photographic catalogues. In the TRC project (Röser and Høg 1993) the AC has to be reduced to the Hipparcos system. This work will be carried out at Astronomisches Rechen-Institut after the Hipparcos catalogue will be available. Magnitude equations and plate distortions will be investigated and removed. GSC 1.0 is known to show serious systematic errors (Taff et al. 1990). Röser et al. (1994) describe a new method to reduce GSC 1.0 to the PPM system. The first author is presently studying magnitude dependent errors in GSC 1.0 by comparing overlapping GSC plates.

References

Bystrov N.F., Polozhentsev D.D., Potter Kh.I., Zalles R.F., Zelaya J.A., Yagudin L.I., 1989, *Astron. Zh.*, **66**, 425.

Congrès Astrophotographique International, 1887, Gauthier-Villars, Paris.

Fricke W., Schwan H., Lederle T. et al., 1988, Veröff. Astronomisches Rechen-Institut Heidelberg No. 32. G. Braun, Karlsruhe.

Heckmann O., Dieckvoss W., Kox H., Günther A., Brosterhus E., 1975, *AGK3*, Hamburg-Bergedorf.

Høg E., Bastian U., Egret D. et al., 1992, *Astron. Astrophys.*, **258**, 177.

Jackson J., Stoy R.H., 1954, Annals of the Cape Observatory, Vols. **17** to **22**, London

Nesterov V.V., Kislyuk V.S., Potter Kh. I., 1990, In IAU Symposium 141, J.H. Lieske and V.K. Abalakin, Kluwer Academic Publishers, Dordrecht, 482

Röser S., Høg E., 1993, In Workshop on Databases for Galactic Structure, eds. A.G. Davis Philip, B. Hauck, A.R. Upgren, L. Davis Press, Schenectady, N.Y.

Röser S., Bastian U., Kuzmin A.N., 1994, in IAU Coll. 148, in press.

Schorr R., Kohlschütter A., 1951, *AGK2*, Verlag der Sternwarte, Hamburg-Bergedorf.

Taff L.G., Lattanzi M.G., Bucciarelli B. et al., 1990, *ApJ*, **353**, L45.

van Altena W.F., 1994, In IAU Symposium 161, H.T. MacGillivray et al., Kluwer Academic Publishers, Dordrecht, 193

Zacharias N., de Vegt C., Nicholson W., Penston M., 1992, *Astron. Astrophys.*, **254**, 394.

1.2 SPACE MISSIONS

PROPERTIES OF THE HIPPARCOS CATALOGUE: WHAT CONFIDENCE CAN WE HAVE IN THE FINAL DATA?[1]

L. LINDEGREN
Lund Observatory
Box 43
S-22100 Lund
Sweden

Abstract. The foreseen complexity of the the ESA Hipparcos mission led to the establishment of two independent scientific consortia (FAST and NDAC) for the parallel processing of all main mission data. The validation of Hipparcos data through internal and external checks and by interconsortia comparisons is outlined. Examples are given based on preliminary solutions. The various checks indicate that the overall accuracy and reliability of the Hipparcos Catalogue will be extremely high.

1. Status of the Hipparcos Catalogue

The Hipparcos satellite delivered high-quality astrometric and photometric data during 37 of the 40 months from December 1989 to March 1993. This information is now being processed by the three 'data reduction consortia' TDAC (for the Tycho mission of about a million stars) and FAST and NDAC (for the main mission of 118,000 selected stars). According to current planning the Hipparcos Catalogue will be released around 1997.0, but selected data will be available a year earlier for approved investigations.

The FAST and NDAC consortia have both completed their independent solutions (here called F30 and N30[2]) based on the first 30 months of data. In this paper I discuss the properties of the final catalogue (HIP) as can be extrapolated from these solutions and the provisional catalogue called H30,

[1]Based on observations with the ESA Hipparcos satellite
[2]Not to be confused with the 'normal system' N30 created by H.R. Morgan in 1952. The Hipparcos N30 catalogue, like F30 and H30, will never be published.

E. Høg and P. K. Seidelmann (eds.),
Astronomical and Astrophysical Objectives of Sub-Milliarcsecond Optical Astrometry, 55–60.

obtained by the averaging and merging of F30 and N30. Table 1 summarizes the main characteristics of H30 and HIP.

TABLE 1. Median formal standard errors in the provisional catalogue H30, and extrapolated to the final catalogue (HIP). The apparent non-improvement from H30 to HIP is due to the assumption that the additional stars all fall above the median accuracy of HIP

Quantity	H30	HIP	Unit
Number of stars (approx)	107,500	118,000	-
Mean epoch, J1900+	91.14	91.24	year
$\sigma_\alpha \cos\delta$	1.17	1.17	mas
σ_δ	0.96	0.95	mas
$\sigma_{\mu\alpha} \cos\delta$	1.65	1.51	mas yr^{-1}
$\sigma_{\mu\delta}$	1.35	1.22	mas yr^{-1}
σ_π	1.45	1.45	mas

The standard errors in Table 1 are *formal* ones obtained from the least-squares solutions of the data reductions. The actual or *external* errors of any experiment are usually larger due to modelling errors, which in general also introduce biases, or *systematic* errors, of the parameters. External and systematic errors are most directly checked by comparison with independent data, but relatively little data of sufficient accuracy is available. Internal checks and interconsortia comparisons can on the other hand be performed on a massive scale, but provide only indirect evidence of data properties. Both kinds of tests are briefly reviewed in Sections 3 to 5.

2. Parallel Reductions

The expected complexity of the Hipparcos science data processing motivated ESA to appoint two independent groups, or consortia, to perform the complete reduction from raw satellite data to the end product. The groups are known by their acronyms FAST (Kovalevsky et al., 1992) and NDAC (Lindegren et al., 1992). Although the general principles of the reductions are the same, each group has developed its own algorithms, software implementation and processing strategy. As for the modelling of the instrument and mission, the goal has been to take into account all known effects that may correspond to a shift of 0.1 mas or more in a single observation.

The main interaction between FAST and NDAC has been to compare intermediate results at various levels of the processing, first using simulated input data and later the real satellite data. As a direct consequence of this activity a number of errors and shortcomings have been identified and corrected on both sides, and some algorithms have been considerably improved.

In retrospect it appears that the strategy of parallel reductions has been very successful, and the most important guarantee that the final Hipparcos Catalogue will fully reflect the high quality of the satellite data.

3. Internal Checks

Within each consortium the consistency between the satellite data and the reduction model can be checked by a number of statistical tests and solution experiments. The simplest tests concern the size and distribution of the residuals and possible correlations with other parameters such as magnitude, colour and position on sky. In general such tests have yielded satisfactory results if allowance is made for a small fraction of outliers. The use of robust estimation techniques is essential at all stages of the reductions.

Another, very powerful data consistency test is to divide the observations into two similar but distinct data sets, which are reduced independently. The differences of the resulting astrometric parameters, divided by the square root of the sum of the variances from the two solutions, should ideally follow the centred unit normal distribution $N(0,1)$. In reality small deviations are found which are used to adjust the *a priori* weights assigned to the input data. By this procedure the formal standard errors can be made consistent with the data on a global scale.

A quite different kind of test is to expand the reduction model by introducing additional *ad hoc* parameters that are estimated along with the 'normal' parameters. Judiciously chosen such parameters may reveal subtle instrumental effects which would otherwise be masked by a much larger observation noise. In the NDAC sphere solution, for example, the temporary introduction of a sixth unknown for each star (in addition to the five astrometric parameters) has allowed a detailed mapping of the instrument chromaticity as function of colour index. This new information will be incorporated in an improved reduction model for the final solution.

Another example of *ad hoc* parameters are the global harmonic coefficients Γ_2 to Γ_{12} introduced to model possible thermally induced periodic variations of instrument parameters (Lindegren et al., 1992). In the N30 solution these coefficients are estimated with formal standard errors below 0.01 mas, and none is found to exceed 0.025 mas in absolute value. This indicates an extremely good short-term stability of the basic angle, and indirectly supports the assertion that the parallaxes are absolute at the 0.1 mas level.

4. FAST–NDAC Comparisons

There are 95,579 stars in common between the two provisional catalogues F30 and N30. After global re-orientation and rotation of the catalogues to a common system the median differences, taken in the sense N30 − F30, are -0.064 ± 0.004 mas in δ, -0.024 ± 0.005 mas/yr in μ_δ and -0.048 ± 0.005 mas in π. (In α and μ_α the median differences are by definition zero after transformation to the common system.) Dividing the material according to hemisphere, magnitude and colour does not produce any larger differences. A significant colour-dependent rotation difference is however noted in the proper motions, possibly caused by inadequate chromatic modelling. Full clarification of this effect is required, if it is still present in the 37 months data, before the results can be accepted as final.

The rms differences are typically about the same size as the formal standard errors in Table 1. Due to the varying precision it is more useful to consider the normalized differences, $\Delta p^* = (p_\mathrm{N} - p_\mathrm{F})/(\sigma^2_{p\mathrm{N}} + \sigma^2_{p\mathrm{F}})^{1/2}$, where p stands for each of the five astrometric parameters. Normal probability plots of the five quantities Δp^* are shown in Fig. 1. The distributions are remarkably close to normal, especially for the parallaxes. The rms values of the normalized differences are less than unity because of the positive correlation of the errors in p_N and p_F. The circumstance that they are still relatively large for the preliminary catalogues (~ 0.7) indicates that there is still room for improvement by iteration and by averaging the consortia results.

5. External Comparisons

Positions. Standard catalogues of stellar positions are totally inadequate for checking the positions in H30. Two sources do however provide a small number of very accurate positions useful for a comparison with Hipparcos.

The USNO Mark III optical stellar interferometer (Hummel et al., 1994) has yielded positions for a small number of bright stars in the 10–20 mas accuracy range. A comparison of arc length differences with respect to H30 for nine stars shows very good agreement within the quoted standard errors, thus confirming the accuracy of both the Mark III results and H30 at the 10 mas level.

VLBI observations of some radio stars relative to quasars already yield astrometric parameters at the sub-mas accuracy level. Preliminary VLBI data for seven radio stars in H30, observed as part of the programme to link Hipparcos to the extragalactic VLBI system (Lestrade et al., in preparation), shows rms residuals of about 2.1 mas. The expected rms residual, from the formal standard errors, is about 1.2 mas. This agreement is quite satisfactory in view of the problematic nature of these stars (binaries, pos-

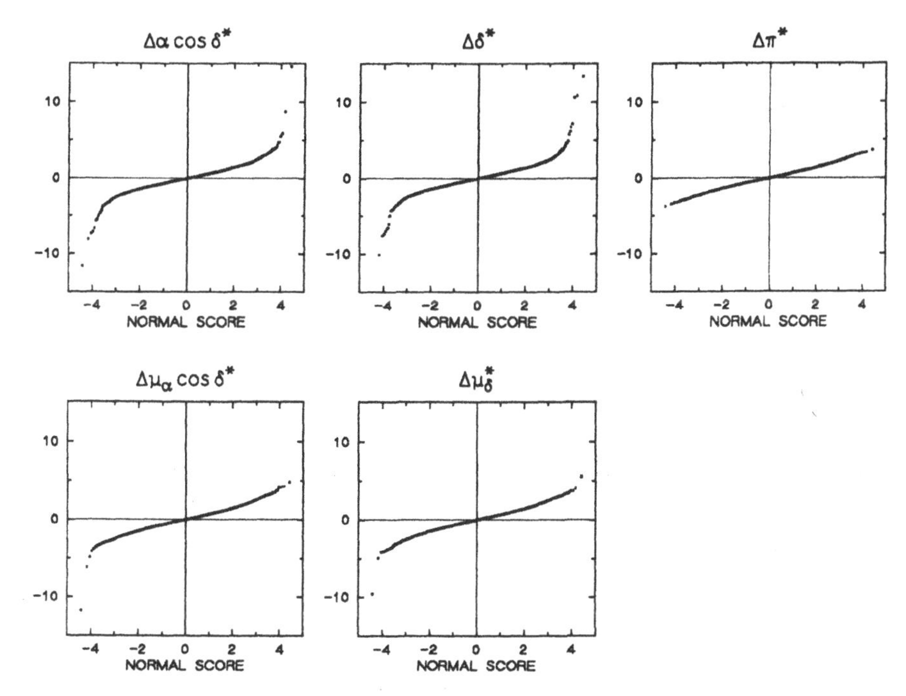

Figure 1. Normal probability plots of the normalized differences Δp^* between N30 and F30 (see text for explanation).

sible offset of visual photocentre from radio centre) and the preliminary state of the data.

Proper Motions. The FK5 proper motions, with estimated standard errors of about 1 mas/yr, should provide a valuable test of Hipparcos proper motions. The distribution of differences is strongly non-gaussian, but there appears to be a 'core' of standard width $\sim$ 2.5 mas/yr, to be compared with the expected rms difference of 1.7 mas/yr. It is likely that systematic (mostly zonal) errors of the FK5 proper motions, and perhaps underestimated individual errors, account for a major part of the discrepancy.

The previously mentioned preliminary VLBI proper motion data for radio stars give a very good agreement, about 1.1 mas/yr, fully consistent with the formal errors.

Parallaxes. The parallaxes offer by far the most substantial confirmation of the quality of the preliminary Hipparcos data. The main reason for this is that photometric and spectroscopic parallaxes become very precise for sufficiently distant stars, and there is a sizable number of distant stars in the Hipparcos programme. An extensive comparison with cluster distances and spectroscopic, photometric and dynamical parallaxes (Arenou et al., in preparation) gives strong evidence that the zero point of the preliminary Hipparcos parallaxes is correct to within 0.1 mas, and that the formal standard errors in H30 are correct to within a few percent.

These conclusions are supported by comparisons with some of the most accurate ground-based parallaxes obtained both by classical optical methods (Harrington et al., 1993) and from the VLBI observations of radio stars (Lestrade et al., in preparation). The external errors of the H30 parallaxes have also been studied from the distribution of negative values (Lindegren, in preparation) resulting in an estimated external/internal error ratio very close to unity.

6. Conclusions

Although the whole set of Hipparcos data is not yet analysed, several kinds of internal and external checks have been applied to the preliminary catalogues. All tests indicate a very high degree of internal consistency and agreement with the best available external data. In particular it appears that the Hipparcos parallaxes are absolute at least to the level of 0.1 mas, and that the standard errors in Table 1 essentially represent the true accuracies of the data.

Acknowledgements

This overview is based on the work of many others. I wish to thank all colleagues who contributed, perhaps without knowing it.

References

Harrington, R.S., Dahn, C.C, Kallarakal, V.V., Guetter, H.H., Riepe, B.Y., Walker, R.L., Pier, J.R., Vrba, F.J, Luginbuhl, C.B., Harris, H.C., Ables, H.D. (1993) U.S. Naval Observatory photographic parallaxes. List IX, *Astron. J.*, **105**, pp. 1571–1580

Hummel, C.A., Mozurkewich, D., Elias II, N.M., Quirrenbach, A., Buscher, D.F., Armstrong, J.T., Johnston, K.J., Simon, R.S., Hutter, D.J. (1994) Four years of astrometric measurements with the Mark III Optical Interferometer, *Astron. J.*, **108**, pp. 326–336

Kovalevsky, J., Falin, J.L., Pieplu, J.L., Bernacca, P.L., Donati, F., Froeschlé, M., Galligani, I., Mignard, F., Morando, B., Perryman, M.A.C., Schrijver, H., van Daalen, D.T., van der Marel, H., Villenave, M., Walter, H.G. (1992) The FAST Hipparcos Data Reduction Consortium: overview of the reduction software, *Astron. Astrophys.*, **258**, pp. 7–17

Lindegren, L., Høg E., van Leeuwen, F., Murray, C.A., Evans, D.W., Penston, M.J., Perryman, M.A.C., Petersen, C., Ramamani, N., Snijders, M.A.J., Söderhjelm, S., Andreasen, G.K., Cruise, A.M., Elton, N., Lund, N., Poder, K. (1992) The NDAC Hipparcos data analysis consortium. Overview of the reduction methods, *Astron. Astrophys.*, **258**, pp. 18–30

TYCHO ASTROMETRY
FROM 30 MONTHS OF SATELLITE MISSION [1]

E. HØG
Copenhagen University Observatory - NBIfAFG
Østervoldgd. 3
DK-1350 Copenhagen K, Denmark

Abstract. The Hipparcos satellite's star mapper gives photon counts in two spectral channels simultaneously, close to Johnson B and V. The transit times and the signal amplitudes for each star across two groups of four slits are derived and used for astrometry and photometry, respectively, and this constitutes the Tycho project. The present paper describes results of Tycho astrometric data processing, leading from the transit times to the astrometric parameters of the Tycho stars.

Some 30 months of Tycho observations, i.e. about 80 percent of the Hipparcos-Tycho mission, have been used to produce a working catalogue of Tycho positions, proper motions and parallaxes of a million stars. The external errors of this preliminary catalogue have been determined by comparison of 98 000 stars common with a preliminary, but much more accurate Hipparcos catalogue. External systematic errors of positions and annual proper motions are less than 0.5 milliarcsecond (mas) and the accidental errors per star are about 30 mas rms at $V = 10.5$ mag, the median magnitude of the catalogue. It is concluded that a satisfactory accuracy has been achieved.

1. Introduction

The Tycho experiment of the Hipparcos satellite (Perryman et al. 1992) supplies astrometric and photometric information. The data are analyzed by the Tycho Data Analysis Consortium (TDAC) as described by Høg

[1]Based on observations made with the ESA Hipparcos satellite

E. Høg and P. K. Seidelmann (eds.),
Astronomical and Astrophysical Objectives of Sub-Milliarcsecond Optical Astrometry, 61–68.

et al. (1992). The work is carried out in cooperation with the two Hipparcos data reduction consortia, NDAC and FAST, and with the Hipparcos Input Catalogue Consortium (INCA). The preparatory tasks of the astrometry processing are carried out at the TDAC sites in Tübingen, Heidelberg, Strasbourg, Cambridge, Copenhagen and Lund, leading to the final astrometric processing at Copenhagen. The astrometric processing is the subject of the present paper.

The photon counts from the B_T and V_T channels of the star mapper are called the Tycho counts or, sometimes more briefly, the B and V counts. The complete counts are subject to a number of processing steps under the responsibility of TDAC. The transit time of a star crossing the star mapper slits (Fig. 1) is the basic astrometric datum obtained. In the astrometry process, the observed transit time for the added $B + V$ counts is compared with the predicted transit times obtained from the satellite attitude, a provisional grid calibration, and the approximate positions of the stars in the Tycho Input Catalogue (the 'Tycho stars'), leading to an identification of the transiting star and subsequently to an updating of its astrometric parameters. The mathematical formulation of the astrometry task is outlined by Høg et al. (1994) where the analysis of the first half of the mission is discussed.

At present, the geometry of the grid has been calibrated by means of a preliminary Hipparcos catalogue and provisional positions, proper motions and parallaxes of stars have been determined. These calculations are based on 'identified transits' from the first 30 months of the Tycho mission data. Final astrometric parameters will be obtained from the complete 36 month mission data after further verification of the processing, based on the combined Hipparcos catalogue, H30, derived by the FAST and NDAC consortia from 30 months of the mission (Lindegren et al. 1994). This will result in positions, proper motions and parallaxes for some 1 000 000 stars with an accuracy of about 30 mas rms for stars of $V = 10.5$ mag, the median magnitude of the catalogue. The standard error increases roughly by a factor of two per magnitude, for magnitudes fainter than $V = 7$, cf. Table 1, discussed in Sect. 4. The Tycho positions will therefore be much more accurate at the epoch of observation than in ground-based catalogues. But the observation error of proper motions and parallaxes is in fact larger than the true values of these quantities, for the majority of faint Tycho stars. Double stars with separations larger than about 2 arcsec will be resolved.

The photometric processing resulting in the Tycho magnitudes B_T, V_T, T, where the T magnitude is derived directly from the $B_T + V_T$ counts, is described by Großmann et al. (1994).

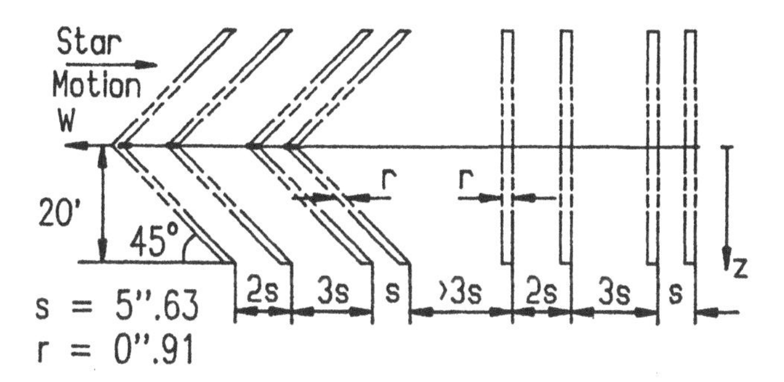

Figure 1. The star mapper slit system consists of the 'chevron' group of four inclined slits, and the group of 'vertical' slits, perpendicular to the motion of the stars.

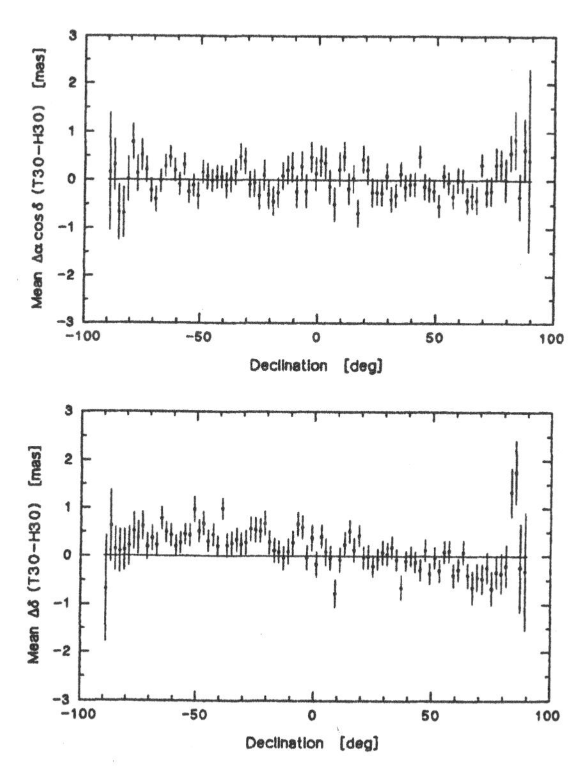

Figure 2. Systematic differences between preliminary Tycho and Hipparcos positions, T30−H30, at the epoch 1991.25 and as function of declination. The differences are generally less than 0.5 mas and are due to errors in the Tycho positions. The systematic trend in the lower plot is significant, but it is expected to disappear in the final Tycho positions. Similar plots versus right ascension (not shown here) appear as non-systematic as the upper plot.

TABLE 1. Expected accuracy of the final Tycho Catalogue. External accidental errors of positions, annual proper motions and parallaxes, are given as function of visual magnitudes, for *bona fide* single stars of $B - V = 0.7$. Systematic errors will be less than 0.5 mas

V	[mag]	2.5-6.0	7.0	8.0	9.0	10.0	11.0	12.0
σ	[mas]	2.5	3.5	5	10	20	50	80

2. Attitude, grid and star catalogues

The satellite attitude data contains the pointing directions of the two optical axes of Hipparcos, as functions of time during the mission. The attitude used in the 'identified transits' was produced by NDAC, based on provisional grid calibration parameters and a Hipparcos catalogue, N18, of positions, proper motions and parallaxes obtained from the first 18 months of the mission.

The grid calibration in Tycho astrometry shall provide corrections to the calibration parameters used in the NDAC attitude so that the corrected identified transits agree with the more accurate Hipparcos catalogue, H30. It is, however, noted that the T30 catalogue discussed here was based on a calibration using the N18 catalogue for positions and proper motions, and only ground-based parallaxes were used. This problem is discussed and probably resolved in Sect. 4.

The star mapper grid consists of accurately etched slits, Fig. 1, on a glass substrate. The light from sky and star(s) passing the slits is collected by two photomultipliers, B and V. The grid geometry, related to the optical axes given by the attitude data, is slowly varying with time due to gradual changes of the mechanical structure of the satellite. In practice, the grid geometry is specified by (a) a table of 'medium scale irregularities', in fact constant for the whole mission, and (b) sets of 11 calibration parameters, each derived from the observations of a period of 24 hours. The calibration parameters are defined as zero-points and orientations of the slit groups, and are in fact corrections to the parameters used in the attitude production.

The final grid calibration shall make use of the H30 catalogue of positions, proper motions and parallaxes of about 100 000 single stars. The calibration provides a data base of parameters covering the mission and supplying the relevant parameters by interpolation for every observation, during the subsequent Tycho catalogue production. This ensures that the positions, proper motions and parallaxes in the final Tycho catalogue are obtained in the coordinate system of the Hipparcos catalogue, with systematic deviations less than about 0.5 mas, as we shall demonstrate in the following.

A number of consecutive or 'iterated' Tycho catalogues have been produced so far:

(1) The Tycho Input Catalogue of 3 million stars (Egret et al. 1992) was obtained from ground based catalogues, with position errors about 1 arcsec;

(2) the Tycho Input Catalogue Revision of one million stars (Halbwachs et al. 1992 and 1994) was obtained from analysis of the first 12 months of the Tycho mission, with position errors about 60 mas;

(3) the T30a of one million stars, obtained from 30 months of the mission and containing positions with typical errors of 30 mas;

(4) the T30b, an iteration of T30a to make sure that all positions have converged;

(5) the T30c, an iteration of T30b, for the first time including proper motions and parallaxes; and

(6) the T30d, the first iteration of proper motions and parallaxes. This is the catalogue discussed presently and it is briefly called T30.

The four catalogues T30a-d are iterative improvements of the Tycho Input Catalogue Revision by means of the first 30 months of observations. Thus, all about 100 million observation equations are calculated in each iteration, corrected for improved astrometric parameters from the previous solution. The observation residual relative to the previous solution is also calculated and is used to reject outlying observations, with residuals larger than about $3\sigma_{\rm obs}$.

3. Comparison with the Hipparcos catalogue, H30

The T30 catalogue has been compared with the H30 catalogue, by way of the 98 000 stars in common. The latter catalogue has standard errors less than 2 mas for all astrometric parameters, i.e., negligible compared to the errors of the T30, except for very bright stars. Therefore, *external* systematic and accidental errors of T30 can be derived directly from the differences. Before differences T30−H30 are formed the T30 positions and proper motions are rotated (by about 60 mas and 1 mas/yr) to coincide with the H30 system.

The differences for positions and parallaxes are plotted versus declination and magnitude in Figs. 2-4. Plots of annual proper motions are numerically very similar to the Fig. 2a of positions, and are therefore not included. Plots of the differences as function of right ascension and plots on two-dimensional sky maps were also studied, but are not included because no new information was obtained. The Hipparcos magnitude, Hp, is related

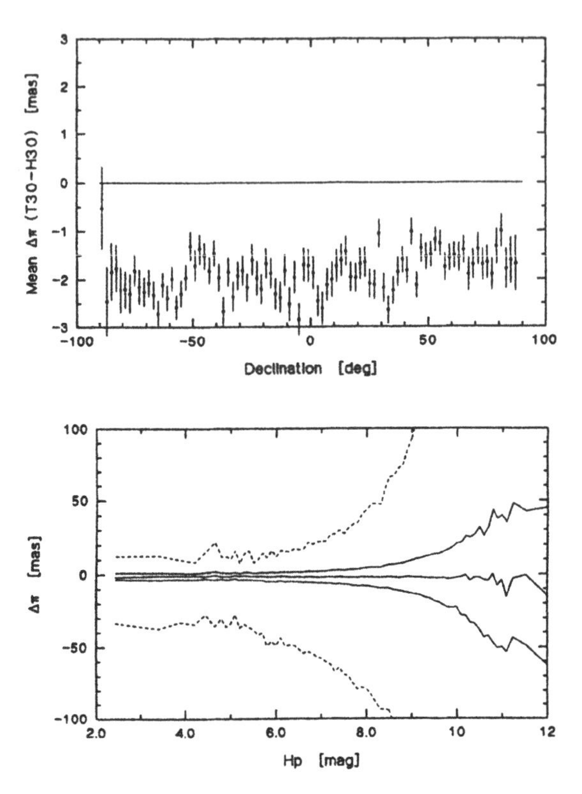

Figure 3. Differences between parallaxes, T30−H30. (a) Systematic differences versus declination. A plot versus right ascension shows the same systematic difference, about −1.9 mas. (b) The curves mark the sextiles, as explained in Fig. 4. The systematic difference seems to be slightly more negative at faint stars.

to the Johnson and Tycho magnitudes as $Hp = V_{\mathrm{J}} + 0.16 = V_{\mathrm{T}} + 0.08$ for a star of the typical colour index $(B - V)_{\mathrm{T}} = 0.7$.

4. Discussion

The external systematic errors of T30 are simply obtained as the systematic difference T30−H30 because the H30 has a superior accuracy. They appear from Fig. 2 to be less than 0.5 mas for positions. They are in fact also less than 0.5 mas/yr for proper motions. The only systematic feature worth noting is the statistically significant trend in $\Delta\delta_{\delta}$ (Fig. 2b) from +0.5 to −0.5 mas. This small trend might perhaps be neglected, but it is expected to disappear in the final Tycho catalogue because a new grid calibration will be used, based on the H30 catalogue, instead of the NDAC 18 month Hipparcos catalogue, used in the present calibration.

The systematic error of parallaxes in T30 appears from Fig. 3a to be

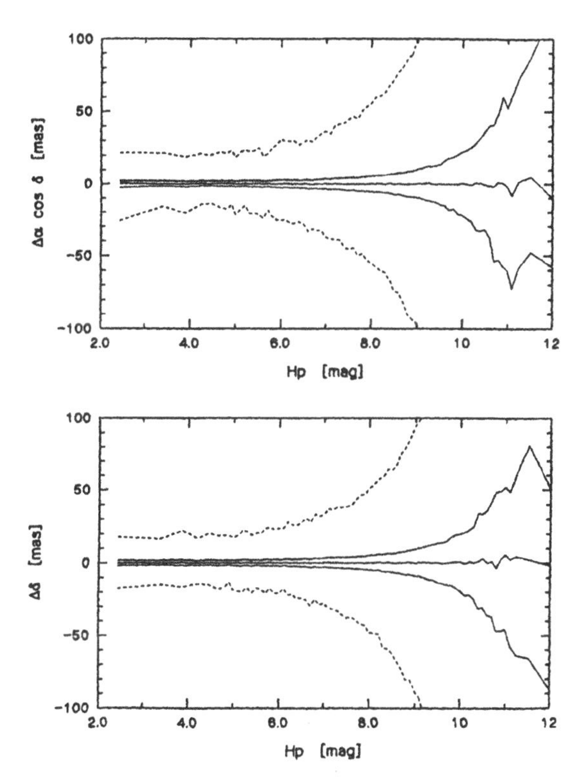

Figure 4. The solid curves mark the 1st, 3rd and 5th sextiles of the differences between preliminary Tycho and Hipparcos positions, T30−H30, of 98000 common stars calculated for bins in Hp of at least 100 stars or of 0.1 mag width (whichever gives the wider bin). The magnitude $V_J \simeq Hp - 0.16$ mag. The dashed curves are the 1st and 5th sextiles in a 10 times expanded scale. *Note*: the 3rd sextile is the same as the median. The 1st and 5th sextiles roughly correspond to −1 and +1 s.d. for a normal distribution.

very significant, equal to −1.9 mas. This error is expected to disappear in the final catalogue because the calculated Tycho parallax will be used to calculate the residual for the subsequent iterative catalogue, and this was not done at present. This means that a zero parallax was in fact assumed, although the parallaxes of the Tycho stars have a median value about +10 mas. This has presumably given a slight bias, resulting in the systematically too small Tycho parallaxes.

The external accidental errors of positions and parallaxes as functions of magnitude can be derived from the 1st and 5th sextiles of the distribution functions in Figs. 3b and 4, as explained in the note to Fig. 4. Corresponding plots for the annual proper motions are nearly identical and are therefore not included. This results in Table 1 giving the expected final Tycho accuracies. The table takes into account the expected improvement from 36 months of observations, 6 months more than in the T30 catalogue. It is noted that the errors are only about 2.5 mas for stars brighter than $V = 6.0$, comparable with the Hipparcos accuracy. The external accidental errors of T30 obtained in this paper agree within less than 10 per cent with the formal standard errors, derived from the normal equations and a statistical model of the photon noise plus empirical values for the attitude noise (about 7 mas for vertical slits and 35 mas for inclined, cf. Høg et al. 1994). The photon noise model is derived from basic statistical principles, as discussed by Makarov and Høg (1994).

It appears from the table that the errors of Tycho proper motions for stars fainter than 8 mag are so large that positions calculated for epochs a few years from the mean epoch 1991.25 will be rather uncertain. This problem will be solved by the proper motions with accuracy about 3 mas/yr, expected in the planned Tycho Reference Catalogue (Röser and Høg 1993) obtained after a new reduction of the AC catalogue plates by means of the Hipparcos catalogue.

Acknowledgements: Many scientists have been involved in the Tycho data reduction and deserve credit for their contributions and encouragements. Those presently contributing to the astrometric part are: F. van Leeuwen (Cambridge), E. Høg, V.V. Makarov, H. Pedersen, C. Petersen (Copenhagen), U. Bastian, P. Schwekendiek (Heidelberg), L. Lindegren (Lund), J.L. Halbwachs (Strasbourg), K. Wagner, A. Wicenec (Tübingen).

References

Egret D., Didelon P., McLean B.J., Russell J.A., Turon C., 1992, A&A 258, 217

Großmann V., Bässgen G., Grenon M., Grewing M., Høg E., Mauder H., Snijders M.A.J., Wagner K, Wicenec A., 1994, A&A in preparation

Halbwachs J.L., Høg E., Bastian U., Hansen P.C., Schwekendiek P., 1992, A&A 258, 193

Halbwachs J.L., Bässgen G., Bastian U., Egret D., Høg E., van Leeuwen F., Petersen C., Schwekendiek P., Wicenec A., 1994, A&A 281, L25-L28

Høg E., Bastian U., Egret D., Grewing M., Halbwachs J.L., Wicenec A., Bässgen G., Bernacca P.L., Donati F., Kovalevsky J., van Leeuwen F., Lindegren L., Pedersen H., Perryman M.A.C., Petersen C., Scales D., Snijders M.A.J., Wesselius P.R., 1992, A&A 258, 177

Høg E., Bastian U., van Leeuwen F., Lindegren L., Makarov, V.V., Pedersen H., Petersen C.S., Schwekendiek, P., Wagner, K., Wicenec, A., 1994, A&A in preparation

Lindegren, L., Röser, S., Schrijver, H., Perryman M.A.C., van Leeuwen, F. et al, 1994, A&A in preparation

Makarov, V.V., Høg, E., 1994, this volume

Perryman M.A.C., Høg E., Kovalevsky J., Lindegren L., Turon C., Bernacca P.L., Crézé M., Donati F., Grenon M., Grewing M., van Leeuwen F., van der Marel H., Murray C.A., Le Poole R., Schrijver J.H., 1992, A&A 258, 1

Röser, S., Høg, E., 1993, 'Tycho Reference Catalogue: A Catalogue of Positions and Proper Motions of one Million Stars.' In: Workshop on Databases for Galactic Structure. Ed.: A.G. Davis Philip, B. Hauck and A.R. Upgren. Van Vleck Observatory Contr. No.13, 137. L. Davis Press, Schenectady, N.Y.

DOUBLE STAR ASTROMETRY WITH THE HIPPARCOS DATA

F. MIGNARD
Observatoire de la Côte d'Azur/CERGA,
Av.Copernic 06130 Grasse, France

Abstract. As we approach the final processing of the observations carried out by HIPPARCOS, in particular for the double and multiple stars, it is possible to provide reliable statistics on the number of such objects detected and on the quality of the relative and absolute astrometry and photometry. About 24 000 stars have been recognized as non-single, including 11 000 already known as double and multiple before the mission and 13000 discovered by Hipparcos. Also, a subset of 16 000 stars among the 24 000 have been successfully solved for their relative coordinates (position angle and separation) with an accuracy in the range of 3 to 30 mas, including 7000 new double stars. I outline in this paper the principle of the internal recognition procedure and present some statistics on the solution.

1. Introduction

The Hipparcos mission was primarily dedicated to the production of an astrometric catalogue for the position, proper motion and parallax of about 118000 stars, with a nominal accuracy of 2 mas and 2 mas/year. The data processing scheme was designed at an early level in the mission definition and was optimized for single stars. However the determination of the parallax of the double stars, in particular for the thousand or so orbital double stars, is of the utmost importance for the determination of the stellar masses. It was then compelling to account for theses complex sources in the data processing. This raised essentially two questions :

- How to be sure that the observed signal is that of a non single star? Although the Hipparcos Input Catalogue includes such information, it was soon recognized that it was too incomplete and that an independent recognition procedure had to be worked out. Scientifically this

E. Høg and P. K. Seidelmann (eds.),
Astronomical and Astrophysical Objectives of Sub-Milliarcsecond Optical Astrometry, 69–76.

proved to be a real bonanza with the detection of thousands of new double stars.

- Once a program star is known to be double or multiple, it must be processed differently from the single stars because of lack of simple relationship between the phase of the signal and the position of the components on the sky. A by-product of this process is the measurement of the separation and position angle of about 16000 double stars, along with the difference of magnitude between the two components.

2. Double star recognition

Observations were carried out over essentially two continuous stretches from December 1989 to July 1992 and then from November 1992 up to March 1993 with a smaller daily coverage in the latter interval due to attitude recovery problems. Altogether this amounts to 37 months of data with a continuous flow of 24000 bits per seconds. On the average a particular star has been observed about 120 times over this period, with considerable variation from star to star with the ecliptic latitude.

The basic signal is described in Kovalevsky et al. (1992), and for double stars in Mignard et al. (1992). The expected signal is properly calibrated for a point source, it is then possible to study the deviation of the actual signal from this ideal and construct statistical tests to reject the hypothesis that the source is single. In particular, one of the tests is based on the photometry computed on the average intensity on one hand, and on the amplitude of modulation on the other hand. The two magnitudes so derived are equal for a single star, whereas the latter is larger than the former for a double star. A plot of the detections is shown in Fig.1 for about 10000 stars as a function of the color. The diagram is very asymetrical, with virtually no negative values, but those accounted for by the photon noise. In the upper part of the diagram ($\Delta m > 0$) all the data points with a magnitude difference larger than 0.025 mag very likely correspond to non single stars, primarily doubles. In this plot one also notes that, thanks to a careful calibration, there is no systematic effect, even for the reddest stars. The statistics of detection are given in table 1, and are split into three categories. A star is categorized as single when the signal has never shown any significant departure from the single star signal over the 120 passages. On the other hand it is said to be double or multiple beyond a certain statistical threshold which rules out it being single. There remains an intermediary class for which no firm decision can be made with the data collected by Hipparcos.

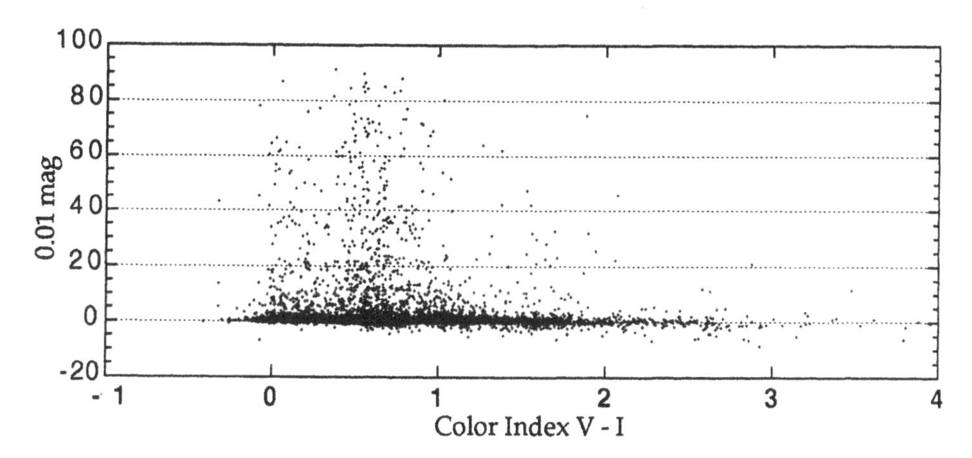

Figure 1. Photometric detection of double stars

TABLE 1. Double star detection

Category	N
Single	83000
Dubious	10500
Double or multiple	24500
Total	118000

3. Double star solution

3.1. THE RELATIVE ASTROMETRY

We cannot in this short paper describe the convoluted track that leads from the reception of the raw data to the astrometric solution. This has been described in several publications, for example in Kovalevsky et al. (1992), Perryman et al. (1992). In the case of double star a fitting of the grid signals, collected over all the available transits, to the double star parameters provides the separation, the position angle and the magnitude difference. Then at a latter stage one can find the absolute position and parallax.

One must stress however that not every double star is seen as such by Hipparcos, in the same way as there are detection limits in ground based observations. Regarding the small separations, the diffraction by the Hipparcos window sets the resolving power at $\rho > 0.10-0.15''$. So stars with smaller separations are termed single, although they might be double, and actually quite often are. For these stars, close to the lower limit of resolution, the magnitude difference is strongly correlated to the separation, and in general cannot be large. On the other hand at the upper limit, the detector allows a detection of the companion, provided it is at a distance from the primary less than 25 to 30 arcsec. This instrumental limitation is due to the using of a image dissector tube (IDT) which diminishes the available light of an image not properly centered on the detector, in such a way that a star lying 20 arcsec off center appears 1.5 mag fainter, nearly 3 mag at 25 arcsec and 6 mag at 30 arcsec. For these stars the detection limit for the magnitude difference is about $\Delta m \approx 4$ mag, including the attenuation effect. So a pair with a separation larger than 25 arcsec generates a signal hardly discernible from that of a single star and is categorized as single on the sole basis of Hipparcos observations.

The statistics of the solution are given in table 2 and refer to the solution before a full synthesis between NDAC and FAST is completed. The figures are then not final, but should be within 10% of the catalogue content. Table 2 refers to the 24500 stars considered non-single in Table 1. A complete solution means that both the relative astrometry and photometry have been obtained from the transit data. It may seem that the solution is made more likely when a star is known to be double from pre-launch observations. This reflects the fact that the range of known visual binaries, which make most of the content of the Input Catalogue double stars, comprises stars rather easily solved with the Hipparcos observations, with separations larger than 0.2 arcsec and magnitude difference less than 3 mag. So out of the 11000 stars detected, about 90 % are properly solved. The situation is very different for the new double stars, which are on the average of smaller separation and/or larger magnitude difference. So they are detected as non single from the signal, but it is more difficult to retrieve the two components.

TABLE 2. Double star solution statistics

	Total	Known DS	New DS
Complete Solution	16500	9500	7000
Photometry only	8000	1500	6500
Total	24500	11000	13500

The distribution of the separations are given in Fig. 2 and 3, respectively for the known and new double stars. The difference is conspicuous, with the large fraction of close binaries belonging to the set of Hipparcos detected

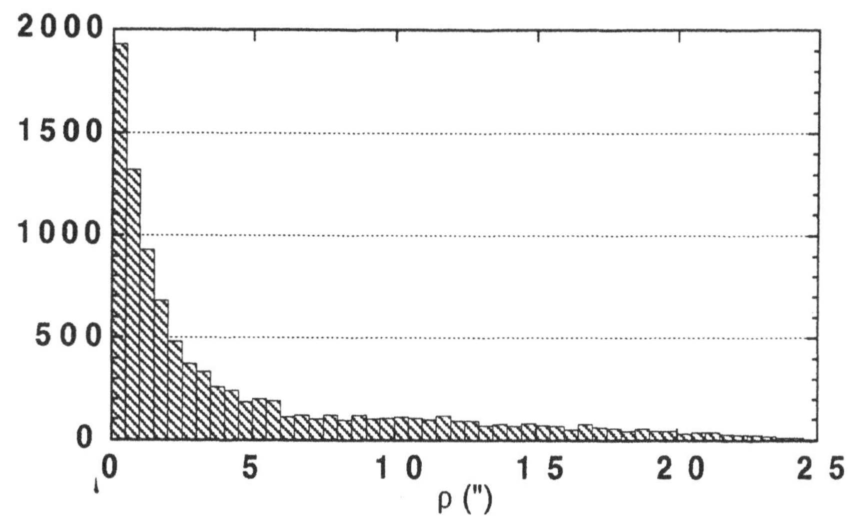

Figure 2. Distribution of the separation for 9000 known double stars

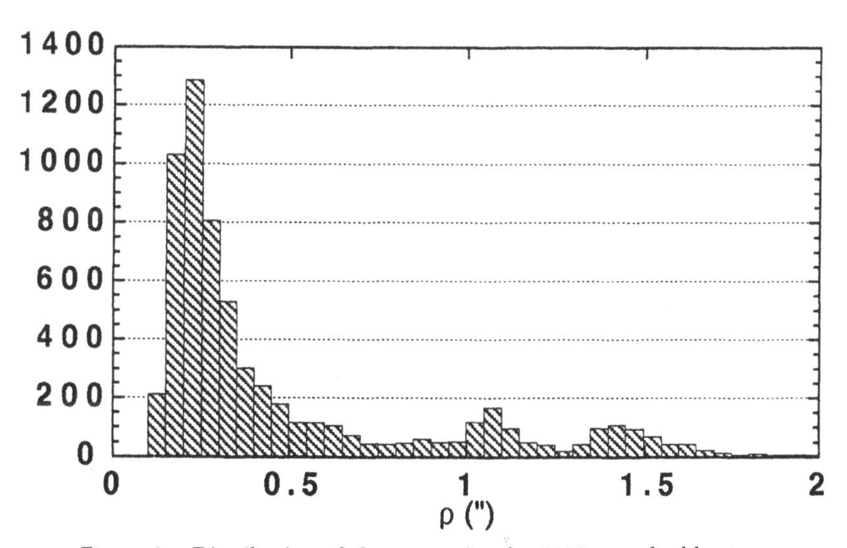

Figure 3. Distribution of the separation for 6000 new double stars

double stars. The two bumps at $\rho \approx 1.2$ arcsec and $\rho \approx 1.4$ arcsec are instrumental effects. The first is caused by grid step errors and is probably not real while the second illustrates a better detection efficiency at the separation of 1.4 arcsec.

Comparison to external data of similar accuracy can be made with the observations of double stars made by speckle interferometry and published

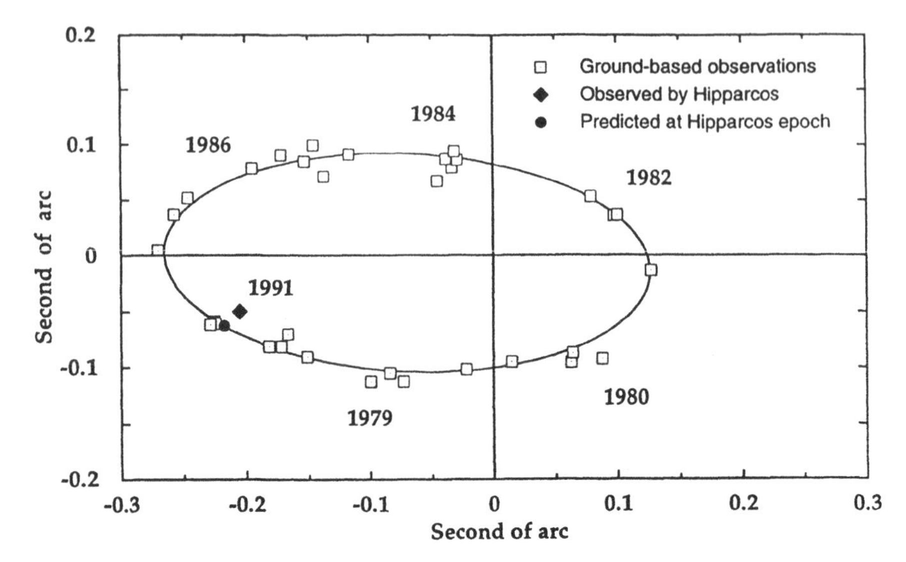

Figure 4. Comparison of an Hipparcos observation to ground based measurements

by McAlister and Hartkopf (1988). One such example is shown in Fig.4 for a star observed regularly over an orbit. The Hipparcos data point matches the predicted position within the expected accuracy of 12 mas on each direction for this object. From our analysis, it seems that the Hipparcos separation are free of systematic error above one mas, at least for separation larger than 0.2 arcsec. For smaller separation this is not yet fully known.

3.2. THE ABSOLUTE ASTROMETRY

The parameters of the double stars are then used to correct the signal from the effect brought about by the duplicity and solve for the absolute astrometry in the same way as for a single star. In an alternative approach used by the NDAC consortium, one fits in a single step both the relative and absolute astrometry and photometry to the grid signal. Thus the methods used by the two consortia differ markedly, to such an extent that the comparison of the results is really meaningful, since the systematic errors that may remain in the solutions are unlikely to be similar. This proved to be a very efficient tool to detect and correct shortcomings in either software. At the end the analysis of the scatter between the two solutions is probably a good indication of the external error. In no case can it supersede the comparison to fully external data, provided they are of comparable accuracy with Hipparcos.

The main result is illustrated in Fig. 5, with the difference in the parallax NDAC - FAST for about 3000 double stars solved by both groups and for

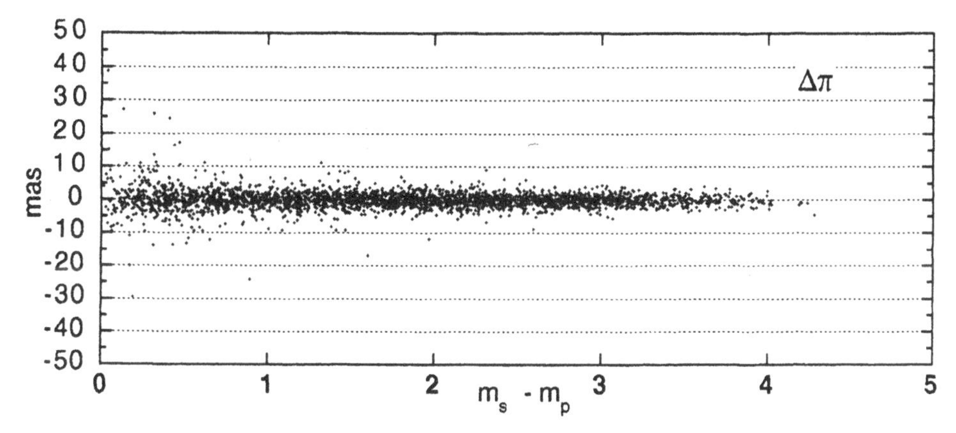

Figure 5. Parallax difference in mas between NDAC and FAST for 3000 double stars

which there is a good agreement in the solution found for the separation and position angle. In general the difference in the parallax is below 3.5 mas with a smaller scatter when the magnitude difference is large, that is to say when the single star conditions are recovered. This plot indicates the level of degradation in the parallax solution between the single stars, with a scatter of about 2 mas and the true double star, with a scatter about 1.5 times that of single stars.

One may fear that the astrometry might be very different in NDAC and FAST for stars processed as doubles by one group and as singles by the other. It turns out that such a situation occurs only for stars at the detection margins, that is to say with a signal not far from the single star signal. So the modeling error is not large on the astrometric parameters. More specifically those stars have usually very small separations and the position produced by Hipparcos will be that of the photocenter, which is exactly what we arrive at by neglecting the duplicity, provided $\rho < 0.35$ arcsec. Experiments have shown that the effect on the parallax is then negligible.

4. Conclusion

The mass processing will continue within each consortium until mid 1995 to produce the final iterated solutions based on the whole data set, including all the single and multiple stars. Then the two independent solutions will be merged into the single Hipparcos catalogue both for the astrometry and photometry.

Acknowledgements

The double star reduction of the Hipparcos data involves many scientists of different countries. They are gratefully acknowledged for their contribution : L. Lindegren, S. Söderhjelm (Sweden), J.L. Falin, M. Froeschlé, J. Kovalevsky, C. Martin (France), M. Badiali, P.L. Bernacca, L. Borriello,D. Cardini, R. Pannunzio, G. Prezioso, A. Spagna (Italy), H. Bernstein (Germany), P. Lampens (Belgium).

References

Kovalevsky J. et al. (1992) The FAST Hipparcos Data Reduction consortium: overview of the reduction software *A&A* ,**Vol. no. 258** , pp. 7–17.

McAlister H.A. and Hartkopf W. I. (1988) Second Catalog of Interferometric Measurements of Binary Stars, Chara Contribution **no. 2**

Mignard F., Froeschlé M., Badiali M., Cardini D., Emanuele A.,Falin J.L., Kovalevsky J. (1992) Hipparcos double star recognition and processing within the FAST consortium,*A&A* ,**Vol. no. 258** , pp. 165-172.

Perryman M.A.C. et al. (1992), In orbit performance of the Hipparcos astrometry satellite, *A&A*, **Vol. no. 258**, pp. 1–6.

Turon C. et al. (1992) , The Hipparcos Input Catalogue , ESA SP-1136.

SCIENTIFIC CONTENTS OF THE HIPPARCOS INPUT CATALOGUE [1]

C. TURON, A.E. GOMEZ, D. MORIN
DASGAL/URA 335 du CNRS, Observatoire de Paris-Meudon
92195 Meudon Cedex, FRANCE
e-mail: turon@obspm.fr

1. Introduction

The Hipparcos Input Catalogue –hereafter called HIC– (Turon *et al.* 1992a, Turon *et al.* 1994) contains the observing programme of the Hipparcos mission: 118 000 preselected stars, 48 minor planets and three satellites of major planets. These objects were selected on the grounds of more than 200 scientific programmes proposed by the world-wide astronomical community, and dealing with a large variety of astronomical and astrophysical topics. It contains the most up-to-date, comprehensive and homogeneous data and information related to these programme stars, collected during the years 1981-1990 by the INCA Consortium (Hipparcos INput CAtalogue consortium).

The accuracy of the astrometric and photometric parameters obtained from the analysis of the first 30 months of Hipparcos data already provides a check the various sources of ground-based data contained in the Hipparcos Input Catalogue. It also provides a statistical estimation of what can be expected for the various types of scientific programmes proposed for observation by Hipparcos, in terms of accuracy of the astrometric parameters with respect to spectral types, luminosity classes, types of objects, etc.

[1] Based on observations made with the ESA Hipparcos satellite, and on work performed within the INCA, FAST and NDAC Consortia.

E. Høg and P. K. Seidelmann (eds.),
Astronomical and Astrophysical Objectives of Sub-Milliarcsecond Optical Astrometry, 77–82.

2. The quality of the astrometric and photometric data in the Hipparcos Input Catalogue

The data collected in the HIC were obtained from extensive cooperative programmes of compilation and new observations or measurements performed by the INCA Consortium during the years 1981-1990. A complete description of the data included in the HIC can be found in the introduction to the printed, tape and CD-ROM versions of the Hipparcos Input Catalogue (Turon *et al.* 1992a, Turon *et al.* 1992b, Turon *et al.* 1994) and in Perryman and Turon 1989.

2.1. POSITIONS

The positions available from the 30-month combined sphere solution (H30) are of an unprecedent accuracy as compared with ground-based positions: median rms of 1.1 and 0.9 mas for ecliptic longitude and latitude respectively. These values are about 1/20 of the rms claimed by the most accurate ground-based catalogues, and confirm that these preliminary Hipparcos results are perfectly suitable for our present purpose. The distribution of differences between H30 and IC8 (version of HIC used for satellite observations) is illustrated in Table 1. Only single entries were considered for the comparisons by source catalogue.

In order to characterise the scatter of these differences, a *width* based on distribution percentiles is used as dispersion estimate instead of a rms scatter, which is too sensitive to heavy tail distributions and outliers.

2.2. PROPER MOTIONS

The provisional proper motions available from the 30-month combined solution show median rms of 1.5 and 1.3 mas/yr for the proper motion in longitude and in latitude respectively. These values will be improved in the final data reduction, including the 37 months of Hipparcos data. Therefore, the results shown here are only tentative. However, for all source catalogues with the exception of the FK5, these preliminary results should be significant. The distribution of differences between H30 and IC8 is illustrated in Table 2.

TABLE 1. Median and width (arcsec) of distribution of differences between H30 and IC8 in $\alpha \cos \delta$ and δ (equinox J2000.0, epoch J1990.0)

	Nb of	$\alpha \cos \delta$		δ	
	stars	median	width	median	width
All stars	107 495	-0.020	0.275	0.070	0.285
Single entries	104 045	-0.020	0.275	0.070	0.280
Joint entries	3 450	-0.060	0.330	0.080	0.310
FK5	1 940	0.005	0.080	0.040	0.095
Meridian	6 283	0.000	0.108	0.040	0.140
IRS *	12 629	-0.030	0.185	0.080	0.200
SSSC	3 851	-0.270	0.255	0.170	0.248
PPM *	41 448	0.030	0.270	0.040	0.285
CPC2 *	9 585	-0.240	0.320	0.160	0.295
Plate meas.	12 199	-0.030	0.355	0.080	0.405

* provisional versions

TABLE 2. Median and width (arcsec) of distribution of differences between H30 and IC8 in $\mu_\alpha \cos \delta$ and μ_δ

	Nb of	$\mu_\alpha \cos \delta$		μ_δ	
	stars	median	width	median	width
single entries	90 619	0.000	0.007	0.001	0.008
FK5	1 940	0.000	0.003	0.000	0.003
Meridian	5 273	0.000	0.002	0.000	0.005
IRS *	7 558	0.000	0.006	0.001	0.006
SSSC	3 440	0.000	0.011	0.007	0.009
PPM *	41 451	0.001	0.006	0.000	0.007
CPC2 *	5 784	-0.005	0.006	0.003	0.005
SAO	11 605	0.000	0.017	0.000	0.017

* provisional versions

2.3. PHOTOMETRY

Within the frame of the Hipparcos Input Catalogue preparation, Hp magnitudes (magnitudes in the wide bandpass of the Hipparcos photometric system, Grenon *et al.* 1992), had to be predicted from ground-based multicolour photometry, in order to optimise the observing time allocation, and to check the star observability. In addition, photoelectric photometry was collected to produce high accuracy standards for the on-orbit calibration of the detection chain. The accuracy of Hipparcos photometric measurements is at the level of a few millimagnitudes. The comparison of these measurements with those included in the HIC is illustrated in Table 3.

TABLE 3. Median and width (magnitudes) of distribution of differences between H30 and HIC in Hp magnitudes

	Number of stars		median	width
all stars	117 610	J	0.000	0.098
photoelectric B and V	44 483		0.000	0.016
photoelectric V	13 298		-0.031	0.064
heterogeneous V *	45 273		0.022	0.240

* colour derived from spectral type and extinction model (Arenou *et al.* 1992)

2.4. CONCLUSION

The astrometric and photometric data included in the Hipparcos Input Catalogue are largely within the original specifications of ESA: ± 0.3 arcsec on 1990 positions (to be compared with a specification of 1.5 arcsec); ± 0.25 mag on Hp magnitudes and ± 0.02 mag for more than a third of the catalogue (to be compared with a specification of 0.5 mag).

3. Expectations with respect to scientific objectives

The 118 000 stars were selected on the grounds of more than 200 scientific programmes proposed by the world-wide astronomical community, and dealing with a large variety of astronomical and astrophysical topics: fundamental astronomy (reference systems; dynamics of the stellar system; astrometry); stellar physics (stellar luminosities; fine structure of the HR diagram; tests of the physics involved in internal stellar structure models; stellar evolution; determination of stellar ages, masses, and diameters; etc); galactic physics (stellar populations; chemical, kinematical and dynamical evolution of our Galaxy; galactic rotation and potential; stellar orbits; etc); cosmic distance scale (new calibrations of photometric and spectroscopic parallaxes, of period-luminosity relations; detailed study of the effects of chemical abundances –metallicity, Helium, Oxygen, etc– on the location of the main sequence or giant branch).

Details on the iterative process which led to selection of the final observing programme are given in Turon *et al.* 1992c. The results obtained from the analysis of the first 30 months of Hipparcos data already give a very good confidence in the possibilities expected from the final Hipparcos Catalogue:

- with respect to fundamental astrometry: the Hipparcos reference frame will be characterised by an accuracy of 1.5 mas for positions at epoch 1991.25 and of 1.5 mas/yr for proper motions; a link to an extragalactic system with an accuracy of 0.5 mas on the orientation R, 0.3 mas/yr on the rotation R'; a stellar density of 3 stars per square degree.
- with respect to stellar physics: the present accuracy already allows a direct determination of luminosities in the range of M_V -4 to 14 mag; about 200 stars with a relative accuracy of the trigonometric parallax better than 1% (a few giants and subgiants, from F2 to K2; dwarfs from A0 to M5; one white dwarf); 3 300 stars with a relative accuracy of the trigonometric parallax better than 5% (a few supergiants; about 200 giants, from B7 to M4; about 300 subgiants, from B2 to K4; about 2 000 dwarfs, from B3 to M8; 10 white dwarfs, 50 orbital systems; a few pulsating stars; etc); 13 000 stars with a relative accuracy of the trigonometric parallax better than 10% (up to B0 stars on the main sequence; about 1 000 giants, from B5 to M5).
- with respect to galactic physics and cosmic distance scale determinations: 13 000 stars with a relative accuracy of the trigonometric parallax better than 10%; 40 000 stars with a relative accuracy of the trigonometric parallax better than 20%; 36 000 stars with a relative

accuracy of the proper motion better than 5%; 82 000 stars with a relative accuracy of the proper motion better than 20%; direct distance determination of about 20 galactic open clusters of various ages and metallicities; direct distance determination of a few pulsating stars, supergiants stars and bright giants, planetary nebulae, novae, etc; discovery of new variable stars.

4. Conclusions

The results obtained with only 30 months of data already allow to deal with all the categories of astrometric and astrophysical programmes submitted by the worldwide astronomical community in 1982 and the Hipparcos Consortia members in 1993. The analysis of the totality of the 37 month mission will still improve the accuracy of the astrometric parameters and the coherence of the whole solution. Moreover, the 'Hipparcos sphere' will be linked to extragalactic reference system, leading to an absolute determination of the proper motions. In parallel, a new reduction of the photometric data is being performed, using an improved set of standard stars and calibrations.

References

Arenou, F., Grenon M., Gómez, A.E. (1992) *A&A*, **258**, 104
Grenon M., Mermilliod, J.C., Mermilliod, M. (1992) *A&A*, **258**, 88
Perryman M.A.C., Turon C. (eds), 1989, ESA SP-1111, Vol. II
Turon, C. *et al.* (1992a) The Hipparcos Input Catalogue, ESA SP-1136.
Turon C. *et al.* (1992b), Bull. Inform. CDS, **41**, 9
Turon, C., Gómez, A., Crifo, F., Crézé, M., Perryman, M.A.C., Morin, D., Arenou, F., Nicolet, B., Chareton, M., Egret, D. (1992c), *A&A* **258**, 74
Turon, C., Morin, D., Arenou, F., Perryman M.A.C. and Priou D. (1994) CD-ROM version of the Hipparcos Input Catalogue, Turon, C. ed

ASTROMETRIC AND ASTROPHYSICAL INSIGHTS INTO THE HIPPARCOS DATA QUALITY[1]

M.A.C. PERRYMAN
Astrophysics Division, European Space Agency
ESTEC, 2200AG Noordwijk, The Netherlands

Abstract. Internal error estimates and external verifications indicate that median errors for the 120000 stars in the Hipparcos Catalogue, for each of the five astrometric parameters, will be in the range 1–1.5 mas. This paper illustrates some of the statistical investigations that have been conducted so far, including comparisons with catalogues of ground-based positions and proper motions, the structure of the Hertzsprung-Russell diagram, results of the reanalysis of meridian circle and photographic plate data using the positions determined from the satellite, and expected results on double and multiple stars and photometric variability.

1. Introduction

The final stages of the Hipparcos Catalogue construction are well advanced; the final catalogue is expected to be completed and distributed to 'internal proposers' at the end of 1995, and to be made fully available to the scientific community at the end of 1996. While astrophysical exploitation or interpretation of the astrometric and photometric data will only commence after the catalogue has been finalised, some statistical comparisons have already been made with existing catalogues, and certain other tests undertaken, with the intention of validating the quality of the preliminary Hipparcos data.

Details of the progress of the data analysis for the Hipparcos and Tycho experiments can be found in Kovalevsky *et al.* (1994) and Høg *et al.* (1994) respectively.

[1]Based on observations made with the ESA Hipparcos satellite

E. Høg and P. K. Seidelmann (eds.),
Astronomical and Astrophysical Objectives of Sub-Milliarcsecond Optical Astrometry, 83–86.

2. Illustrations of the Hipparcos Data Quality

A detailed comparison of the preliminary Hipparcos Catalogue with positions and proper motions from the FK5 and PPM Catalogues can be found in Lindegren *et al.* (1994). The most rigorous external verifications of the astrometric parameters (Lindegren 1994) are given by comparisons with (a) the USNO (optical) Mk III interferometer and VLBI interferometry (for positions); (b) the FK5 Catalogue (for proper motions; and (c) a variety of possible zero-point determinations (for parallaxes). These investigations suggest that the formal standard errors derived in the sphere solutions by the FAST and NDAC Data Analysis Consortia are likely to be very close to the true external errors of the derived astrometric parameters. In the following, other tests or applications of the Hipparcos data which provide further evidence for the overall data quality are reported.

2.1. MERIDIAN CIRCLE ANALYSIS

Réquième *et al.* (1994) report the re-analysis of meridian circle observations (from the Bordeaux and Carlsberg instruments) using preliminary positions from the 18-month Hipparcos Catalogue, H18, compared with reductions made using the FK5 Catalogue. A significant decrease in the residuals is found in both α and δ. The form of the residuals in α as a function of declination differ between the Carlsberg and Bordeaux instruments, and amount to as much as 30 mas, suggesting that the origin of the differences cannot lie within the preliminary Hipparcos Catalogue, but arises rather from small defects in the instruments or their calibration, masked up until now by the errors in the astrometric reference positions. Again, different signatures in δ of up to 50 mas suggest that modelling of refraction may be improved once better reference star positions become widely available.

2.2. PHOTOGRAPHIC PLATE ANALYSIS

A variety of photographic plate reductions have already been carried out using the preliminary Hipparcos positions; these confirm the previously-held suspicions that the limited precision of the available reference catalogues has compounded the difficulties of determining the proper choice of plate model. For example, Platais *et al.* (1994) have completed a preliminary analysis of the plates from the Yale/San Juan Southern Proper Motion program, using the preliminary 30-month Hipparcos Catalogue, H30, to provide a reference system with negligible random errors. They have inferred the presence of a consistent magnitude equation and certain significant cubic terms, concluding that the Hipparcos positions offer a very powerful tool for detecting systematic errors in wide-field photographic astrometry.

2.3. THE HERTZSPRUNG-RUSSELL DIAGRAM

One of the primary goals of the Hipparcos mission was to furnish high quality trigonometric parallaxes for tens of thousands of stars, in order to refine the detailed structure of the observational Hertzsprung-Russell diagram and to extend the determination of absolute magnitudes to stars significantly more luminous than $M_v \sim 0$ mag. With a significant fraction of parallaxes having standard errors below 1 mas, and systematic errors at 0.1 mas or better, distance estimates to many tens of thousands of stars in the Hipparcos Catalogue within 100 pc will have an accuracy of better than 10%. A dramatic indication of the quality of the parallaxes is given by the HR diagram constructed from the preliminary 30-month Hipparcos Catalogue, H30. A presentation and discussion of this diagram is given by Perryman *et al.* (1994).

3. Miscellaneous Applications of the Hipparcos Reference Frame

3.1. THE OPTICAL/RADIO EMISSION FROM SN1987A

Preliminary Hipparcos positions have been used by Reynolds *et al.* (1994) to allow registration of high-resolution optical and radio images of SN1987A to the 100 mas level. The significance of the problem is illustrated by the radio-optical overlay published by Staveley-Smith *et al.* (1992), which shows a 0.5 arcsec displacement between the radio and optical centroids. In a careful succession of reference frame links, Reynolds *et al.* (1994) have been able to show that this mis-registration was the result of an inadequate (optical) astrometric reference frame, at least in the vicinity of SN1987A; they are thus able to conclude that the radio emission originates from the interaction between the whole of the expanding shock wave and the surrounding medium.

3.2. EPHEMERIDES OF ASTEROIDS AND COMETS

While the use of the preliminary Hipparcos Catalogue has been largely restricted to 'internal' tests and verifications, preliminary positions have been circulated to groups requiring timely availability to improved astrometric data—thus the Hipparcos results have been used by ESO observers to improve the prediction of the time of impact between Comet Shoemaker-Levy 9 and Jupiter (West & Hainaut, private communication), and were used to assist navigation of the Galileo satellite for its encounter with the asteroid Ida (Owen & Yeomans, 1994; see also ESA Bulletin No. 77, p143).

3.3. METRIC DETERMINATION

The Hipparcos data have been reduced within a relativistic framework, including accounting for gravitational light-bending by the Sun (and Earth). Unlike previous determinations of light-bending, either in the optical or in the radio (see, e.g., Soffel 1989) the regions over which the effect is significant for Hipparcos are no longer restricted to a few solar radii, but extend to most of the celestial sphere. Within NDAC, L. Lindegren (private communication) has determined a value of $\gamma = 0.9893 \pm 0.014$ from the 12-month Hipparcos sphere solution. This preliminary result suggests that a precision on γ of better than 0.5% will be available from the final sphere solution.

4. Double/Multiple Stars and Photometry

This summary only provides space to underline the fact that the Hipparcos mission will provide a wealth of data related to double and multiple and stars—roughly 10,000 known systems were contained within the Hipparcos Input Catalogue, and a similar number of newly-discovered systems with $\Delta\rho > 0.1$ arcsec and $\Delta m < 3-4$ mag will have their astrometric and photometric characteristics tabulated within the final Hipparcos Catalogue. The photometric data available from Hipparcos and from Tycho will yield light-curves on an unprecedented scale, permitting characterisation of variability over the entire HR diagram.

References

Høg, E. *et al.* (1994) Tycho Astrometry from Half of the Hipparcos Mission, *Astr. Astrophys.*, in preparation.

Kovalevsky, J. *et al.* (1994) Construction of the Intermediate Hipparcos Astrometric Catalogue, *Astr. Astrophys.*, in preparation.

Lindegren, L. (1994) Properties of the Hipparcos Catalogue, this volume.

Lindegren, L. *et al.* (1994) Properties of the Intermediate Hipparcos Astrometric Catalogue, *Astr. Astrophys.*, in preparation.

Owen, W.M. & Yeomans, D.K. (1994) The Overlapping Plates Method applied to CCD Observations of 243 Ida, *Astronomical Journal*, in press.

Perryman, M.A.C. *et al.* (1994) The Hertzsprung-Russell Diagram from the first 30 months of Hipparcos Data, *Astr. Astrophys.*, in preparation.

Platias, I. *et al.* (1994) A Study of Systematic Positional Errors in the SPM Plates, *Astr. Astrophys.*, in preparation.

Réquième, Y. *et al.* (1994) Meridian Circle Reductions using Preliminary Hipparcos Positions, *Astr. Astrophys.*, in preparation.

Reynolds, J.E. *et al.* (1994) Accurate Registration of Radio and Optical Images of SN1987A, *Astr. Astrophys.*, in preparation.

Soffel, M.H. (1989) Relativity in Astrometry, Springer-Verlag.

Staveley-Smith, L. *et al.* (1993) Radio Emission from SN1987A, *Nature*, **366**, p136.

BINARY STAR ASTROMETRY WITH THE HUBBLE SPACE TELESCOPE: ONE MILLISECOND OF ARC ACCURACY AND BEYOND

O. FRANZ, K.J. KREIDL, L.H. WASSERMAN,
A.J. BRADLEY, G.F. BENEDICT, R.L. DUNCOMBE,
P.D. HEMENWAY, W.H. JEFFERYS, B. McARTHUR,
E. NELAN, P.J. SHELUS, D. STORY,
A.L. WHIPPLE, L.W. FREDRICK and W.F. van ALTENA

Abstract. We briefly review the concept of double star measurement with HST Fine Guidance Sensors (FGS) in the Transfer Function (TF) Scan mode and give results for three calibration binaries observed with FGS3. Agreement among multiple observations indicates an astrometric precision of 1 millisecond of arc (mas) per observation. We compare measured angular separations with ephemeris values from orbits based entirely on speckle observations. This comparison shows that the accuracy of binary-star astrometry with FGS3 in the TF-Scan mode is 1 mas per observation. Multiple observations can be expected to produce relative positions of binary components at sub-millisecond of arc accuracy.

E. Høg and P. K. Seidelmann (eds.),
Astronomical and Astrophysical Objectives of Sub-Milliarcsecond Optical Astrometry, 87.

HUBBLE SPACE TELESCOPE:
A GENERATOR OF SUB-MILLIARCSECOND PRECISION PARALLAXES

G. BENEDICT, W. JEFFERYS, B. MCARTHUR, E. NELAN, A. WHIPPLE
Q. WANG, D. STORY, P. HEMENWAY AND P. SHELUS
McDonald Observatory
University of Texas
Austin, TX 78712

W. VAN ALTENA
Astronomy Department
Yale University
New Haven, CT 06511

O. FRANZ
Lowell Observatory
Flagstaff, AZ 86001

R. DUNCOMBE
Aerospace Engineering
University of Texas
Austin, TX 78712

AND

L. FREDRICK
Astronomy Department
University of Virginia, Charlottesville, VA 22903

Abstract. Hubble Space Telescope Fine Guidance Sensor 3 can generate sub-milliarcsecond precision parallaxes in eighteen months. We discuss the internal precision and external accuracy of our observations of Proxima Centauri and Barnard's Star. For some classes of targets Hubble Space Telescope will remain the parallax tool of choice for years to come. It can offer 0.5 mas precision. It will remain useful by satisfying urgent needs for quick results, by offering a 13 magnitude dynamic range, and by providing an unparalleled binary dissection capability.

E. Høg and P. K. Seidelmann (eds.),
Astronomical and Astrophysical Objectives of Sub-Milliarcsecond Optical Astrometry, 89–94.

1. Introduction

We will discuss the precision and accuracy of parallaxes and proper motions acquired with Fine Guidance Sensor 3 (FGS 3) aboard the Hubble Space Telescope (HST). The FGS is a white light interferometer. A description of the FGS, its data products, and data processing can be found in Bradley *et al.* (1991). The FGS is operated in either TRANS or POS mode. TRANS produces a scan across a star, which results in a complete interferometer response (transfer) function. POS mode provides a string of position error signals, generated by the FGS as it attempts to track the zero-crossing of the transfer function. Parallax studies utilize POS mode.

Benedict *et al.* (1992) showed that FGS 3 had the best performance characteristics for astrometry of any of the three FGS aboard HST, then demonstrated (Benedict *et al.* 1994) that FGS 3 had the precision and stability required for long-term astrometric studies, such as parallax work. We obtain 1.5 milliarcsecond (mas) precision per axis per observation. No other ground- or space-based device can beat this noise figure for the very faint stars we study. Details of our Optical Field Angle Distortion (OFAD) calibration can be found in Jefferys *et al.* (1994). While stable enough for astrometry, FGS 3 does change over time. See Whipple *et al.* (1994) for a description of the character of these changes and of our OFAD monitoring strategy.

We now have data to determine parallaxes for Proxima Centauri and Barnard's Star. We chose these two targets for an astrometric search for planetary companions. The results of the Proxima Cen planet search will be reported elsewhere. The Barnard's Star planet search continues.

2. Proxima Centauri

• Background - Proxima Cen = GL 551 is the nearest star to us other than the Sun. It is a known flare star (V645 Cen), with quiescent V = 11.22. We have used HST FGS 3 in POS mode to monitor the position of Proxima Cen in an effort to detect perturbations due to very low mass companions. We have 48 data sets (orbits) from 23 March 1992 to 4 May 1994. The time coverage is not uniform, with gaps due to solar constraint and equipment difficulties. We secured many observations quickly because gas giants orbiting M stars may have short periods (Black and Scargle 1982). If one scales the temperature distribution in the solar preplanetary nebula around the far less luminous Proxima Cen, gas giants could form with orbital periods as short as 50 days.

A key element in planet searches is the determination of the proper motion of the star and the elimination of the effects of the earth's orbit. Anyone who has looked for planets around other stars has found that the earth orbits the Sun! Another key element, and one germane to parallax work, is the demonstrated random noise level of 1.5 mas per axis. This level is crucial to establishing upper limits for companion masses and for detecting actual perturbations.

• Results: Internal Precision - We analyse these data as described in Benedict *et al.* (1994), first showing what can be obtained with many more observations than any one proposer is likely to receive from the HST Time Allocation Committee. Given that constraint, we show the precision levels obtainable with far fewer (and, hence, more likely granted) observations.

Entire Data Series - A solution including all 48 observation sets yields a formal error (1-σ) for the parallax of 0.4 mas. For the proper motion (μ) we obtain 0.5 mas precision. See Table 1 for the parallax value. We reduce the relative parallax to absolute parallax (based on the faintness and galactic latitude of the reference frame, $V_{av} \simeq 14$) $\pi_{abs} = \pi_{rel} + 0.7$mas (van Altena, 1994).

A 1.5 year subset - We carried out an extension of a previous analysis (Benedict *et al.* 1994) with a 15 observation subset of the original 48 observations. This subset spanned 18 months and provides a formal error for the parallax of a ($V = 15$) reference star of 0.5 mas, and 1 mas for μ. We then reanalyzed Proxima Cen for this same subset and achieved similar results (Table 1). Note that in going from 15 to 48 observation sets we do not quite achieve a $\sqrt{n}$ reduction in the parallax error. This is probably due to not having 48 observation sets independent in time. There are many tight groupings of observations.

3. Barnard's Star

• Background - To date (8/94) we have 15 observation sets (from 2 February 1993 to 4 May 1994) of Barnard's Star, $V = 9.54$. The time interval between observations over this 14 months is not uniform. In particular, we have not sampled both extremes of the parallactic ellipse.

• Results: Internal Precision - Even though our study duration is strikingly less than a typical parallax series, and we have not sampled the entire parallactic ellipse, our precision is good. The (1-σ) precision for parallax is 0.9 mas and for μ is 1.3 mas. Our absolute par

allax ($\pi_{abs} = \pi_{rel} + 0.8$mas) is given in Table 1.

4. External Accuracy

Ideally, we should check against results with similar precision, but there are none. Rather, we compare against compilations of many less precise determinations, such as the Yale Parallax Catalog (van Altena *et al.* 1991, 1994). Our comparison is given in Table 1.

TABLE 1. Comparison with Yale Parallax Catalog

	Reduction	HST	YPC(1991)	YPC(1994)
1	Proxima Cen (48 obs)	0.7699±0.0004	0.7718±0.0041	0.7699±0.0004
2	Proxima Cen (15 obs)	0.7701±0.0005	0.7718±0.0041	0.7699±0.0004
3	Barnard's Star (15 obs)	0.5420±0.0009	0.5467±0.0008	0.5463±0.0013
4	Barnard's Star (USNO μ)	0.5442±0.0028	0.5467±0.0008	0.5463±0.0013

The last line in the table is a parallax value obtained by constraining μ of Barnard's Star to a previously determined value (Harrington & Dahn, 1980). Much of the difference between the HST and the YPC parallax values (line 3) is likely due to poor sampling of the parallactic ellipse by HST. We cannot yet separate parallax from proper motion. Finally, it should be noted that the ecliptic latitude is 27° for Barnard's Star and -44° for Proxima Cen. Thus, there is parallax information in two directions. A parallax for a target near the ecliptic might require more than 1.5 years.

5. The Future of HST Parallax Work

The only other program routinely providing milliarcsecond precision parallaxes at optical wavelengths is the ground-based CCD work of Monet *et al.* (1992). HST parallax precision is somewhat better than they typically achieve, and is attained in one half to one third the time. Once large array optical interferometers are in routine use (e.g. Simon *et al.* 1991), the list of potential parallax targets for HST may shrink, but not vanish. In this section we summarize the strengths of HST. We then point out a few examples of the kinds of objects for which HST is ideally suited. We end with a brief discussion of areas requiring further work.

5.1. WHY USE HST TO OBTAIN PARALLAXES?

• Precision - Our demonstrated precision for two stars in the Proxima Cen field is 0.5 mas. Given the precision we attained for Barnard's Star under less than ideal circumstances (§3), it is probable that we will reach 0.5 mas for most targets.

• Field of View - Often a parallax target is unmeasurable because reference stars are not located within the field of view of the measuring device, be it photographic plate or CCD. For the data discussed above sufficient reference stars existed within the 3.5 arcmin diameter center of the FGS field of view. If a larger field of view is necessary for a parallax series, one can trade study duration for field of view. Observations could be spaced by six months. The FGS field of view would then flip 180°. The paraboloid-shaped region in common to the two orientations has a short axis = 3.5 and a long axis = 14 arcmin.

• Dynamic Range - FGS 3 can obtain POS mode position measures for stars in the range $4 \leq V \leq 17$. This large dynamic range is provided by a neutral density filter that reduces the magnitude of bright stars by 5 mag. The unfiltered range is $9 \leq V \leq 17$. There will be a small but unknown shift in position (due to filter wedge) when comparing the positions of the bright star to the faint reference frame. The shift is constant in direction (relative to FGS 3) and size, since the filter does not rotate within its holder. However, the shift can become a nuisance parameter in our model, because it represents another annual term, like parallax. The FGS field of view slowly rotates as the HST solar arrays are kept normal to the Sun throughout the year. To fully exploit this dynamic range, it is essential that the magnitude and direction of the filter wedge shift be determined by a cross-filter calibration.

• Binary Stars - HST can obtain precise parallaxes for close binaries. In POS mode, the FGS will lock on a null position which is generated by two closely overlapping s-curves. A nearly simultaneous TRANS mode observation gives the relative positions of the two components. One can determine the null position with an accuracy and precision of 1 mas relative to the two components, once the s-curves from the two stars are deconvolved (Franz *et al.* 1994a). POS and TRANS mode measurements provide the null position and component positions relative to reference stars.

• Timeliness of Result - One no longer need wait 3-6 years for the parallax of astrophysically interesting objects. The distance to a sufficiently interesting and important object can be obtained on the same time scale as other astrophysical information. *Without distances, no one does astrophysics.* If an object or class of objects is interesting now, a theory can be tested now. A caveat; accuracy will require complete sampling of the parallactic ellipse.

5.2. TARGET EXAMPLES

• Targets requiring the dynamic range and precision of HST include the classical cepheid variable, SU Cas and the defining member of the class, RR Lyrae.

• As an example of the need for speed, consider asteroseismology, a particularly active research area. It may be possible to measure distances to very bright (young) white dwarf stars by photometric study of their vibrational modes (Kawaler and Bradley 1994). The technique requires calibration. These stars are all quite distant, with very poorly determined distances. As soon as HST can provide a more precise distance for (e.g.) PG 1159-035, astronomers may have a new distance scale tool.

• We have discovered many previously unknown faint binaries in the Hyades (Franz *et al.* 1994b). Precise mass determinations require precise distances. Given the nearness and depth of this cluster, we must obtain parallaxes for each binary, rather than rely on an average cluster distance.

5.3. FUTURE IMPROVEMENTS

After working with FGS 3 for several years, we feel that improvements are still possible. Why do we need 6 coefficients in our overlapping plate model? Avenues of investigation include the spatial resolution of our OFAD calibration and color effects (FGS has refractive elements). In December 1993, NASA serviced HST. This introduced some rather large and, as yet, unexplained changes in our OFAD. To fully exploit the dynamic range of FGS astrometry, we will push for a cross filter calibration. The precision, field of view, dynamic range, and speed of HST/FGS POS mode astrometry demand that we press for the best possible treatment of the data.

References

Benedict, G. F. *et al.* (1992) *PASP*, 104, 958

Benedict, G. F. *et al.* (1994) *PASP*, 106, 327

Black, D. and Scargle, J. (1982) *ApJ*, 263, 854

Bradley, A. *et al.* (1991) *PASP*, 103, 317

Franz, O. G. *et al.* (1994a) "Binary Star Astrometry and Photometry from Transfer-Function Scans" in *Calibrating Hubble Space Telescope*, ed. by J. C. Blades and S. J. Osmer, STScI

Franz, O. G. *et al.* (1994b) *BAAS*, 26, 929

Harrington, R.S. and Dahn, C.C. (1980) *AJ*, 85, 454

Jefferys, W. H. *et al.* (1994) "Optical Field Angle Distortion of FGS 3", in *Calibrating Hubble Space Telescope*, ed. by J. C. Blades and S. J. Osmer, STScI

Kawaler S. and Bradley, P. (1994) *ApJ*, in press

Monet, D. G. *et al.* (1992) *AJ*, 103, 638

Simon, R. S. *et al.* (1991) "Imaging Optical Interferometry" in *IAU Coll. 131* ASP Conf. Ser. Vol 19, 358.

van Altena, W. F., Lee J. T., and Hoffleit E. D. (1991) *The General Catalogue of Trigonometric Parallaxes, Preliminary Version*, in Astronomical Data Center CD-ROM Selected Astronomical Catalogs, Volume 1, L. E. Brotzman, S. E. Gessner, J. M. Mead and M. E. Van Steenberg, eds., Goddard Space Flight Center, Greenbelt.

van Altena, W. F., Lee J. T., and Hoffleit E. D. (1994) *The General Catalogue of Trigonometric Parallaxes*, Yale University Observatory, New Haven.

van Altena, W. F. (1994), private communication

Whipple, A. L. *et al.* (1994) "Maintaining the FGS 3 OFAD Calibrations with the Long-Term Stability Test", in *Calibrating Hubble Space Telescope*, ed. by J. C. Blades and S. J. Osmer, STScI

FRINGE INTERFEROMETRY IN SPACE: THE FINE STRUCTURE OF R136A WITH THE ASTROMETER FINE GUIDANCE SENSOR ABOARD HST

M.G. LATTANZI
Space Telescope Science Institute,
Affiliated with the Astrophysics Division,
SSD, ESA; on leave from Oss. Astr. di Torino

R. BURG
Johns Hopkins University

J.L. HERSHEY, L.G. TAFF AND S.T. HOLFELTZ
Space Telescope Science Institute

AND

B. BUCCIARELLI
Osservatorio Astronomico di Torino

Abstract.

We report on the highest angular resolution observation to date of the bright core, R136a, of the massive star cluster R136 within the 30 Doradus complex in the LMC. This visual observation was obtained with the interferometric fringe mode of operation of Fine Guidance Sensor No. 3 (FGS3) on board HST. Crowding and strong diffuse background from nebular emission make this a challenging observation.

The giant HII region 30 Dor has provided some of the best candidates for the most massive stars, like R136a1. We provide evidence for a new component, R136a1B, within the previously known R136a1-a2 system with a separation of 80 mas (or $\approx$ 4000 au from a1 at the distance of the LMC), and $\Delta V = 1.1$ mag fainter than the brightest component a1. Estimates from current evolutionary models of massive stars based on the new FGS photometry predict, after subtraction of a1B, that the present mass of R136a1 is 30 $M_{\odot}$ with a main sequence progenitor of 60 $M_{\odot}$. To date, this is the *lowest* direct estimate of the mass of R136a1.

E. Høg and P. K. Seidelmann (eds.),
Astronomical and Astrophysical Objectives of Sub-Milliarcsecond Optical Astrometry, 95–100.

The success of this difficult observation adds a new, unique feature to FGS3 and gives a much expanded, astrophysically very rewarding, role to the interferometer.

1. Introduction

We discuss a 10–minute TRANSfer Mode observation with FGS3, taken as part of that portion of the STScI FGS Cycle 3 Calibration program designed to explore *new* and *unique* ways of using the interferometric capabilities of this instrument. TRANS mode samples the visibility fringe (or Transfer Function, TF) produced by the Koester's prism–based interferometers inside FGS3.

The choice of the target, the core (R136a) of the massive star cluster R136 in 30 Dor, was motivated by its fundamental importance as the site of some of the most massive stars known. Indeed, the establishment of an upper mass limit for individual stars remains an open question of great importance in astronomy. Its resolution will influence the studies of stellar evolution, HII regions and galaxy evolution. In particular, R136a had for many years been unresolved and its structure and photometry are still not fully explored.

The challenges for the FGS in this never–attempted–before observation were: a) crowding (R136a has the appearance of a relatively dense open cluster), and b) the high background level from nebular emission (Burg et al. 1994).

In this paper, emphasis is on the technical details of this difficult interferometric observation and its interpretation (sections 2 and 3). Some of the astrophysical implications of the results are discussed in section 4.

2. The Observation

The FGS measurements consist of ten, identical position angle, consecutive "Transfer Scans", each 2."1 long with 0.8 mas step size, through R136a. As described in the FGS Instrument Handbook (Version 4.0), and in recently published papers (Bernacca, Lattanzi et al. 1993; Lattanzi et al. 1994) each scan samples the interference fringe produced by the Koester's prism interferometer, the "heart" of the Fine Guidance Sensor. The Koester's prism works with an *afocal* polarized beam. There are two Koester's prism interferometers in each FGS. These are fed by a beam–splitter (which provides the required polarization) to give sensitivity in two orthogonal directions, usually referred to as X and Y axes. When a scan is executed, the FGS 5" x 5" instantaneous field–of–view scans across the target at a fixed 45°

angle to the X and Y axes. Therefore, each FGS3 scan produces two fringes which can then be independently analyzed for signatures other than those characteristic of the FGS3 X and Y TFs of the standard single star UP69; which is observed with the same instrument setup as for the science observation. Indeed, the deviations from the single star TFs are used to measure multiple objects (see next section).

We can regard the X (Y) scan as driving a slit of the size of the resolution limit in X (Y) and 5" wide in the Y (X) direction along the scan path. Another object appearing anywhere in the rectangular slit has the *same* projected X (Y) coordinate, within the resolution limit, as that of the primary target. The resulting fringe (given as counts from the two PMTs for the given axis, i.e. $C_A - C_B$) will have an amplitude which is the sum of the two individual fringe amplitudes.

In the configuration used, the FGS can be considered a *white light* interferometer with a central wavelength of 583 nm and a bandpass (FWHM) of 234 nm. Therefore, the fringe sampled at each scan corresponds to the Fourier transform of the "real" central fringe which would form on the focal plane of the 2.5m, filled, aperture (the HST primary mirror).

The PMT integration time per pixel during each scan was 0.025 sec. To increase the signal–to–noise ratio (S/N) the 10 scans were added together for a total exposure time of 10.3 min. Each scan has a S/N of $\simeq 3$. Co–adding the scans increased the S/N to ~ 11, consistent with the expected improvement of $\sqrt{10}$.

The stability of the platform, i.e. the accurate control of the telescope pointing during FGS scans through a source, is of utmost importance for the success of this kind of observations. This is addressed in the work by Bradley (1994), which shows that a pointing stability of ≈ 2 mas residual jitter is routinely achieved.

3. Interferometry of Multiple Stars

The observed R136a region is almost coincident with that of the high-resolution FOC F/288 image discussed by Weigelt et al. (1991). For details on the scan orientation and the field see Lattanzi et al (1994).

The presence of several point sources in the FGS FOV during the scan with separations larger than the FGS resolution limit will generate the complex fringe in Eq.(1). There, S(X) is the *normalized* single–star TF [i.e., $(C_A - C_B)/(C_A + C_B)$, $S(X + \Delta X_i)$ is the same as S(X) but displaced along the X-axis by ΔX_i, the projected separation of the i–th component from R136a1, Δm_i is the magnitude difference between the two stars, and nc is the total number of point like sources in the scan. There is an analogous expression, M(Y), for the multiple star Y–axis fringe.

$$M(X) = \frac{S(X) + \Sigma_{i=1}^{nc}\ 10^{-0.4\Delta m_i}\ S(X + \Delta X_i)}{1 + \Sigma_{i=1}^{nc}\ 10^{-0.4\Delta m_i}} \quad (1).$$

According to Eq.(1), the presence of companion stars decreases the fringe visibility (its peak–to–peak value) of the bright star; the worst case being that of components of the same magnitude as that of the "primary" star. Because of the denominator, $C_A + C_B$, in the expression of S(X), the effect of the very high level of background in this field is also a reduction of the peak–to–peak amplitude of S(X) in Eq.(1) by a factor of ten as compared to average background. The observed fringe is represented by the solid curve in Fig.1. For reference, a typical peak–to–peak (single star in average background) is $\simeq 1.1$.

From the FGS data themselves, and using the technique described in Bernacca, Lattanzi et al (1994), the background from the nebular emission is estimated at V=15 mag arcsec^{-2}. This value is actually consistent with the contributions from Hα, Hβ, [O III] $\lambda\lambda$ 4959, 5007 emissions and the spectral response of FGS3 in the configuration used here.

4. Measuring a Multiple Star

Although the standard algorithms used with double stars must be modified for the multiple star case, the basic technique remains unaltered. Simply, the measurement of separations and magnitude differences of multiple component objects is performed, as for double stars, by measuring the departures of the corresponding TFs from the template TFs of our reference single star Upgren 69 in NGC 188 (Lattanzi et al 1994, and references therein).

Fig. 1 shows the result for the X–axis scan obtained via the correlation method, which synthesizes the best possible model from the template scans (i.e., those of the standard star UP69) and compares that to the observed scans via Eq.(1). The solid curve is the X–axis observed TF and the dashed line superimposed is the residual curve. The locations and relative amplitudes, proportional to the relative brightnesses, of the 'spikes' appearing on the abscissas represent the solutions of our adjustments. The dashed curve shown in the bottom part of Fig. 1 represents the result of applying our alternative deconvolution technique to the OBSERVED X–axis TF (Hershey 1992). There is quite good agreement between this independent method and the correlation solution. As expected, the deconvolved TFs generally overlay the locations and the relative amplitudes of the set of point source positions deduced from the synthetic TF method. There are also cases of two spikes too close together for the deconvolved TF to show two *resolved* peaks, as it is the case for the peak which overlays the a1 and a1B X–axis spikes (Fig. 1). In these cases the shapes of the peaks exhibit significant

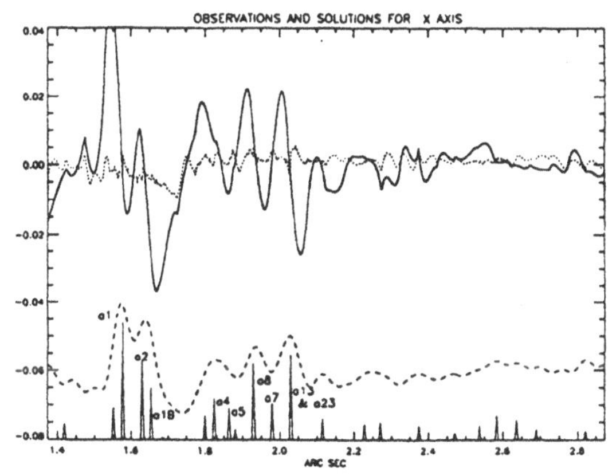

Figure 1. The results of interpreting the 30 Dor signal on the X axis. The thick curve is the normalized observed TF, the dotted line is the residual curve, the dashed curve is the profile generated from the deconvolution of the TF. The position and relative amplitude of the spikes on the abscissa represent the individual stellar components from our adjustments. The amplitude of the spikes is proportional to the relative brightness of the components. Only the sharpest features are identified (from Lattanzi et al 1994).

deviations from that of a fully resolved point source. Note that the projected separations and the magnitude differences are estimated at the same time. Also, the M(X) and M(Y) best–correlation fits yield two independent estimates of the Δm of each component.

The unambiguous identification of components a1, a2, and of the new component a1B from our solution is simple, since they are the closest, most luminous objects in the field. For a detailed discussion of the star identifications presented in Fig. 1 we refer to Lattanzi et al (1994). It is shown there that the FGS has "seen" all of the classical components (a1 through a8) already found by Weigelt and collaborators via speckle techniques (Weigelt and Baier 1985), plus some more previously discovered with the HST PC and FOC (pre–COSTAR) cameras.

5. The Case of R136a1 in The LMC

The component a1B is about ΔV=1.1 mag fainter than a1 (V $\sim$ 12.75) and 80 mas away from it. The X–axis projected separation shown in Fig. 1 is $\sim$ 75 mas. Notice also that a1B is only 23 mas from a2. The estimated (internal) errors of those measurements are 0.1 mag and better than 10 mas, respectively. Ground based spectroscopy tends to classify a1 as WN6. This classification, after the subtraction of the apparent luminosity of the newly resolved component and using Maeder's models (1990), yields a present–day mass of 30 $M_\odot$ for a1 and a main–sequence progenitor of 60 $M_\odot$.

As for R136a1B, it is a very early spectral type star as well, probably an O7 V of $\sim$ 20 $M_\odot$, given the magnitude difference measured by the FGS.

However, the luminosity class is uncertain as the FGS cannot provide any color information. It could be a much later class III giant, although this would pose a serious problem in dating the system, if one assumes that a1B is associated with the R136a complex and coeval star formation within the cluster.

6. Discussion and Conclusions

The results discussed in the preceding sections clearly demonstrate that the fringe mode of FGS3 is viable, in terms of both high angular resolution and accurate photometry, even in crowded fields with strong background as 30 Dor. The effect on the measurements (limiting resolution and magnitude error) of these "adverse" observing conditions appears negligible given that some of the S/N lost can be regained through the technique of co–adding consecutive scans. This new, unique, use of FGS3 gives this instrument a much expanded, astrophysically very rewarding, role.

7. Acknowledgements

This paper is based on observations with the NASA/ESA Hubble Space Telescope, obtained at the Space Telescope Science Institute, which is operated by the Association of Universities for Research in Astronomy, Inc., under NASA contract NAS5–26555. This work was supported in part by NASA grants NAGW–2597 and CW–0016–92. R.B. is partially supported by NASA grant NAGW–2509

We wish to thank F. Macchetto and P. Stockman for their careful reviews during the preparation of the calibration observations. We are also thankful to the members of the HST Astrometry Team (led by W. Jefferys) for stimulating discussions during their regular meetings, which affected several aspects of this work. Our warm appreciation goes to Linda Abramowicz-Reed (HDOS) for her precious help in establishing the photometric properties of FGS3 in the configuration used in this work. Finally, the useful comments of an anonymous referee are gladly acknowledged.

References

Bernacca, P.L., et al. 1993, *A&A*, **278**, L47.
Bernacca, P.L., et al. 1994, *A&A*, in press.
Bradley, A. 1994, private communication.
Burg et al. 1994, *STScI Newsletter*, **11**, No.1,p.7).
Hershey, J.L. 1992, *PASP*, **104**, 592.
Lattanzi, M.G., et al. 1994, *ApJ*, **427**, L21.
Maeder, A. 1990, *A&AS*, **84**, 139.
Weigelt, G., and Baier, G. 1985, *A&A*, **150**, L18.
Weigelt, G. et al. 1991, *ApJ*, **378**, L22.

ASTROMETRY WITH THE HST PLANETARY CAMERA

T. M. GIRARD, Y. LI AND W. F. VAN ALTENA
Dept. of Astronomy, Yale University
P.O. Box 208101, New Haven, CT 06520-8101

J. M. NUNEZ
Univ. of Barcelona, Spain

G. F. BENEDICT, R. L. DUNCOMBE, P. D. HEMENWAY,
W. H. JEFFERYS, B. MCARTHUR, J. MCCARTNEY, E. NELAN,
P. SHELUS, D. STORY AND A. L. WHIPPLE
Univ. of Texas, USA

O. G. FRANZ AND L. W. WASSERMAN
Lowell Obs., USA

AND

L. W. FREDRICK
Univ. of Virginia, USA

Abstract.

The astrometric capability of the Hubble Space Telescope Planetary Camera (WF/PC1) is investigated, motivated by a study of the internal velocity distribution of globular clusters. The astrometric accuracy of the HST PC will be determined ultimately by 1) the accuracy to which the aberrated images can be 'centered', and 2) the accuracy to which the distortions across the PC field can be modeled. A series of overlapping exposures of two clusters, NGC 6752 and M15, are utilized to examine these issues.

We have made use of maximum-likelihood image reconstruction to address the first issue, with good success. Reconstruction improves both the detectability and precision of the image centers. A preliminary exploration of the second issue, that of modeling the distortion across the PC field, is also presented, using positions derived from the multiple overlapping exposures.

E. Høg and P. K. Seidelmann (eds.),
Astronomical and Astrophysical Objectives of Sub-Milliarcsecond Optical Astrometry, 101–106.

1. Introduction

Among the Guaranteed Time Observing programs undertaken by the Space Telescope Astrometry Team, is a project designed to determine the distribution of internal velocity dispersion within several globular clusters. This is to be accomplished by measuring the relative proper motions of a sufficient number of stars over a range of mass and distance from the cluster core. This, in turn, will require the astrometric determination of precise relative positions of a large sample of stars at two separate epochs.

The primary astrometric instrument on board the Hubble Space Telescope is the Fine Guidance Sensor, whose superb astrometric performance is described elsewhere in this volume, (Benedict *et al.* 1994). A fundamental limitation of the FGS, however, is its 'serial' mode of operation which makes it unsuitable for measuring large numbers of stars. As an alternative, we have chosen to use the Planetary Camera for this task. A number of first-epoch exposures have been obtained using the WF/PC 1 in Planetary Camera mode. Second-epoch exposures are to be taken with the WF/PC 2. It has yet to be conclusively demonstrated that the required astrometric precision can be attained with the PC data. We describe here our present efforts at investigating the astrometric capabilities of the PC with specific attention given to our proposed study of the proper-motion distribution within globular clusters.

2. Motivation

Even with its aberrated images, the HST Planetary Camera allows imaging of individual main-sequence stars very near the cores of globular clusters. It is our goal to use the HST PC as an astrometric instrument to determine stellar proper motions within several globular clusters. The proper motions would yield:

1) internal velocity dispersions for these clusters,
2) virial mass estimates for these clusters,
3) kinematic distance estimates (statistical parallax), and
4) for a subgroup of the more nearby clusters; *i)* the radial and azimuthal velocity dispersions as a function of distance from the cluster center, and *ii)* the degree to which energy equipartition exists between various stellar mass groups within the clusters.

The clusters to be observed are 47 Tuc, M13, M22, NGC 6752, M15, M30, NGC 1851, and PAL 1. We have thus far obtained first-epoch exposures for the first five of these clusters.

2.1. OBSERVATIONS

The Planetary Camera (WF/PC 1) consists of four 800x800 CCDs with a scale of 0.043 "/pixel. We adopt a nominal expected precision for relative stellar positions within individual frames of ± 1 mas, i.e. 1/40 of a pixel. This would allow second-epoch exposures to be taken after five years. The resulting proper-motion errors would be ± 0.3 mas/yr per 'plate pair', which corresponds to 7 km/s for a cluster at 5 kpc distance. Thus, the component due to measuring error of the observed velocity dispersion would be 3 ± 0.3 km/s, again assuming d=5 kpc. This value assumes five 'plate pairs' per star and 50 stars per velocity bin. Since the intrinsic velocity dispersions of globular clusters are typically 6 to 10 km/s, a measuring dispersion of 3 ± 0.3 km/s is small enough and determined accurately enough to be confidently 'subtracted' from the total observed dispersion to yield the intrinsic cluster dispersion.

In order to calibrate the distortion across the PC field and to cover a large enough area to ensure a sufficient number of cluster star images, a strategy of using multiple-exposure patterns has been adopted. Two patterns are used, one a 5x pattern, the other, a 10x pattern. Within each pattern, the exposures are offset by 100 pixels in the x,y system of the WF/PC. The cores of the program clusters are observed with 25-second and 100-second exposures using the 10x and 5x patterns, respectively. The clusters are also observed with 1000-second exposures of the 5x pattern at positions corresponding to five and to ten times the cluster core radius. Observations are made in two passbands using the F555W and F785LP filters. The range in exposure times will hopefully allow us to measure stars roughly two magnitudes below the main-sequence turnoff for most of the clusters.

3. Image Centering

As stated earlier, the final astrometric precision of the relative positions derived from PC images will depend on the accuracy to which individual stellar images can be centroided and the degree to which the distortion across the PC field can be modeled. We have addressed the issue of optimum image centering with a variety of tests involving raw and reconstructed PC images, which we now describe in detail. Throughout the following discussion, the actual centroiding itself is done with a two-dimensional Gaussian fitting of the intensity profiles.

3.1. RAW IMAGES

Previous tests, (Girard and van Altena, 1990), using synthetic PC frames and a limited number of early-observation frames of the R136 region showed that 2 mas relative positional accuracy could be achieved for well-exposed images using Gaussian fitting of the raw intensity profiles. The accuracy quickly fell off for fainter images and in crowded regions.

Although we are interested in an application involving crowded fields with low signal-to-noise images, (the globular cluster program), it is worth mentioning the centering precision we have recently achieved with a set of high signal- to-noise, isolated PC stellar images. This series of PC frames, taken in conjunction with FGS observations intended to monitor the long-term stability of the FGS instrument frame, was taken with the same pointing on consecutive orbits. The PC fields contain from two to three well-exposed stars. We have fit the 'raw' intensity profiles of the stars with our 2-D Gaussian centroider and estimated the centering precision from the deviations in the separation of star pairs. Based on 22 separation measures, the single- coordinate centroiding precision is 0.6 mas, (0.014 pixels)!

It must be stressed that this high level of repeatibility is thus far demonstrated only over consecutive HST orbits and with nearly identical pointing, (the individual star positions varied by $\approx$ 0.06 to 0.20 pixels).

3.2. RECONSTRUCTED IMAGES

Image reconstruction of the aberrated PC images can make substantial improvements in qualitative appearance, and possibly recover faint star images from within the extended wings of nearby bright images. The question remains as to whether or not image reconstruction can retain or improve the astrometric information to be extracted from a PC image. We have performed a series of tests to help answer this question.

We have done so using the maximum likelihood image reconstruction code of Nunez and Llacer, (1990). Synthetic psf's generated by the Tiny Tim software package were used, and the frames were reduced assuming a uniform psf, (i.e., not a spatially varying psf across the PC field).

The first of these tests involved two non-program PC frames of the globular cluster NGC 6752, (a 40-second and 500-second exposure with F675W). Fifty-five (uncrowded) stars from PC-5 were selected by hand. The entire 800x800 pixel frame was restored using the maximum likelihood version of the Nunez and Llacer code. The reconstruction process is iterative, and was halted after 100 iterations. The restored intensity profiles were centered and the long exposure positions were transformed into those of the short exposure to determine the unit weight measuring error. The results are given in Table 1.

TABLE 1. Long-to-short exposure reductions.

	σ_x (mas)	σ_y (mas)	No. Stars	Mag. Range.
raw	1.4	2.1	33	2.3
psf @ (200,600)	1.4	1.2	39	2.9
psf @ (400,400)	1.3	1.0	38	2.8
psf @ (600,200)	0.8	0.8	42	3.3

It should be noted that the 'raw' solution required the removal of many 'outliers' both bright and faint, thus the unit weight errors may be underestimated. The original target list consisted of 55 stars. Several of these were so saturated they did not center in any reductions. In addition, 8 to 10 very bright stars were eliminated due to a non-linear magnitude effect present.

These results suggest that 1 mas positional precision may be obtained with image reconstruction, for an uncrowded field of relatively well-exposed images.

3.3. SUBSAMPLED RECONSTRUCTIONS

Synthetic psf's may be generated on arbitrarily fine grids, allowing the reconstruction to be performed on a grid spacing smaller than that of the detector, one that better samples the psf. A subsampled reconstruction should be much less affected by the centering bias discussed above. We have used subsampled reconstructions, at subsampling factors of 1 through 5, to explore its effect on the final centered positions.

Two additional PC exposures of NGC 6752 have been used for this part of our study, 100-second exposures in F555W and F785LP . A 128x128 subframe containing roughly 40 faint stellar images was extracted from PC 5. Tiny Tim psf's were calculated, in both filters, at the center of the subframe, and for each subsampling factor from 1 to 5. The subframes were then processed through 80 iterations of the reconstruction code.

Twenty-five stars successfully centered on each of the 1x processed frames. The differences between the fractional pixel coordinates from the 1x and 5x reconstructions for these stars showed no indication of a systematic bias in the 1x fractional pixel coordinates, relative to the presumably accurate 5x coordinates, giving us somewhat more confidence in the error estimates quoted in Table 1.

In an attempt to determine just how the image coordinates "converge" from their 1x to their 5x values, linear transformations of the Nx positions were transformed into the 5x positions. Unexpectedly, the standard error does not decrease monotonically but is a minimum for the 3x transformation. If instead the transformations are performed into the 4x coordinates, the minimum appears for the 2x data! Clearly there is an artifact of the subsampling/reconstruction process rearing its ugly head here. This effect is currently being explored.

4. Distortion

The second of the two sources of astrometric uncertainty to be considered, that of distortion across the PC, has just begun to be explored. The positional displacements due to classical cubic distortion can be written

$$\Delta x = D\, x\, (x^2+y^2), \text{ and } \Delta y = D\, y\, (x^2+y^2).$$

Comparing two exposures with a relative offset $(\delta x, \delta y)$, the differential effect of the cubic distortion becomes quadratic in form

$$\Delta x_2 - \Delta x_1 = 3\, D\, \delta x\, x^2 + 2\, D\, \delta y\, xy + D\, \delta x\, y^2, \quad \text{and}$$
$$\Delta y_2 - \Delta y_1 = 3\, D\, \delta y\, y^2 + 2\, D\, \delta x\, xy + D\, \delta y\, x^2.$$

Therefore, a general, quadratic, least-squares transformation between a pair of offset frames can yield the distortion coefficient. We have used five reconstructed, core exposures of M15 to test this method. Our preliminary results indicate that PC-5 of the WF/PC 1 exhibits classical cubic distortion of magnitude, $D = -1.8 \pm 0.1 \times 10^{-5}$ arcsec^{-2}.

References

Benedict, G. F., McArthur, B., Nelan, E., Story, D., Whipple, A. L., Wang, Q., Jefferys, W. H., Duncombe, R. L., Hemenway, P. D., Shelus, P. J., van Altena, W. F., Franz, O. G., and Fredrick, L. W. (1994) 'HST, a generator of sub-millisecond of arc parallaxes' (This volume)

Girard, T. M. and van Altena, W. F., (1990) *BAAS* **22**, 1277.

Nunez, J. M. and Llacer, J. (1990) *Astrophys. and Space Sci.* **171**, 341.

1.3 EXTRAGALACTIC REFERENCE FRAME

THE DYNAMICAL REFERENCE FRAME

E. M. STANDISH
JPL/Caltech
301-150; Pasadena, CA 91109; USA

Abstract. Planetary and lunar ephemerides continue to improve in accuracy as they continue to be adjusted to newer and more accurate observational data. An additional improvement will be that of the orientation of the ephemerides; in the future, the ephemerides produced at JPL will be based upon the reference frame of the radio source catalogues. Recent planetary observations have been made directly with respect to the radio reference frame, and these observations have shown a satisfying degree of absolute accuracy and internal consistency; they enable the automatic orientation of the ephemerides onto the radio reference system during the ephemeris adjustment process.

1. Introduction

Lunar and Planetary Ephemerides continue to improve for a number of reasons.

- Newer and more accurate observations increase the internal accuracy of the motions.
- Most of the relevant astronomical constants are now accurately known from other sources.
- Recent and future positional measurements of the planets provide a direct tie between the planetary and radio source reference frames.

In the past, the ephemerides of the inner planetary system, Mercury through Mars including the Moon, have been very well determined in a relative sense, due to the highly accurate ranging observations. The orientations of the systems, however, have been less certain by orders of magnitude (Standish & Williams 1990).

E. Høg and P. K. Seidelmann (eds.),
Astronomical and Astrophysical Objectives of Sub-Milliarcsecond Optical Astrometry, 109–116.

The paper discusses the observational data set involved in the creation of the ephemerides, concentrating upon reference frames and how the data have influenced the orientation of previous ephemerides produced at JPL. The recent progress of determining frame-ties is described, and it is shown how it is now possible to automatically orient the ephemerides onto the radio reference frame by fitting to recent VLBI measurements of the planets. Arguments for and against basing the ephemerides onto the radio reference frame are presented.

2. Orientations of Past Ephemerides at JPL

Lunar and planetary ephemerides are adjusted to various types of observational measurements in a least squares sense. The accuracy of the ephemerides relies primarily upon the accuracy of these data. Similarly, the accuracy with which the ephemerides are oriented onto any particular reference frame depends upon both the accuracy of the measurements of the planets with respect to that frame and the accuracy of that frame itself. If there is more than one frame involved, frame-ties must be applied in order to have the data sets be consistent among themselves.

Table 1 lists the various types of observations to which the ephemerides are adjusted and shows both the approximate accuracy of the observations and the frames to which they are referenced. The planetary ranging observations are virtually independent of outside reference frames; they tend to refer the position of the measured planet with respect to the orbit of the earth. The lunar laser ranging (LLR) data are sensitive to the earth's orientation also; but, if the locations of the LLR telescopes are not constrained, that sensitivity is alleviated. Therefore, of the earlier data types, only the optical observations are referenced to an outside reference frame. Consequently, the earlier JPL ephemerides (up to and including DE130) were oriented automatically by the optical observations and thus were aligned onto the origin of the FK4 reference frame. The optical observations, however, as shown in the table, are two orders of magnitude less accurate than the other data types. (The ranging measurements are comparable to millisecond observations, since one kilometer at one AU subtends an angle of $0\overset{''}{.}0014$.) It is now known that systematic errors remain in the FK4/FK5 optical systems amounting to tenths of arcseconds; as such, the past orientations onto the optical frame represented one of the least accurate features of the ephemerides.

Starting with DE200, the first J2000-based ephemeris from JPL, attempts were made to orient the ephemerides onto the mean equator and mean dynamical equinox of J2000 (Standish 1982). These attempts were not entirely successful, however. First, there is more than one definition of

TABLE 1. The observations to which the ephemerides are adjusted.

Observation	Accuracy	Reference Frame
Radar-ranging	2 km → 100 m	Earth Orbit
s/c ranging	100 m → 7 m	Earth Orbit
LLR	30 cm → 3 cm	Earth Orientation and Orbit
Transits	0."5 → 0."2	Optical
Astrolabe	0."5	Optical
Astrometry	0."5	Optical
Occultations	0."1	Optical
Thermal Emission	0."03	Radio
VLBI : Magellan	0."002	Radio
VLBI : Phobos	0."001	Radio
VLBI : Ulysses	0."003	Radio
CCD Astrometry with QSO's	0."03	Radio

the ecliptic (Standish 1981); secondly, the determination of the mean of a non-periodic quantity is poorly-defined; and, thirdly, the precession of the equator, from the mean weighted epoch of the observational data to J2000, was inaccurate due to the now-known error (-0."3/century) in the value of the J2000 precession constant.

For the ephemerides, there is now a preferable reference frame. Table 1 shows that VLBI observations of spacecraft near to planets now exist at the milliarcsecond level. When used in conjunction with the spacecraft's orbit about the planet, these measurements refer the planet to the reference frame of the radio source catalogue. Of course, these observations are inconsistent with the optical observations, since the two corresponding reference frames are not coincident. However, since the VLBI measurements are so much more accurate than the optical ones, they tend to dominate the orientation of the ephemerides. To correctly use the two types of observations in the same adjustment, though, a frame-tie between the two frames should be introduced.

3. Ties Between Different Reference Frames

For an optimum adjustment of the ephemerides, all of the available observational data must be referenced onto the same reference frame. Consequently, the relations between the different reference frames must either be known or be determined as part of the adjustment process. Fortunately, enough progress has been made in this respect, so that the necessary relations now seem sufficiently well-determined. In particular, there have been determinations of the ties between the IERS radio catalogue and the dynamical reference frames (as represented by the JPL ephemerides), between the IERS catalogue and the FK5, and between the IERS catalogue and the Hipparcos reference frame. These frame-ties are discussed here.

3.1. PLANETARY EPHEMERIS - IERS FRAME-TIE

Folkner et al. (1994) have determined a tie between the IERS radio catalogue and the JPL planetary ephemerides by doing a joint analysis of VLBI and LLR observations. Basically, processing VLBI observations with the IERS catalogue determines a set of coordinates for the VLBI stations in the radio frame; similarly, processing the LLR data, using a given ephemeris, determines a set of coordinates for the LLR telescopes in the ephemeris frame. Since the two sets of coordinates don't involve the same stations, a geodetic survey between the two sets is used for the comparison. The differences in the comparison define the rotation matrix which produces optimum agreement; that rotation is the frame-tie, expressed by three differential angles about the x-, y-, and z- axes, respectively.

An alternative determination of the same frame-tie may be produced by processing the VLBI measurements listed in Table 1 and solving for the rotation of the ephemeris which gives the best fit to the data; again, the resultant rotation is the frame-tie.

Comparison of the two methods for DE200 shows agreement at the milliarcsecond level:

(-0."002, -0."012, -0."006) [±0."004] for the former method;
(-0."002, -0."003, -0."002) [±0."002] for the latter.

Importantly, DE200 is very close to the IERS reference frame, differing by only 0."01; thus, the transformation from the DE200 reference frame onto that of the IERS will not represent a major change.

3.2. OPTICAL - RADIO SOURCE FRAME-TIES

A comparison by Ma et al. (1990), using 28 sources visible in both optical and radio wavelengths, provides a frame-tie between the FK5 and a radio source catalogue aligned using the adopted value of 3C 273. Their result,

($+0\overset{''}{.}030$, $+0\overset{''}{.}053$, $+0\overset{''}{.}023$) [$\pm 0\overset{''}{.}020$], is still being modified by later studies, now using 50 sources (Johnston 1994). A similar study is reported by Kumkova et al.(1994). The eventual uncertainty of these frame-ties, however, will probably not improve much below the quoted value of $0\overset{''}{.}020$, owing to the inherent uncertainties of the FK5 itself.

Similarly, measurements of 11 radio stars with respect to both the Hipparcos reference frame and the IERS radio frame are enabling a frame-tie to be determined between those two (Lestrade et al., 1994). Kovalevsky (1994) mentions a similar effort by the Hipparcos Science Team. Zacharias et al. (1994) discuss the links of both optical systems to the radio source reference frame.

4. The Orientation of Future JPL Ephemerides

With VLBI measurements now relating the inner planet system to the radio reference frame, the present intention is to base future JPL ephemerides on that frame. This seems especially advantageous since the radio frames, the IERS frame itself or a similar one specified by the IAU, have stable orientations at the sub-milliarcsecond level.

The changing of the frame of the ephemerides onto that of the radio catalogue will involve modifications to the ephemeris adjustment process. However, these are relatively minor.

- Observations which are based upon reference frames other than that of the radio sources will have to be transformed before processing. The obvious examples are the optically-based observations, which form the major part of the observations of the Jovian planets.
- The equations of motion for the lunar librations are derived under the assumption that the reference frame is that of the dynamical ecliptic and equinox. The resultant formulation will require pre- and post-applications of the frame-tie rotation.
- The orientation of the earth, to which the LLR data is sensitive, should now be done in the IERS reference frame using the IERS formulation and earth orientation parameters. However, since the IERS determinations do not extend back to the time of the beginning of the LLR data (1969), the earth orientation will not be strictly homogeneous for the LLR data.

There are distinct advantages to changing the frame of the ephemerides onto that of the radio source catalogue.

- The VLBI measurements will orient the ephemeris frame with unprecedented accuracy.
- With the advent of CCD astrometry, the capability now exists of measuring the outer planets and/or their satellites along with the positions

of a number of QSO's (Stone and Dahn, 1994). Since positions of the QSO's are well-known in the radio catalogues, the whole CCD reduction process can be done in that frame.
- The reference frame will remain stable between different ephemerides at the milliarcsecond level; in the past, it was not unusual to see a difference of 0.''04 between two ephemerides.
- The orientation of the earth will be done directly in the frame of the ephemerides themselves.
- If the VLBI observations eventually span a decade, they will begin to contribute to the determination of the inertial mean motions of the inner planet system.

5. Ephemeris Mean Motions

The possibility of determining mean motions with VLBI observations is significant. The accurate ranging measurements, especially those of the Mariner 9 orbiter (30 m, 1971-2) and of the Viking Lander (10 m, 1976-82), enabled the mean motions of Mars and the earth to be determined with unprecedented accuracy; however, the mean motions of the ephemerides do not remain accurately-known over extended periods of time. The uncertainties remain small only for times which are relatively near to the measurements themselves; away from those times, the accumulated effects of the perturbations of the asteroids begin to accumulate.

To account for the forces of asteroids affecting the ephemerides, the gravitational effects of Ceres, Pallas and Vesta are modeled directly: the orbits of these asteroids are approximated using mean keplerian orbits, and their forces upon the sun, moon, and planets are included directly in the equations of motion. Corrections to these three masses may be determined during the adjustment process.

For nearly 300 other asteroids, the following procedure is used: the density of each is assigned according to the taxonomic class (C, S, or M); the volume is computed from the estimated diameter; using the derived masses, the summed forces of the three taxonomic classes upon the Moon, Earth and Mars are computed for each day and stored upon a file; and the file is interpolated during the integrations. Corrections to the density of each taxonomic class may be determined during the adjustment process.

Even with such attempts to account for the asteroid perturbations, large uncertainties still exist. The mass of Ceres is uncertain at the 5-10% level; those for Pallas and Vesta, 10- 20%; for the rest, the masses may be as much as 100% uncertain, due to large uncertainties in the estimations of both the densities and the diameters. The net effect is that uncertainties in the mean motions of the earth and Mars (the best-known of all the planets) can grow to 0."01/century or more (Williams 1984).

6. Conclusions

The orientation of all of the planets in the solar system, done accurately with respect to the same, stable reference frame, will represent a major milestone in planetary ephemeris improvement. Many factors now make it advisable to pursue this goal: to base the future JPL ephemerides upon the radio source reference frame.

- The radio frame is accurate, stable, accessible and well-defined.
- Frame-ties between the radio frame and the previously-used reference frames now exist.
- Earth orientation is now done with respect to the IERS frame.
- VLBI observations of the inner planets with respect to the IERS frame have now being taken, so that the ephemeris may be automatically aligned onto that frame.
- CCD measurements of the outer planets and their satellites, in conjunction with measurements of QSO's, are now able to refer those positions onto the radio frame.

Acknowledgements

The research in this paper was carried out by the Jet Propulsion Laboratory, California Institute of Technology, under a contract with the National Aeronautics and Space Administration.

References

Folkner,W.M., Charlot,P., Finger,M.H., Williams,J.G., Sovers,O.J., Newhall, XX, & Standish,E.M. (1994) Determination of the extragalactic frame tie from joint analysis of radio interferometric and lunar laser ranging measurements, *Astron. Astrophys.*, **287**, 279-289.

Johnston,K.J. (1994) private communication.

Kovalevsky,J. (1994) The Hipparcos Extra-Galactic Link, *IAU Symposium 166*, The Hague.

Kumkova,I., Tel'nyuk-Adamchuk,V., Babenko,Yu., & Vertypolokh,O. (1994) CONFOR Program: Determination of Relative Orientation Parameters Between VLBI and FK5 Reference Frames, poster paper, *IAU Symposium 166*, The Hague.

Lestrade,J.-F., Phillips,A.E., Jones,D.L., & Preston,R.A. (1994) VLBI Astrometry of Radio Stars, *IAU Symposium 166*, The Hague.

Ma,C., Shaffer,D.B., de Vegt,C., Johnston,K.J., & Russell,J.L. (1990) A Radio Optical Reference Frame. I. Precise Radio Source Positions Determined by Mark III VLBI: Observations from 1979 to 1988 and a Tie to the FK5, *Astron. J.*, **99**, 4, 1284-1298.

Standish,E.M. (1981) Two Differing Definitions of the Dynamical Equinox and the Mean Obliquity, *Astron. Astrophys.*, **101**, L17-18.

Standish,E.M. (1982) Orientation of the JPL Ephemerides, DE200/LE200, to the Dynamical Equinox of J2000, *Astron. Astrophys.*, **114**, 297-302.
Standish,E.M. & Williams,J.G. (1990) Dynamical Reference Frame in the Planetary and Earth-Moon Systems, in *Inertial Coordinate System on the Sky* (J.H.Lieske & V.K.Abalakin, eds.) Kluwer Academic Publishers, Dordrecht, 173-181.
Stone,R.C. & Dahn,C.C. (1994) CCD Astrometry, *IAU Symposium 166*, The Hague.
Williams,J.G. (1984) Determining Asteroid Masses from Perturbations on Mars, *Icarus*, **57**, 1-13.
Zacharias,N., Fey,A.L., Russell,J.L., Johnston,K.J., Archinal,B., Carter,M.S., de Vegt,C., Eubanks,T.M., Florkowski,D.R., Ma,C., McCarthy,D.D., Reynolds,J.E., & Sovers,O. (1994) RORF - A Radio-Optical Reference Frame, poster paper, *IAU Symposium 166*, The Hague.

THE EXTRAGALACTIC
OPTICAL-RADIO REFERENCE FRAME

C. de VEGT, K.J. JOHNSTON

Abstract. The new extragalactic reference frame will be based primarily on the positions of a globally selected number of about 500 compact radio sources, which almost all display optical counterparts, mainly quasars or BL Lac objects. Precise radio positions of these objects on the mas level have been determined by VLBI techniques and will be monitored and updated on a regular basis. In parallel an extensive optical observing program is underway to determine precise optical positions of these objects. The present optical reference frame is based on the FK5 fundamental catalog, containing 4652 brighter stars. The transfer of the VLBI radio frame to the optical domain will be achieved now by the Hipparcos stellar net, containing about 120,000 stars with comparable accuracy. The paper will address details and consequences of the implemenation of the new system for astrometry and general astronomical applications.

E. Høg and P. K. Seidelmann (eds.),
Astronomical and Astrophysical Objectives of Sub-Milliarcsecond Optical Astrometry, 117.

VLBI ASTROMETRY OF RADIO-EMITTING STARS

J.-F. LESTRADE
Observatoire de Paris-Meudon, arpeges ura-cnrs 1757
F92195 - Meudon - Principal Cedex - France

D.L. JONES, R.A. PRESTON
Jet Propulsion Laboratory, Calfornia Institute od Technology
4800 Oak Grove, Pasadena, CA, 91109, USA

R.B. PHILLIPS
Haystack Observatory, Massachussets Institute of Technology
Westford, Mass, 01886, USA

AND

J. KOVALEVSKY, M. FROESCHLE, F. MIGNARD
Centre d'Etudes et de Recherche en Géodynamique et Astronomie
F06130 - Grasse - France

Abstract. High-accuracy astrometric VLBI observations of 7 radio stars are presented with applications to the connection between the Hipparcos and VLBI extragalactic reference frames, to the identification of the radio emitting region in the ternary system Algol and the detectabity of Jupiter-size planet orbiting the radio star σ^2 CrB.

1. Introduction

Very Long Baseline Interferometry (VLBI) observations of radio-emitting stars conducted over the last decade have yielded their positions, annual proper motions, trigonometric parallaxes with a precision of 1 milliarcsec (mas) or better. The applications of such high-accuracy astrometric parameters include the improvement of the astronomical distance scale through trigonometric parallax measurements, the connection of the radio and optical radio celestial reference frames, the attempt to detect Jupiter-size companions orbiting radio-emitting stars, the determination of the orientation (node) and angular separation of binary system orbits for dynamical studies of these systems, the determination of the crucial proper motion of

E. Høg and P. K. Seidelmann (eds.),
Astronomical and Astrophysical Objectives of Sub-Milliarcsecond Optical Astrometry, 119–126.

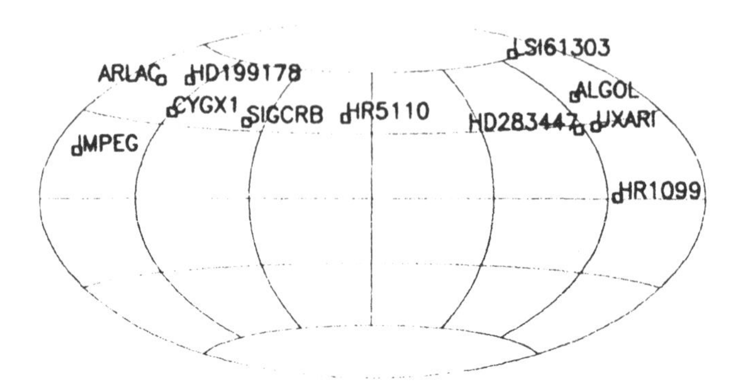

Figure 1. Sky distribution of the VLBI link stars for Hipparcos.

the reference star in the Gravity Probe B mission to test the Lense-Thirring effect of General Relativity.

There are about 30 stars detectable with the state of the art VLBI data acquisition system and there might be 100 detectable by the end of the decade with the new VLBI capabilities being developed in the US and in Europe.

Our VLBI astrometric program of 11 radio-emitting stars (Figure 1) started in 1982 and its initial motivation was to connect the Hipparcos optical reference frame to the VLBI (radio) extragalactic reference frame at the milliarcsec level. The flux densities of these 11 stars are variable between a few milliJansky and a few tens milliJansky, i.e. 100 to 1000 times weaker than the extragalactic radio sources usually observed by VLBI. We had to resort to the phase-referencing VLBI technique both to enhance sensitivity with multi-hour integrations and to achieve high astrometric accuracy through use of the differential inteferometric phase between the target star and an angularly nearby VLBI extragalactic source. The details of this technique are described in Lestrade *et al* (1990). We shall summarise the state of the Hipparcos/VLBI connection based on our program and present two additional results as spin-offs of these VLBI observations.

2. Preliminary link of the Hipparcos and VLBI reference frames

We have carried out a comparison between the VLBI and Hipparcos astrometric parameters of seven radio-emitting stars of our program. Both techniques (VLBI and Hipparcos) have provided positions, annual proper motions and trigonometric parallaxes of these stars with comparable formal uncertainties at the 1 mas level or better. We have found that the systematic discrepancies between these two sets of astrometric parameters can be removed by performing a single global rotation between the extragalactic and Hipparcos reference frames. The three angles and annual rates of this rotation are determined at better than the milliarcsec level as shown in Table 1. The Hipparcos parameters used for this determination are from the Hipparcos reduction consortium FAST but a similar result has been found with the other consortium NDAC.

TABLE 1. VLBI/Hipparcos-FAST connection: the 3 angles (A_1, A_2, A_3) and the 3 rates of rotation ($\dot{A}_1$, $\dot{A}_2$, $\dot{A}_3$) between the Hipparcos and VLBI reference frames have been determined with 7 link stars.

Weighted Least-squares-fit solution
Rotation angles at epoch 1991 April 1 :
A_1 = -26.87 ± 0.46 mas
A_2 = -12.64 ± 0.55 mas
A_3 = 22.99 ± 0.46 mas
Rotation rates :
$\dot{A}_1$ = 0.60 ± 0.48 mas/yr
$\dot{A}_2$ = 0.06 ± 0.46 mas/yr
$\dot{A}_3$ = 1.35 ± 0.43 mas/yr

The robustness of the solution has been tested by splitting the 7 FAST link stars into two independent subsets, one with 3 stars (HR1099, HR5110, AR Lac) and one with 4 stars (UX Ari, σ^2 CrB, CygX1, IM Peg). Rotation angles and rates have been solved independently for with these two subsets and compared. The differences are no more than the quadratically combined uncertainties of the two solutions (Table 2).

The three angles of rotation are consistent with the expected values since the FAST Hipparcos coordinate system is aligned upon a *quasi-FK5* reference frame and the FK5 is aligned with the VLBI reference frame to within 70 mas. The measured rotation rates are consistent with the uncertainty of about 0.5 mas/yr in Table 1, although the $\dot{A}_3$ rate reaches 3σ. This means that the procedure used in the Hipparcos data reduction to stop the rotation of the Hipparcos sphere is efficient.

The angles of rotation found are relative to a VLBI extragalactic reference frame that is defined by the International Earth Rotation Service (IERS) VLBI coordinates of the extragalactic reference sources (Arias, Feissel, Lestrade 1991) used for the differential VLBI measurement of the link stars. The post-fit residuals of the star coordinates and proper motion com-

TABLE 2. Differences between the angles and rates of rotation determined after splitting the 7 VLBI link stars into two independent subsets of 3 and 4 stars. The symbol σ is the quadratically combined uncertainties of the 2 solutions.

Rotation parameters	Differences between solutions [mas]
A_1	$+0.53 \sim 0.5\sigma$
A_2	$-0.48 \sim 0.5\sigma$
A_3	$+0.08 \sim 0.1\sigma$
$\dot{A}_1$	$+1.17 \sim 1.0\sigma$
$\dot{A}_2$	$+0.91 \sim 1.0\sigma$
$\dot{A}_3$	$+0.54 \sim 0.5\sigma$

ponents after the adjustment of the global rotation indicate that the consistency between the Hipparcos and VLBI astrometric techniques is at the milliarcsec level. This is the first cross-check between these two astrometric techniques of comparable precision. The level of agreement found is consistent with the expected accuracy of the technique, even though the astrometric parameters used are from preliminary reductions of the data for both techniques. A complete account of this preliminary link is in Lestrade *et al* (1994a).

3. VLBI Astrometric Identification of the Radio-emitting Region in Algol

Spectroscopy, photometry and speckle interferometry have already provided accurate values for almost all the geometric and orbital parameters of the ternary system Algol (e.g. Söderhjelm 1980). Consequently, it is known that the mass ratio M_A/M_B between the primary (main-sequence) and secondary (subgiant) stars of the close binary of this system is ~ 4.5 and thus significantly different from unity. This makes the motion of the radio source quite different depending on whether it is associated with the subgiant or the main sequence star of the close-binary (astrometric measurements exclude that the third component of the system could be responsible for the radio emission). The maximum displacement of the radio source projected on the sky occurs between the quadrature points of the close binary orbit. This means between orbital phase $\phi = 0.25$ and $\phi = 0.75$, if $\phi = 0$ is the primary eclipse time when the cool subgiant is in front of the bright B8V star. The magnitude of this displacement would be $0.0036''$ (from $2 \times \frac{M_A}{M_A+M_B} \times a$) if the emitting region is centered on the subgiant but it

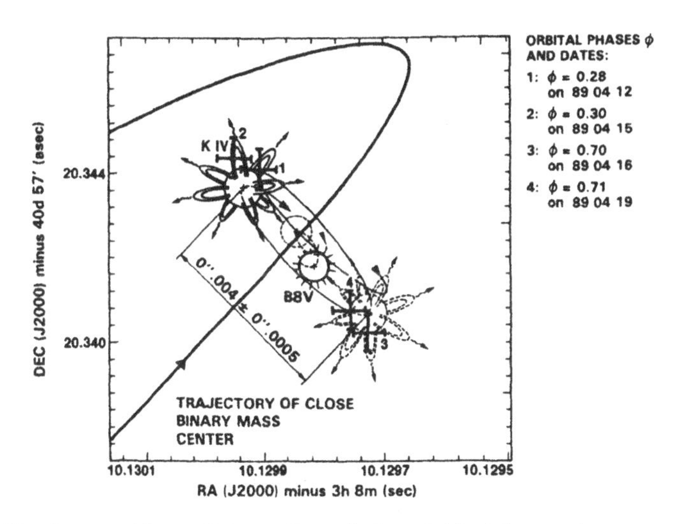

Figure 2. The four positions of the radio emission of Algol (crosses) measured by VLBI are superposed on the configuration of the close binary during the observations at orbital phases 0.25 and 0.75. A portion of the trajectory of the close binary mass center orbiting the third star in 1.86 yr is displayed.

would be 0.0008″ (from $2 \times \frac{M_B}{M_A+M_B} \times a$) if centered on the more massive main sequence star B8. a is the semi-major axis and is known. The displacement would be intermediate if the emitting region is associated with the transient flow of gas between the two stars.

In April 1989, we made four VLBI astrometric observations of the Algol radio emission. The observations took place on April 12th, 15th, 16th and 19th, allowing 2.5 orbital revolutions of the close binary to be sampled over these 7.5 days. These four epochs were chosen so that the observations took place when the two stars of the close binary were as close as possible to the orbital phases $\phi = 0.25$ and $\phi = 0.75$. A single observation at each of these two orbital phases would have sufficed, but two additional observations were scheduled because of possible technical failures during the observations and for redundancy of the measurement. All four observations were successful.

We measured the displacement of the radio source in Algol of 4±0.5 mas between two orbital phases 0.25 and 0.75 over two consecutive orbital revolutions of the close binary (see Figure 2). The *magnitude* of the displacement unambiguously indicates that the less massive star of the close binary, the K subgiant, is the star responsible for the non-thermal radio emission of the system. This is consistent with the idea that the radio emission in Algol is related to the strong magnetic activity of the subgiant.

The *orientation and sense* of the displacement on the sky that are di-

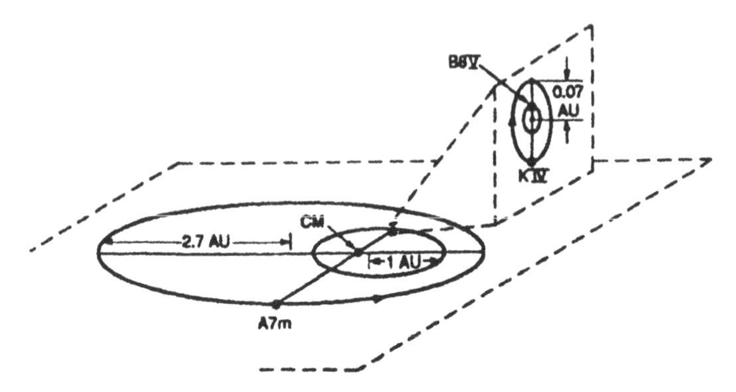

Figure 3. Configuration of the two orbital planes in the Algol ternary system.

rectly deduced from our astrometric VLBI observations imply 1) that the orbital plane of the close binary is at p.a. = $+52 \pm 5°$ (p.a.= 0° is North and p.a. = 90° is East) and 2) that the sense of circulation of the close binary is clockwise, as seen on the sky. Thus, the long-period and close binary orbital motions are almost orthogonal and counter-revolving (Figure 3). The orientation and sense of circulation of the close binary determined by our VLBI observations are identical to the findings of Rudy (1979) and Kemp *et al* (1981, 1983) obtained less directly with optical polarisation observations. This intriguing configuration is difficult to reconcile with our present idea on the formation, evolution and dynamical stability of multiple stellar systems, since it is difficult to conceive that the angular momentum of the primordial cloud of this ternary system was split along two orthogonal axis. For a full account of these observations see Lestrade *et al* 1993.

4. Astrometric detectability of a Jupiter-size planet orbiting the radio-emitting star σ^2 CrB

The motion of a single planet in a circular orbit around a star causes the star to undergo a reflexive circular motion around the star-planet barycenter. When projected on the sky, the orbit of the star appears as an ellipse with angular semimajor axis θ given by $\theta = \frac{m_p}{M_*}\frac{a}{d}$ (*eq.* 1) where θ is in arcsec when the semimajor axis a is in AU, the mass of the planet (m_p) and the mass of the star (M_*) are in solar masses and the distance d is in pc. For example, observing the solar system from a distance of 10 pc, the presence of Jupiter would be revealed as a periodic circular displacement in the Sun's position, with a diameter 2θ of 1.0 mas and a period of 11.9 years.

Ground-based optical astrometry has generally been limited to a precision of a few tens of milliarcsec although the best measurements are at the 1 mas level now (Gatewood et al 1992). VLBI astrometric observations of

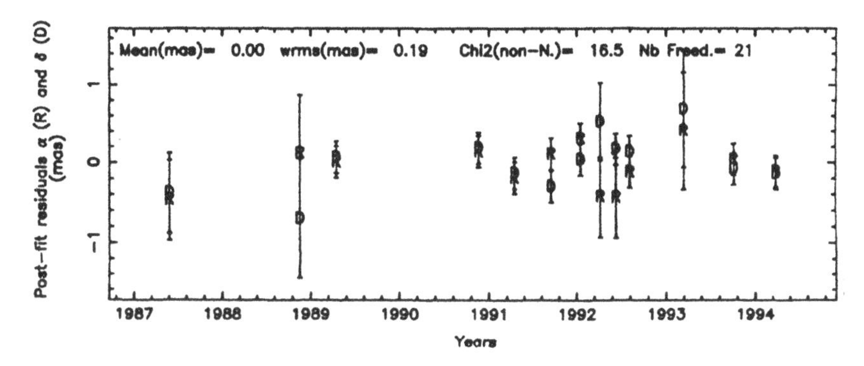

Figure 4. Post-fit coordinate residuals for σ^2 CrB. The 5 astrometric parameters of σ^2 CrB (coordinates, proper motion components and trigonometric parallax) are adjusted to the coordinates measured by VLBI at 13 epochs.

σ^2 CrB conducted by us at 13 epochs between 1987 and 1994 have yielded its position, proper motion and trigonometric parallax and the resulting post-fit coordinate residuals are characterised by an rms as small as 0.2 mas (see Figure 4). The formal uncertainties for the 5 fitted astrometric parameters are 0.08 mas for the relative position between σ^2 CrB and the reference source 1611+343, 0.04 mas/year for the proper motion and 0.08 mas for the trigonometric parallax (Lestrade *et al* 1992, 1994b).

The lack of a clear sinusoidal signature in the post-fit coordinates residuals of Figure 4 sets a limit on the presence of a planet around σ^2CrB for the 7 years observation span. The rms of the post-fit residuals (0.2 mas in Figure 4) is an upper limit on any systematic departure from linear motion of the star. Eq (1) can be used to exclude a range of planetary perturbations by taking $2\theta = 0.2$ mas, $M_* = 2.26\ M_\odot$ and d = 22.7 pc for σ^2CrB. The log-log representation of eq (1) with these parameters is in Figure 5 and in Lestrade *et al* (1994b). The diagonal line of constant astrometric signature in Figure 5 follows eq (1) for $2\theta = 0.2$ mas. We assume that a full orbital period of the planet must be sampled during the observations to separate the sinusoidal planetary signature from the fitted linear proper motion. In these conditions, the maximum semimajor axis a of the orbit of a planet corresponds to the total observation span through the third Kepler law. This lower limit on a is 4.8 AU for the 7 years of observations and is indicated by the vertical dashed line in Figure 5. Finally, the shaded area indicates the planet parameter space (a, m_p) that are excluded by the present observations, i.e. any planetary perturbations in the shaded area is too large to be consistent with the post-fit coordinate residual rms of 0.2 mas measured over 7 years by VLBI. Note that for $a = 4.8$AU, the upper limit on the planet mass m_p is 0.0012 $M_\odot$, ($\sim$ 1 Jupiter mass) as derived

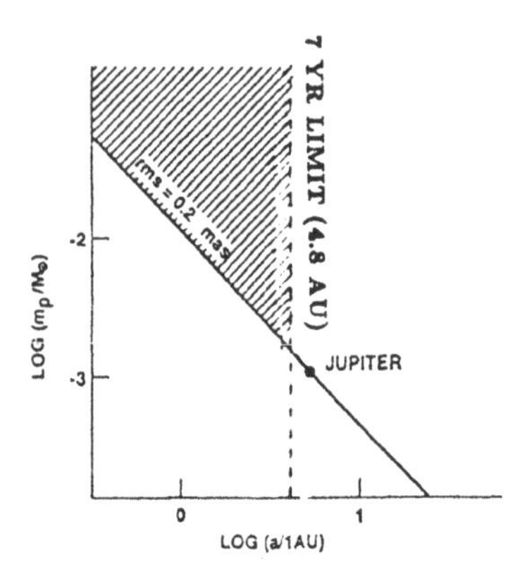

Figure 5. Log-Log representation of eq (1) for the rms of the post-fit coordinate residuals (0.2 mas) of σ^2 CrB. The shaded area is the parameter space (semi-major axis a, mass m_p) that are excluded by our 7 years of VLBI observations for a planet around this star.

by eq.(1) for σ^2 CrB and $2\theta = 0.2$ mas.

This limit will become more stringent if the period of observation is extended. The ultimate goal of our program is now to extend these 7 years of astrometric VLBI observations of σ^2 CrB to complete a full 12 year cycle typical of a Jupiter-size planet that might orbit this star. The goal is also to improve the level of precision of our astrometric measurement since this level is not presently SNR-limited. It might reach 20 **microarcseconds** if all systematic errors could be removed by an improved strategy of observations and data analysis.

References

Arias, F., Feissel, M., Lestrade, J-F., 1991, The IERS extragalactic Reference Frame and its Tie to Hipparcos, IERS Technical Note 7, December 1991, Observatoire de Paris, France.

Gatewood, G., Stein, J., Joost Kiewiet de Jonge, T., Persinger T., Reiland, T., 1992, Astron. J., 104, 1237

Lestrade J.-F., Rogers A.E.E., Whitney A.R., Niell A.E., Phillips R.B., Preston R.A., 1990, *Astron. J.*, **99**, 1663-1673.

Lestrade J.-F., Phillips R.B., Preston, R.A., D. C. Gabuzda, 1992, Astron. Astroph., 258, 112-115

Lestrade J.-F., Phillips R.B., Hodges, M.W., Preston, R.A., 1993, Astroph. J., 410, 808-814.

Lestrade J.-F., Jones, D.L., Preston, R.A., Phillips R.B., Titus M.A., Kovalevsky J., Lindegren L., Hering R., Froeschle M., Falin, J-L, Mignard, F., Jacobs, C.S., Sovers, O.J., Eubanks, M., Gabuzda, D., 1994a, Astron. Astroph. (submitted)

Lestrade J.-F., Jones, D.L., Preston, R.A., Phillips R.B., 1994b, Astroph. and Space Science, vol. 212, Kluwer Academic Publishers, p.251-260.

Rudy, R.J., 1979, MNRAS, 186, 473

Kemp, J.C., Barbour, M.S., McBinney, R.E., Rudy, R.J., 1981, Astroph. J., 243, 557

Söderhjelm, S., 1980, Astron. Astroph., 89, 100.

THE HIPPARCOS EXTRAGALACTIC LINK

JEAN KOVALEVSKY
Observatoire de la Côte d'Azur/CERGA,
Av. Copernic 06130 Grasse, France

Abstract. In conformity with the IAU resolutions on reference frames adopted in 1991, the Hipparcos catalogue will represent, in the visible spectrum, the celestial reference system defined by fixed positions of extragalactic radio-sources. This will be realized by the strongest possible link between the IERS celestial reference frame with positions and/or proper motions of the largest possible number of Hipparcos stars determined also with respect to extragalactic objects. The data which will be used must be available before April 1995. It will include the following: positions and proper motions of radio stars observed by VLBI, VLA and MERLIN; photographic positions in fields including quasars; proper motions with respect to galaxies of the Lick, Yale, and Kiev programs, proper motions derived from pairs of photographic plates taken at large time intervals; and possibly data acquired by Hubble Space Telescope and from Earth's rotation data. The organization of the tasks within the working group is briefly described. The final accuracy of the link is expected to be of the order of, or better than, half a milliarcsecond.

1. Introductory remarks

The set of recommendations that constitute the IAU resolution A4 (Bergeron, 1992) on celestial reference systems and frames states that the new space coordinate grids will have no global rotation with respect to a set of distant extragalactic objects. The conventional celestial reference frame which will represent this system will consist of positions of about 600 extragalactic radio sources the list of which is being presented at the present IAU General Assembly by the working group set up to this effect and chaired by C. de Vegt (see resolution N°B6, this volume). A preliminary realization of

E. Høg and P. K. Seidelmann (eds.),
Astronomical and Astrophysical Objectives of Sub-Milliarcsecond Optical Astrometry, 127–132.

this frame already exists: it has been prepared by IERS. However, note 5 to recommendation VII states that as long as the relationship between the optical and the extragalactic radio frames is not sufficiently accurately determined, the FK5 catalogue shall be considered as a provisional realization of the celestial reference system in optical wavelengths.

Now that it is assessed that the Hipparcos catalogue will be considerably more precise and denser than the FK5, and that it will be in the future years the main accurate astrometric reference catalogue, it was resolved by the Hipparcos Science Team that this catalogue should be linked as tightly as possible to the fundamental reference frame in radio wavelengths so that it satisfies the demands of the IAU resolutions and hence becomes the realization of the new celestial reference system in optical wavelengths.

This task is undertaken by a working group of the Hipparcos Science Team consisting of P.D. Hemenway, K.J. Johnston, J. Kovalevsky, J.L. Lestrade, L. Lindegren, R.A. Preston, W.F. Van Altena and C. de Vegt.

2. Method

The principle of the link was described by Froeschlé and Kovalevsky (1982). Let us call $\mathbf{X_V}$ and $\mathbf{X_H}$ respectively the unit vectors of the direction of a star as observed in the extragalactic and in the Hipparcos reference frames. Similarly, $\mathbf{X_V}'$ and $\mathbf{X_H}'$ are the proper motions of a star in these systems. One has

$$\mathbf{X_H} = R\mathbf{X_V} \tag{1}$$

$$\mathbf{X_H}' = R\mathbf{X_V}' + R'\mathbf{X_V} \tag{2}$$

From both indices, one has

$$\mathbf{X}\begin{cases} \cos\delta\cos\alpha, \\ \cos\delta\sin\alpha, \\ \sin\delta, \end{cases} \qquad \mathbf{X}'\begin{cases} -\mu_\alpha\cos\delta\sin\alpha - \mu_\delta\sin\delta\cos\alpha, \\ \mu_\alpha\cos\delta\sin\alpha - \mu_\delta\sin\delta\sin\alpha, \\ \mu_\alpha\cos\delta. \end{cases}$$

where the rotation matrices are

$$R = \begin{vmatrix} 1 & \gamma & -\beta \\ -\gamma & 1 & \alpha \\ \beta & -\alpha & 1 \end{vmatrix} \qquad \mathrm{R}' = \begin{vmatrix} 0 & \gamma' & -\beta' \\ -\gamma' & 0 & \alpha' \\ \beta' & -\alpha' & 0 \end{vmatrix}$$

The work consists in writing as many observation equations (1) and (2) as possible and then solving them for R and R'. Then, applying the rotations R

and R' to all positions and proper motions of the Hipparcos catalogue, one will obtain a new version of the catalogue which will be in the extragalactic reference system with an accuracy to be derived from the accuracy of the conventional extragalactic reference frame used and the variance-covariance matrix of the determination of the matrices R and R'.

3. Data available

In order to determine R, it is necessary that the positions of the Hipparcos stars be directly determined with respect to the extragalactic radio sources belonging to the fundamental list of objects. The most precise source of such data is VLBI astrometry. A total of 14 stars are expected to be observed by the Deep Space Network and the Australian VLBI system. A sub-milliarcsecond precision is expected. However, some of these stars are double and the necessary corrections for orbital motion to represent the motion of the radio source may not be so accurately determinable. Less precise, but more numerous, are the positions of stars determined by VLA or Merlin interferometers. About 40 star positions have been obtained since 1984 with a precision of the order of 30 mas (see for instance, Florkowski et al., 1985). Some of them will be reobserved and will provide proper motions.

Another source of information are photographic plates in which positions of Hipparcos stars are determined with respect to quasars whose positions are known in the extragalactic reference frame. Some 300 such plates are available and are being measured and reduced by C. de Vegt. Although the precision of each plate measurement does not permit to determine star positions to better than $0\overset{''}{.}1$, an accuracy of 3 mas is expected for the link using these data alone.

The most important quantity requested from the link is R'. Although the best possible coincidence of the origins of the coordinate systems at epoch is very desirable, it is even more important for all kinematic and dynamical studies which will be performed with the catalogue, that the residual rotation of the coordinates be reduced to a minimum. It is to be remarked that it is not necessary that the proper motions be referred to extragalactic objects of known positions. It is sufficient that the proper motions of stars are determined with respect to extragalactic objects. So several types of data are available.

The VLBI observations provide absolute proper motions with a sub-milliarcsecond per year precision for the 14 stars already referred to. Reobservation by VLA and Merlin of some 10 other radio sources will provide also accurate proper motions.

Pairs of photographic plates of the same region separated by a large time interval and reduced with respect to quasars assumed to be fixed, will provide absolute proper motions. This will be the case for about 50 pairs of Schmidt plates taken in Tautenburg at about 25 year intervals, while a dozen of pairs of plates separated by intervals of time as long as 90 years are available in Bonn Observatory. It is to be noted that a discussion of residuals of the de Vegt quasar photography program as a function of the date may give some information on the time-dependent rotation.

A very important contribution to the link are the results of large surveys whose main objective was to measure absolute stellar proper motions with respect to an extragalactic reference. Although they were generally designed for faint stars of astrophysical interest, there is also a large number of brighter stars which are in the Hipparcos catalogue. The most important is the Lick Northern Proper Motion Program (Klemola et al., 1987). Among the 300 000 measured stars, about 7000 are in the Hipparcos catalogue. A smaller program, the Yale Southern Proper Motion Program (van Altena et al., 1986), has the advantage of concentrated observations around the Southern pole. It has some 1200 stars in common with Hipparcos. Finally, the Kiev survey for the northern hemisphere has about 2500 common stars with Hipparcos.

There are also some hope that despite a late start in the program, some radial components of proper motions from quasars can be determined from Hubble Space Telescope observations. If they are not ready in time, they will provide an a-posteriori check of the link.

Another type of link was proposed by Vondrak during this general assembly of the IAU. Earth orientation parameters obtained by optical methods (astrolabes and transit instruments) during the last years will be re-reduced using positions and proper motions of the Hipparcos catalogue and then compared with those deduced VLBI observations. The differences will be interpreted as a rotation and a time dependent rotation of the Hipparcos reference system in the VLBI extragalactic reference system.

4. Status of the work

For all the data described above, reduction software is written by the teams responsible for the various observation techniques (radio, photographic plates, surveys). A provisional catalogue of reference radio sources had been provided by IERS. The positions and proper motions from the Hipparcos 30 month combined solution were provided to all the teams. Except in the case of radio observations, only single stars will be included in the test solution which should be examined and discussed in early 1995. There are two reasons for this decision. While the position of the radio emission in a couple is well identified within a couple of mas, the point representative of a close binary may not be the same in Hipparcos and on a photographic plate. In addition, the proper motion determined in a short time interval may not be well corrected for orbital motion in the present version of the catalogue.

The final version of the combined Hipparcos catalogue will be available in April 1995, so that the final link should be ready by mid-1995. At present, it is not possible to assess what will be the accuracy of the link. From the very partial results obtained and from the accuracy evaluations of various techniques (see table 1), one should guess that half a milliarcsecond in the components of both matrices R and R' could be achieved. An optimistic view is that the accuracy could be even significantly better for R'.

TABLE 1. Link of Hipparcos catalogue to the extragalactic reference frame - Preliminary expectations

Technique	σR (mas)	$\sigma R'$ (mas/yr)
VLBI	0.5	0.5
VLA/MERLIN	3	1
HST		1.5
QSO + Photo	3	?
Photo Bonn		0.7
Photo Potsdam		1
Survey		0.3
Earth's rotation	?	?

Adopting this optimistic approach and assuming an error of the link of the order of 0.25 mas per year, one can compare the evolution of the error with time as compared with the error evolution of the major catalogues and of the random errors of Hipparcos (Fig. 1). One can see from this figure how much better the Hipparcos frame will be in comparison with the Hipparcos catalogue. A major task for future astrometry will be to extract from the Hipparcos catalogue a number of stars whose positions and proper motions will be improved in order to materialize as accurately as possible this frame.

5. Acknowledgements

Fifteen different institutes are contributing to this very diversified task. These are, first of all, ESA, European Space Agency, responsible of Hipparcos programme and, in alphabetic order : Astronomical Institute (Czeck Republic), CSIRO/ATNF (Australia), Hamburg Observatory (Germany), Haystack Observatory (USA), JPL (USA), Kiev Observatory (Ukraine),

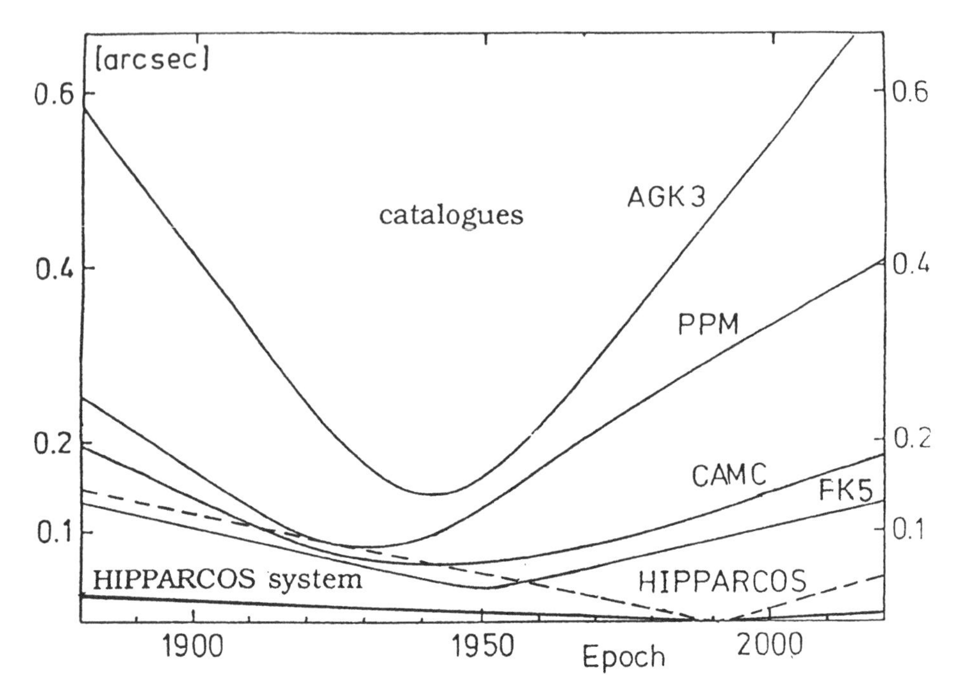

Figure 1. Evolution of the precision of various catalogues and of the Hipparcos system as a function of time

Lund Observatory (Sweden), Observatoire de la Côte d'Azur/CERGA - (France), Observatoire de Paris (France), Observatorium Hoher List (Germany), Potsdam Observatory (Germany), US Naval Observatory (USA), University of Texas (USA), Yale Observatory (USA). Their contribution is greatfully acknowledged.

References

Bergeron, J. (ed.), 1992, *Proceedings of the twenty-first General Assembly of the IAU*, Kluwer Acad. Publ., pp 41–64.

Florkowski, D.R., Johnston, K.J., Wade, C.M. and de Vegt, C., 1985, Astron. J., **90**, pp 2381–2386.

Froeschlé, M. and Kovalevsky, J., 1982, A & A, **116**, pp 89-96.

Klemola, A.R., Jones, B.F. and Hanson, R.B., Astron. J., **94**, pp 501-515.

van Altena, W.F., Girard, T., Lòpez, C.E., Klemola, A.R., Jones, B.F. and Hanson, R.B., 1986, *Highlights of Astronomy*, J.P. Swings (ed.), **7**, pp 89-92.

2. CURRENT AND FUTURE NEEDS FOR VERY ACCURATE ASTROMETRY

2.1 STELLAR ASTROPHYSICS

ACCURATE PARALLAXES AND STELLAR AGES DETERMINATIONS

Y. LEBRETON, E. MICHEL, M.J. GOUPIL, A. BAGLIN, J. FERNANDES
DASGAL, URA CNRS 335, Observatoire de Paris, Section de Meudon
92195 MEUDON Cedex, FRANCE

Abstract. Uncertainties on stellar ages due to the physical description of the stellar material entering the models and to measurements of global parameters and chemical composition are estimated in the case of A/F stars.

With combined efforts on atmosphere modeling (to improve global parameters), on asteroseismology (to improve the physical description of the stellar interior) and on distance determination at the level of the sub-milliarcsecond (to improve luminosity), an accuracy on age determination of about 10% is foreseeable.

1. Introduction

Precise stellar ages determinations are now required for a variety of topics, as galactic evolution or cosmological time scale. The "primary" determination of ages relies on comparisons of models of internal structure with the best available data on individual stars, or stellar groups.

Though the principle of the method is general, its application to different types of stars requires specific developements. We consider the determination of the ages of individual stars and we focus on late A or F spectral type stars on the main sequence, which age range (several 10^6 to a few 10^8 yr) is appropriate for galactic evolution studies.

Stellar age is currently derived from global observable parameters (luminosity and effective temperature), matching the position of a real star and of models in the HR diagram. Uncertainties on models are often ignored. We will show here that, in order to take advantage of the forseeable accuracy on the determination of the global parameters, this simple (first degree) method is inappropriate. One needs to constrain seriously other parameters, with the help of other variables, i.e. asteroseismologic quantities.

2. The "Model-Observation" Confrontation

Stellar interior models are built by integration of the hydrodynamical equations with initial and boundary conditions, assuming spherical symmetry of the star and neglecting rotation and magnetic field. Mass M and chemical composition, i.e. helium content Y and metallicity Z, have to be specified as initial conditions. A physical description of the material is needed, it is assumed here to be globally correct but for some adjustment

E. Høg and P. K. Seidelmann (eds.),
Astronomical and Astrophysical Objectives of Sub-Milliarcsecond Optical Astrometry, 135–142.

parameters (P_i). The solution also depends on time, hereafter called "age" A. Consequently any output global characteristic Q_j of the model, such as observable quantities like effective temperature or luminosity, is obtained as a function Ψ of these inputs, i.e. $Q_j = \Psi(A, M, Y, Z, P_i)$.

In order to obtain an age one has to find the model which reproduces the observed quantities. It requires in principle to inverse Ψ, which is generally non linear, and sometimes not bijective. We consider here A/F stars on the main sequence, in a region of the HR diagram where the Vogt-Russell theorem applies and where the star position is very sensitive to age. The correct model and its age are obtained by iteration, starting from an approximate model and linearizing about this solution to get the necessary derivatives. A "differential" formalism writes:

$$dQ_j = D\Psi(dA, dM, dY, dZ, dP_k) \tag{1}$$

where $D\Psi$ is the linear operator built with the partial derivatives of the observable quantities with respect to the variables V_i (i.e. initial conditions, "physical" description and age). If the number of accessible observable quantities is larger than the number of parameters a χ^2 minimization technique can be used to estimate the parameters and their uncertainties. Brown *et al.* (1994) have proposed to solve this problem in the case of solar type stars through a singular value decomposition of $D\Psi$.

However, in the "classical" problem, luminosity, effective temperature and metallicity Z are the only observable quantities and the age is estimated by comparison in the HR diagram for a "given" chemical composition. All the variables cannot be determined directly which means that age determinations will rely on the choice of the additional parameters entering physical description.

3. Evaluation of $D\Psi$

Numerical computations of a "reference" sequence and of a set of sequences where one variable is modified with respect to the reference variable V_{j0} are performed to calculate:

$$\partial Q_k / \partial V_i \;\; \text{at} \;\; V_j = V_{j0} \;\; \text{for all j} \neq \text{i} \tag{2}$$

3.1. THE "REFERENCE" MODEL $A_0, M_0, Y_0, Z_0, P_{K0}$

The models are calculated with the CESAM code (Morel, 1993, 1994). The reference model is built with the most standard updated physics and a solar chemical composition. We took the nuclear reaction rates from Caughlan and Fowler (1988) and used the third generation of OPAL opacities (Iglesias *et al.*, 1992). Convection is treated with the mixing-length theory. We considered that an overshooting process extends the size of the convective core over a distance $d_{Ov} = O_v H_p$ where H_p is the pressure scale height. The reference value $O_v = 0.2$ comes from Schaller *et al.* (1992) who derived it by comparison of the observed main sequence width of clusters with that given by theoretical isochrones. The solar mixture and metal content ($Z_\odot = 0.019$) are from Grevesse (1991). The constraint that the solar model must yield at solar age the observed luminosity and radius gives an initial helium content $Y_\odot = 0.287$ and a mixing-length parameter $\alpha_0 = 1.67 H_p$.

3.2. THE UNCERTAINTIES

3.2.1. *The Physical Terms P_k*

Many aspects of the physical description of stellar structure remain unknown. We focus here on the current uncertainties in the range of mass considered.

* The $^{14}N(p,\gamma)^{15}O$ reaction is the slowest reaction in the main CN cycle and controls the energy generation in the CNO cycle. The uncertainty on the cross-section S_0 is of about 17% as discussed by Parker and Rolfs (1991).

* Recent OPAL opacities have been validated by different facts, as the coherent modeling of double mode cepheids. To estimate the effects of the remaining uncertainties we use a linear parameter O_p. The O_p-value is 1 if the OPAL tables are used and 0 for the Los Alamos tables (Huebner *et al.*, 1977).

* The question of the universality of the mixing-length parameter α is still open (see i.e. Neuforge and Fernandes, 1994). We assume here an uncertainty of 20% around the solar value corresponding to our standard physical ingredients.

* There has been an already long debate on whether overshooting should be taken into account. Some authors do not consider this process while Napiwotski *et al.* (1993) give $d_{Ov} = 0.15\ H_p$ and Schaller *et al.* (1992) propose $d_{Ov} = 0.20\ H_p$.

3.2.2. *The Chemical Composition Terms*

They have to be determined for each object.

* The helium content is difficult to observe in A/F stars, and also to constrain. One has then to rely on "prejudices" to estimate Y and its uncertainty. We consider (Fernandes *et al.*, 1994) that the enrichment law, as constrained by the width of the main sequence, is valid here, i.e. $(Y - Y_p)/Z \subset [2,5]$, where Y_p is the primordial value taken at 0.228.

* The metallicity can be determined by observations. For simplicity, we assume that all objects have a solar mixture and that only Z varies.

3.3. THE DIFFERENTIAL OPERATOR $D\Psi$

Derivatives in (2) are estimated as finite differences. The most appropriate estimate is obtained using increments in the variables of the order of the probable uncertainties. Since derivatives strongly depend on mass and on the stage of evolution, we have chosen two "characteristic" models , both slightly evolved ($X_c \sim 0.25$): $M1$ is a $1.4 M_\odot$ star at an age of 2.3 10^9 yr and $M2$ is a $2 M_\odot$ star at an age of 0.81 10^9 yr. The positions in the HR diagram are shown on Fig. 1. and derivatives are given in Table 1 with the increments used for this evaluation.

4. The Leading Factors in Age Determinations

4.1. THE CLASSICAL CASE

Once $D\Psi$ is known, the accuracy on the model parameters are in principle obtained by solving the linear equation $\Delta Q_i = D\Psi \Delta V_j$, where ΔQ_i represent the relative errors on the observable quantities. This inversion "mixes" the role of the various parameters, as also discussed by Brown *et al.* (1994). In the "classical case" we are dealing with (comparison in the HR diagram), there are less observables than unknown quantities and we treat physical terms as parameters varying in a "reasonable" interval and the "chemical

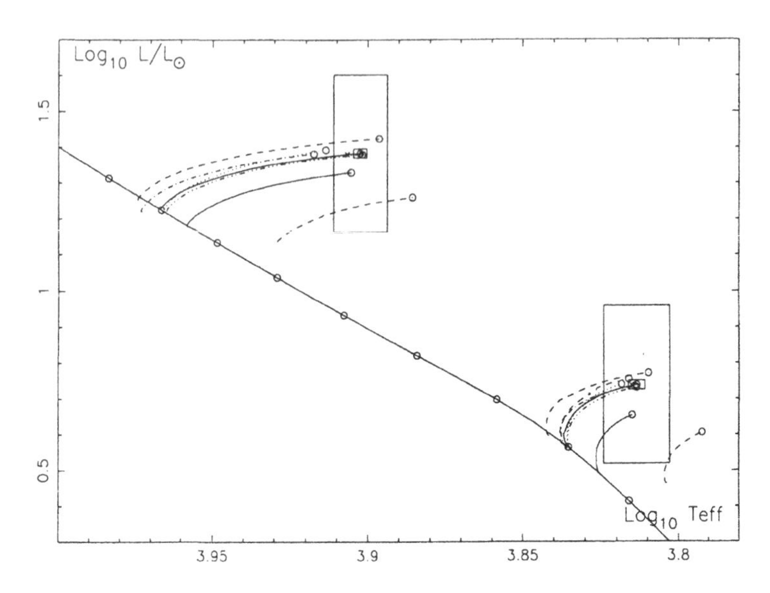

Figure 1. HR diagram showing the evolutionary sequences of the two models $M1$ and $M2$, as computed with the different parameters. The error boxes are given in both cases: $Si1$ (the large one) and $Si2$ (the small one).

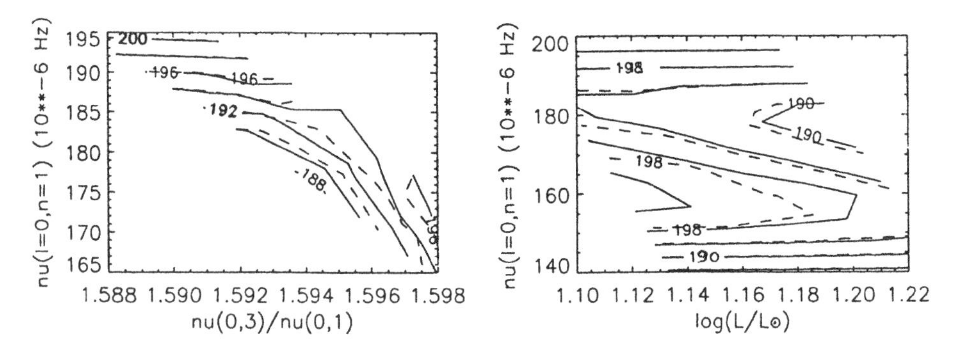

Figure 2. Isocontours of ν_{G1} for models between 1.75 and 1.85 $M_{\odot}$ along the main sequence, (a) in the plane ($\nu_{0,3}/\nu_{0,1}$, $\nu_{0,1}$), (b) in the plane ($logL/L_{\odot}$, $\nu_{0,1}$). Dashed lines refer to models with $O_v = 0.20$, and solid lines to $O_v = 0.19$, other physical and chemical parameters being the same.

TABLE 1.

	model	∂LnA	∂LnM	∂LnS	$\partial LnOp$	$\partial Ln\alpha$	$\partial LnOv$	∂LnY	∂LnZ
increment		0.015	0.05	0.17	1	0.33	0.1	0.0013	0.011
∂LnL	M1	0.25	5.3	-.098	-0.0077	0.046	0.012	1.63	-0.74
∂LnL	M2	0.56	4.9	-0.052	-.0049	0.0048	0.051	2.02	-0.62
∂LnT_{eff}	M1	-0.16	-0.086	0.0014	0.0016	0.12	0.028	-0.19	-0.075
∂LnT_{eff}	M2	-0.21	-0.26	-0.011	0.034	0.0025	0.063	-0.31	-.080

composition terms" as determined through observations. We extract ΔA and ΔM from the uncertainties ΔL and ΔT on $logL$ and $logT_{eff}$, and ΔP_i, ΔZ, ΔY on the parameters.

For simplicity we write $dQ_{Vi} = |\partial Q_k / \partial V_i|$ at $V_j = V_{j0}$ for all $j \neq i$.

$$\Delta A = (dT_M\, \Delta L + dL_M\, \Delta T + \Sigma_i (dL_{Pi}\, dT_M + dL_M\, dT_{Pi}) \Delta P_i) / |dL_A\, dT_M - dT_A\, dL_M| \quad (3)$$

where the summation in (3) has to be understood as a r.m.s.

4.2. MAGNITUDE OF THE UNCERTAINTIES

At the primary level discussed here, luminosity determinations rely on distance and bolometric correction BC. We assume that the targets are studied with the most accurate methods and that the temperature indicator is directly the effective temperature, i.e. that no preliminary calibration is needed. Bolometric corrections are computed from an adjustment of a model atmosphere, based on extended measurements of the flux at a wide range of wavelength. Metallicity is given by observations and is supposed to be derived from detailed analysis. We examine a present situation, situation 1 ($Si1$), that includes the HIPPARCOS enormous progress on distance measurements and a future one, situation 2 ($Si2$) that includes a sub-milliarcsecond mission and the foreseeable progresses in stellar atmosphere modeling.

Presently, around 10 000K uncertainties of 200K on effective temperature and 0.1 magnitude on bolometric correction are reachable. With future progress we can expect to get an accuracy of 40K for T_{eff} and 0.02 magnitude for BC, at least for a number of reference objects.

For chemical composition, the present uncertainty on Z is of the order of 0.006 (Edvardsson *et al.*, 1993) and one can hope to reduce it by a factor 3 to 5. The uncertainty on Y depends on the Z-uncertainty as well as on the value of the enrichment factor for the particular star considered. The present uncertainty is estimated at 0.04, and a reduction by a factor 3 is assumed for situation 2.

For luminosity, since these age indicators are quite far, we choose a typical distance $r = 200pc$, necessary to handle a sufficient sample of targets. In the present situation, distances cannot be obtained directly, even with HIPPARCOS, and one has to rely on photometric calibrations. A direct determination will be possible with the next astrometric mission, if it reaches an accuracy 20 times higher than HIPPARCOS. Table 2 summarizes these estimates.

For the "physical " terms, we have chosen: $S = 0.17\ S_0 = 0.54$ keVb, $\Delta O_p = 0.2$, $\Delta\alpha = 0.3$, $\Delta O_v = 0.15$.

TABLE 2.

	$\Delta log T_{eff}$	ΔBC	$\Delta r/r$	$\Delta log L$	ΔZ	ΔY
Si1	0.01	0.1		0.08	0.006	0.04
Si2	0.002	0.02	0.02	0.025	0.002	0.015

Table 3 gives the values of the different contributions for the two models $M1$ and $M2$, in the two situations $Si1$ and $Si2$, and a typical object at 200 pc. We note, for instance, ΔA_T the contribution of the term ΔT in equation (3).

Presently, age determinations between $1.4 M_\odot$ and $2 M_\odot$ (i.e. ages from 0.7 to 3 10^9 yr) cannot reach an accuracy better than 30 to 40%.

For stars around $1.4 M_\odot (M1)$, the situation is quite complex. Age determinations are sensitive to both the observational terms of the atmospheric analysis and the physical terms from the hydrodynamical processes, as outer convection and overshooting. For the $2 M_\odot (M2)$ star, the relative importance of the different terms changes from $Si1$ to $Si2$. In situation 1, distances determinations contribute significantly to the uncertainty on age. For situation 2, the overshooting term clearly becomes the leading term. To get a maximum advantage of an increased accuracy on atmosphere modeling and distance determinations, one is bound to fix it more precisely.

TABLE 3. (expressed in %)

		ΔA_T	ΔA_L	ΔA_{SO}	ΔA_{Op}	ΔA_α	ΔA_{Ov}	ΔA_Y	ΔA_Z	ΔA *total*
Si1	M1	12	2.5	0.02	0.2	14	12	21	16	35
Si1	M2	11	7	1.2	4	0.3	24	22	19	40
Si2	M1	3	0.8	0.02	0.2	14	12	10	5	21
Si2	M2	3	2.5	1.2	4	0.3	24	8	5	25
$Si2^*$	M1	3	0.8	0.02	0.2	5	3	5	5	10
$Si2^*$	M2	3	2.5	1.2	4	0.1	6	4	5	10

* including the improvements due to seismology on O_v, α, Y, see $ 5.

5. A Need for Seismology

The advent of helio- and asteroseismology has brought new observable quantities, the eigenfrequencies, which can be used to built a model for a given object. Table 1 can then be complemented by several lines, associated with seismological quantities.

Oscillations are presently observed in several δ Scuti stars (A/F stars) in the low frequency domain which contains low order modes (Michel *et al.*, 1992). Some of these modes (Goupil *et al.*, 1992;Dziembovsky and Pamyatnykh, 1991) are very sensitive to the structure of the inner regions, in particular to the μ gradient layer produced by the recession of the convective core.

As an example, Table 4 gives the derivatives of the frequencies of some low order modes, for a model of $1.8M_\odot$, slightly evolved ($X_c = 0.44$), at an age of 767 10^6 yr. This demonstrates the sensitivity of the "mixed mode", G_1-mode ($l = 1, n = 1$, usual nomenclature), to the overshooting parameter O_v, as compared for instance to a radial mode $(0,1)$. Fig. 2a shows how three well chosen modes can determine O_v. It illustrates nicely the nonlinearity of these indicators, confirming that a preliminary guess of the model is necessary. With three frequencies $\nu_{0,1}$, ν_{G1} and $\nu_{0,3}$, three variables can be determined: M, A and O_v, assuming that the other parameters are precisely known.

The accuracy on frequencies measurements can be very high: in self excited modes, it is only limited by the total duration of observations. If we choose an uncertainty of $0.1\mu Hz$ corresponding to an observing run of 40 days, we can expect to determine O_v at $\sim$ 20 %, i.e. $\sim 0.03H_p$. This reduces its contribution to the error on age to 3 % for $M1$ and 6% for $M2$ and brings it to a level comparable to the other terms. The remaining parameters have then to be adjusted using other good quality observables. Fig. 2b shows that, at the level of accuracy proposed for the sub-milliarcsecond mission, luminosity will be sufficiently precise ($\sim$ 2 %) to become as constraining as the frequencies for stellar modelling.

TABLE 4.

mode	$\nu(\mu Hz)$	$\partial Ln\nu/\partial LnA$	$\partial Ln\nu/\partial M$	$\partial Ln\nu/\partial d_{O_v}$	$\partial Ln\nu/\partial Y$
$\nu_{0,1}$	172.78	-.985	-.682	.0889	-2.9
$\nu_{0,2}$	223.66	-.990	-.685	.0899	
$\nu_{0,3}$	275.85	-.970	-.661	.0875	
ν_{G1}	192.06	.342	.259	-.856	.7
$\nu_{0,3}/\nu_{0,1}$	1.595	-.0153	-.0210	.0013	-.06

These desired eigenmodes will probably not be excited and measurable in all objects for which we would like to have a primary determination of age. However, it is reasonable to expect that in the region of the HR diagram that we have selected (Fig. 1), which includes the instability strip, enough objects will oscillate in these modes. These objects will serve to assess the value of the overshooting parameter and tell whether it can be used as an "universal" value, applicable to non oscillating stars. Moreover if a larger number of modes were observed, a parallel treatment, not discussed here, could be done to reduce ∂A_α and ∂A_Y.

6. The Sub-milliarcsecond Mission Contribution

Once the overshooting parameter is fixed at $\sim$ 20 % and the errors due to α and Y are reduced, the contribution of the physical terms remains around 6%, i.e. slightly below the foreseen level for the "chemical composition terms". With the present accuracy on distance, luminosity has to be determined through absolute magnitude calibrations and remains one of the major source of error on age. In situation 2, the luminosity term is similar to the others, as given on the two last lines of Table 3. The hypothesis we made on future progress in atmosphere modelling lead to comparable error terms on luminosity due to errors on distances and to errors on bolometric correction. If the sub-milliarcsecond mission was to reach a level of accuracy higher than 20 times the HIPPARCOS one, the contribution of the distance terms would fall below the "atmospheric ones".

The other "physical terms" will probably be improved by the study of a significant sample of objects, as physical effects are "universal". On the contrary, the atmosphere analysis has to be performed on individual objects, and in this primary phase of scaling, it will be difficult to accept general trends, except for groups. The strong dependence of age on Y remains a difficulty which has to be solved by seismology.

We have shown here that a sub-milliarcsecond mission will significantly participate to the reduction of the uncertainties in age determinations below 3 10^9 yr, accompanying the efforts in other fields, to reach an age accuracy of $\sim 10\%$.

References

Brown T.M., Christensen-Dalsgaard J., Weibel-Mihalas B., Gilliland R. (1994) *ApJ* **427**, 1013

Caughlan G.R., and Fowler W.A. (1988) *Atomic Data Nuc. Data Tables* **40**, 284

Dziembovsky W.A., Pamyatnykh A.A. (1991) *A&A* **248**, L11

Edvardsson B., Andersen J., Gustafsson,B., Lambert D.L., Nissen P.E., Tomkin, J. (1993) *A&A* **275**, 101

Fernandes J., Lebreton Y., Baglin A. (1994) submitted to *A&A*

Goupil M.J., Michel E., Lebreton Y., Baglin A. (1993) *A&A* **268**, 546

Grevesse N. (1991) in *Evolution of stars: The Photospheric Abundance Connection*, eds. Michaud G., Tutukov A. , p. 63

Huebner WF., Mertz A.L., Magee Jr. N. H., Argo M. F (1977), *Astrophysical Opacity Library*, Los Alamos Scientific Lab., Report LA-6760-M

Iglesias C.A., Rogers F.J., Wilson, B.G. (1992) *ApJ* **397**, 717

Michel E., Belmonte J.A., Alvarez M., Jiang S.Y., Chevreton M., Auvergne M., Goupil M.J., Baglin A., Mangeney A., Roca Cortes T., Liu Y. Y., Fu J.N., Dolez N. (1992) *A&A* **255**, 139

Morel P. (1993), in *Inside the stars*, IAU coll. **137**; eds. W.W. Weiss and A. Baglin, p. 44

Morel P. (1994), submitted to *A&A*

Napiwotski R., Rieschick A., Blöcker T., Schönberner D., Wenske V. (1993), in *Inside the stars*, IAU colloquium **137**, eds. W.W. Weiss and A. Baglin, p. 461

Neuforge C., Fernandes J. (1994), *A&A* in press

Parker P.D. MacD., Rolfs C.E. (1991), in *Solar Interior and Atmosphere*, eds. A.N. Cox, W.C. Livingstone, M.S. Matthews, Space Sci. Ser., p. 51

Schaller G., Shaerer D., Meynet G., Maeder A. (1992) *A&AS* **96**, 269

HIGH-PRECISION ASTROMETRY AND THE MODELLING OF STELLAR SPECTRA

BENGT GUSTAFSSON
Uppsala Astronomical Observatory
Box 515
S-751 20 Uppsala
Sweden

ABSTRACT. The impact of high-precision astrometry on modelling of stellar spectra and determination of fundamental stellar parameters for single stars is discussed. It is found to be considerable, but improvements in the understanding of stellar atmospheres, and in particular of the physical background of luminosity (i.e., surface-gravity) criteria will be needed in order to match the possibilities of stellar parallaxes with accuracies on the order of a few percents or better.

1. Introduction

It is an interesting fact that the organisers of this symposium have planned a discussion of the possibilites of modelling stellar atmospheres and spectra. This demonstrates the connections between stellar positional astronomy and astrophysics, revitalized by the remarkable developments discussed, in particular by the possibility of getting accurate parallaxes for a very great number of stars. I shall restrict this presentation to single stars, since Johannes Andersen will later discuss the situation for binary stars.

2. Direct determination of stellar radius and mass

A first task is to use the parallaxes, π, together with photometry to derive stellar radii, R, according to

$$\log R = -\log \pi - 2 \log T_{eff} - 0.2\, V + 0.2\, A_V - 0.2\, B.C. + \text{const.} \quad (1)$$

Obviously, if we wish to match an accuracy of, e.g., 5% in the parallaxes, we need to establish the effective temperatures T_{eff} to about 2%. This can be achieved, both for late- and early-type stars today, using model-atmosphere techniques. For the late-type stars, one may use colours, properly calibrated (see, e.g. Edvardsson et al. 1993, Buser and Kurucz 1992, Bell et al. 1994), or the "infrared flux method" (see, e.g. Blackwell et al. 1991, Bell

E. Høg and P. K. Seidelmann (eds.),
Astronomical and Astrophysical Objectives of Sub-Milliarcsecond Optical Astrometry, 143–148.

and Gustafsson, 1989) which is less model dependent, or profiles of hydrogen lines (Cayrel et al. 1985, cf. also Fuhrmann et al. 1993). For the hot stars fits to model atmosphere spectra of suitable spectral line profiles seem to give the most satisfactory results (cf. Kudritzki and Hummer, 1990). (An interesting question is whether the effective-temperature determination could be made with an accuracy of about 1% - a figure claimed as an internal error in several contemporary studies. This seems indeed possible, but further work is needed before such an accuracy can be regarded as well established in any temperature interval.) In addition, interstellar extinction (A_v) and bolometric correction (B.C.) need to be known to an accuracy of about $0.^m1$. The latter quantity is also derived from model atmospheres, however, for all stars but the hottest and the coolest ones there should not be any severe problem to estimate it with this accuracy.

Knowing the radius, one could estimate the stellar mass M from the spectroscopically determined surface gravity, $g = GM/R^2$ (G being Newton's contant of gravity). I.e.,

$$\log M = \log g - 2 \log \pi - 4 \log T_{eff} - 0.4\ V + 0.4\ A_v - 0.4\ B.C. + const. \quad (2)$$

The dominating problem here is to estimate the surface gravity. For this one uses pressure-sensitive spectral features, such as the strength of spectral lines from ions relative to neutral atoms, molecular lines, wings of pressure-broadened strong lines, or pressure-sensitive features in the continuous flux distribution, such as the Balmer discontinuity in F star spectra. These different methods have been studied and compared in recent work, e.g. by Bonnell and Bell (1994). We note in passing that, in fact, stellar spectra give knowledge about the gravity scaled by a function of the helium abundance Y (cf. Strömgren et al. 1982). This dependence is, however, so weak for a realistic variation of Y that one may hardly hope for interesting bounds on Y resulting from this effect - see, however, the comments on the "Hyades anomaly" below.

A general conclusion from these studies is that it is very difficult to reach an accuracy better than about 0.2 in log g determinations from spectral lines for late-type stars, while for early-type stars an accuracy of about 0.1 in log g can be reached. The problems are due to different reasons, one being the non-LTE effects affecting ionization and possibly molecular equilibria and often more difficult to model in late-type stars, another being the lack of adequate atomic and molecular data (such as damping constants), a third being the strong temperature sensitivity of molecular equilibria, which leads to severe effects from uncertainties in effective temperatures and temperature structures, a fourth being crowded spectra which makes it difficult to trace extended line wings.

An interesting group of stars for which more accurate log g determinations are possible are the main-sequence stars of spectral types F - early G. Edvardsson et al. (1993) have shown that the Balmer discontinuity index δc_1 of the Strömgren four-colour photometry may be used with a semiempirical calibration for deriving log g for these stars with errors in the interval 0.05 - 0.10 dex. A prevailing problem here is the "Hyades anomaly" (Strömgren et al. 1982, Nissen 1988) which indicates that some as yet unknown effects on the δc_1 index set a limit on the accuracy obtainable. It seems that the good fit between stellar models and the observed colour-magnitude diagram and Li abundances for the Hyades rule out a severe He abundance anomaly as a reason for this phenomenon (cf. Swenson et al. 1994).

We conclude that one cannot expect to derive masses for individual stars from Eq (2) with an accuracy better than about 0.2 dex for general late-type stars in the field; for early-type stars the similar error is about 0.1 dex, provided that accurate parallaxes may be obtained. For solar-type dwarfs, one may get even somewhat more accurate masses than that.

3. Parallaxes and checks on spectroscopic estimates

The difficulties mentioned in spectroscopic gravity determination for late-type stars make, however, other applications of accurate parallaxes interesting. For certain types of stars, e.g., the very cool giants and dwarfs, the art of determining spectroscopic gravities is very difficult or undeveloped. This causes trouble in spectroscopic analyses. E.g., in our study of carbon stars, based on highly resolved infrared spectra (Lambert et al. 1986), we found that errors in log g, estimated to 0.5 dex, were of major significance for the uncertainties in the resulting abundances. In this case the gravities were not estimated from the spectra but from mass estimates, based on different astrophysical arguments, and from estimated absolute magnitudes. With direct and relatively accurate parallaxes for these stars (we note that the proposed Roemer mission would produce parallaxes better than 1% for hundreds of carbon stars), one could reduce the uncertainty in the spectroscopic analyses to that reflecting the uncertainty in the masses.

In this connection one should note that spectra of late-type M and N giants and supergiants in fact contain not only information on mass and radius in the combination $g = GM/R^2$ but also, which is a consequence of the spherical atmospheric structure, on M and R individually, as was first noted by Schmid-Burgk et al. (1981). Attempts to determine both these parameters from spectroscopy have, however, not given any reliable results yet, due to the complexity of the situation, including the fact that the extension of these tenuous atmospheres is not only the result of radition transfer and hydrostatic equilibrium - velocity fields are probably also of decisive importance for the structure. Here, accurate parallaxes will overdetermine the parameter problem; in principle radii will be known and that should help in the mass determination and give a useful consistency check. Also, for stars with measured angular diameters for different wavelengths, the accurate parallaxes give checks on the limb-darkening predictions from the model atmospheres and the absolute extension of the atmospheres.

For hot stars with radiatively driven and reasonably stationary winds Kudritzski and collaborators (cf. Kudritzski and Hummer 1990) have shown that the observed terminal wind velocities can be used for estimating the escape velocity (α $(M/R)^{1/2}$) and thus, with the spectroscopically determined surface gravity (α M/R^2) both M and R may be determined spectroscopically. This can then be used for deriving spectroscopic parallaxes, offering a method to determine distances to nearby galaxies. Again, comparisons with accurate trigonometric parallaxes for hot stars, in our Galaxy, may give further checks on the method and thus on the basic assumptions behind it.

4. When evolutionary tracks are also used

What is said above is, as regards astrophysics, in principle only related to our understanding (or lack of understanding) of stellar atmospheres. However, we also have considerable knowledge about stellar structure and evolution. Naturally, a good deal more may be learnt about the stellar parameters from parallaxes and spectra if this knowledge is also added. I will not go deeply into that, since it will be covered in several other contributions at this symposium - just add a few remarks on the significance and shortcoming in the understanding of stellar atmospheres in this respect.

In principle, the radius and luminosity of a star are functions of its mass, age, chemical composition (Y, Z, ...) and possibly angular momentum. Measuring the stellar luminosity (via the parallax), radius (via T_{eff} and Eq. (1)), and estimating log g and Z spectroscopically, one may ask with which accuracy age and, say, helium abundance may now be derived. Clearly, the answer is very different for different regions in the HR diagram. For dwarfs of low mass, presumably close to the ZAMS, the helium abundance (Y) may be directly determined from luminosity and effective temperature, provided that the heavy-element abundance is satisfactorily known. In fact, Z should be determined from stellar spectra with an accuracy of about 0.1 dex in order to enable a determination of Y with an accuracy better than 0.03 for disk stars (cf. Perrin et al. 1977). Similarily, Teff must be known to 1% in order to allow an accuracy of 0.03 in the Y determination. We see that the problem of estimating Y to an interesting accuracy puts rather heavy demands on the quantitative interpretation of stellar spectra, at least if not a strictly differential approach can be used for minimizing the errors.

For solar-type stars in the main-sequence band but at some distance from the ZAMS (in order to avoid the "degeneracy" caused by the early evolution almost along the ZAMS in the log L-T_{eff} diagram) one may derive evolutionary ages with an accuracy of about 20%, using uvby photometry with a semiempirical calibration (Edvardsson et al. 1993), provided that a helium abundance is adopted. Good parallaxes give an interesting check on these ages. In order to trace effects corresponding to the Hyades anomaly, however, they have to be accurate to at least 3%.

One may ask whether it would be possible to separate the effects of age from those of the He abundance for individual stars and thus determine both quantities from L, T_{eff} and log g. Using the evolutionary tracks of Hejlesen (1980) I have found that the very high accuracy in log g of about 0.03 would be needed to bring down the error in Y to 0.03. Obviously, for single stars we are very far from reaching the accuracy in various parameters that one may obtain for visual binaries.

An interesting group of stars for which stellar ages and masses may be determined relatively accurately from good parallaxes, photometric data and evolutionary tracks are the subgiants with relatively low masses at temperatures so high (T_{eff} > 5500K) that they are still not on the giant branch where the evolutionary tracks merge. Thus, with parallaxes (giving luminosities by means of photometric data) accurate to 5% one may derive ages with an accuracy better than 20% for these stars, provided that the metal abundances are determined to 0.1 dex. This also assumes that the helium abundance Y is known within

0.05 and that the convection parameter (l/H_p) is known or constant with an accuracy of about 0.5. This was estimated from the standard model sequences; if, instead convective overshoot is allowed (using the tracks of Dowler and VandenBerg 1994, kindly supplied by the latter) similar conclusions may be drawn as long as the degree of overshoot behaves regularily (i.e. is constant or smoothly varying with the stellar fundamental parameters: mass, age and chemical composition). These ages and masses are not very dependent on the effective-temperature estimate, since the evolutionary tracks are almost horizontal in this part of the HR diagram. Thus, the distribution of stellar ages, masses and metallicities could be determined from Hipparcos parallaxes for these stars, without relying on stellar spectroscopy or narrow-band photometry except when estimating the metallicities.

Also for red giants the Hipparcos data, and parallaxes with higher accuracy from possible future missions for stars that are more rare, will contribute important but rather indirect statistical evidence on the stellar physical properties. Using Eq. (1) and an adopted mass distribution, e.g. derived from an observationally derived Initial Mass Function and an assumed star-formation rate, one may empirically calibrate measures of surface gravities. These may then be compared with theoretical calibrations, based on model atmospheres and synthetic spectra, and conclusions may be drawn concerning the adequacy of the modelling or concerning stellar mass loss. The study of binary stars or cluster giants offer similar, less statistical, tests. In passing, we note that current mass loss estimates for cool giants, based on observations of CO millimetre line or IR dust emission, are strongly dependent on the distance estimate (cf. Olofsson et al. 1993); in this case accurate parallaxes will considerably improve the situation.

5. Conclusions

We have found that accurate parallaxes for nearby single stars may significantly improve our determinations of stellar radii, contribute to better determinations of stellar masses and surface gravities, will improve stellar age estimates and even enable rough helium-abundance estimates for some types of stars. However, when the errors in parallaxes decrease below about 5%, basic uncertainties in the interpretation of stellar fluxes and spectra will often limit the degree to which the astrometric accuracy may be fully used for deriving fundamental stellar parameters. Instead, such parallaxes may in some cases contribute to our basic knowledge about the physics of stellar atmospheres.

ACKNOWLEDGEMENTS. Bernard Pagel is thanked for valuable comments on an earlier version of this manuscript, Don VandenBerg for supplying new evolutionary tracks and Giusa Cayrel de Strobel for interesting discussions.

REFERENCES

Bell, R.A., Gustafsson, B.: 1989, MNRAS, 236, 653
Bell, R.A., Paltoglou, G., Tripicco, J.: 1994, MNRAS, 268, 771
Blackwell, D.E., Lynas-Gray, A.E., Petford, A.D.: 1991, A&A, 245, 567
Bonnell, J., Bell, R.A.: 1993, MNRAS, 264, 334
Buser, R., Kurucz, R.L.: 1992, A&A, 264, 557
Cayrel, R., Cayrel de Strobel, G., Campbell, B.: 1985, A&A, 146, 249

Dowler, P., VandenBerg, D.: 1994, in preparation
Edvardsson, B., Andersen, J., Gustafsson, B., Nissen, P.E., Lambert, D.L., Tomkin, J.: 1993, A&A, 275, 101
Fuhrmann, K., Axer, M., Gehren, T.: 1993, A&A, 271, 451
Hejlesen, P.M.: 1980, A&AS, 39, 347
Kudritzki, R.P., Hummer, D.G: 1990, ARA&A, 28, 303
Lambert, D.L., Gustafsson, B., Eriksson, K., Hinkle, K.H.: ApJS, 62, 373
Nissen, P.E.: 1988, A&A, 199, 146
Olofsson, H., Eriksson, K., Gustafsson, B., Carlström, U.: 1993, ApJS, 87, 267
Perrin, M.-N., Hejlesen, P.M., Cayrel de Strobel, G., Cayrel, R.: 1977, A&A, 54, 779
Schmid-Burgk, J., Scholz, M., Wehrse, R.: 1981, MNRAS, 194, 387
Strömgren, B., Olsen, E.H., Gustafsson, B.: 1982, PASP, 94, 5
Swenson, F.J., Faulkner, J., Rogers, F.J., Iglesias, C.A.: 1994, ApJ, 425, 286.

STELLAR ASTROPHYSICS WITH SUB-MILLIARCSECOND OPTICAL INTERFEROMETRY

PH. STEE, D. BONNEAU, D. MOURARD

AND

P. LAWSON, F. MORAND, I. TALLON, F. VAKILI
Equipe GI2T
Observatoire de la Côte d'Azur
2130 route de l'Observatoire, Caussols,
06460 St Vallier de Thiey, France.
email: stee@rossini.obs-nice.fr

1. Introduction

Although stellar interferometers are capable of measuring the angular diameters of stars, with longer baselines they may also be used to measure brightness variations across a star's surface and to provide constraints on models of stellar envelopes. In this paper we will look at the interpretation of visibility data and some of the more exciting prospects within the reach of current interferometers.

2. Measurements of Giant Stars

With sub-milliarcsecond angular resolution and a simultaneous spectral resolution better than 10000, it will be possible to measure limb

darkening and small-scale structures on or near the surface of giant stars. The brightest yellow and red giant stars have photospheres with diameters of several milliarcseconds, sufficiently large to be well resolved in the visible and infrared using baselines of a few tens of meters. This allows us, using longer baselines, to study details on the surface of the star itself and to test astrophysical models.

E. Høg and P. K. Seidelmann (eds.),
Astronomical and Astrophysical Objectives of Sub-Milliarcsecond Optical Astrometry, 149–155.

2.1. LIMB DARKENING

Limb darkening can be detected through measurements of the first sidelobe of the visibility function, and to properly map the intensity across the disk an interferometer must be able to measure fringes whose contrast is less than about 13%. The observations require baselines of about twice those which resolve the star's angular diameter.

Usually, limb darkening of a stellar disk is predicted by modeling the emission in the continuum or photospheric lines. An important test of these predictions is the comparison with interferometric measurements of the star's intensity distribution.

Up until now, only a few attempts have been made to measure limb darkening. These include observations of Sirius with the Narrabri Intensity Interferometer (Hanbury Brown et al., 1974) and observations by Michelson interferometry of the stars α Boo, R Lyr, β Peg, and ρ Per using the I2T interferometer (Di Benedetto & Foy 1986). Figure 1 illustrates a measurement of limb darkening of the star α Boo, made by Di Benedetto & Foy (1986) with I2T. The corresponding fit is indicated. To perform the same measurements with stars of smaller angular diameter, longer baselines are necessary. For α Uma (KOIII, $\phi = 7.2$mas) a baseline of about 30m is required, and for o Uma (G5 III, $\phi \sim 2.6$mas) an 80 m baseline is necessary at visible wavelengths. These baselines are scaled and overlaid on Fig. 1.

2.2. SURFACE STRUCTURE

To observe a star's surface features, one must be able to detect fringes that have low contrast $\sim 5\%$, and which lie close to, or beyond the first zero of the visibility function for the stellar disk. An angular resolution better than 10 or 100 times that necessary to resolve the stellar diameter is necessary to resolve small structures.

Theoretical studies of convection suggests that a small number of convective elements, 50 or less, would be located at the surface of a red giant or supergiant star. These would have a temperature contrast of about ± 1000K and evolve with timescales of 100 to 300 days (Schwarzchild 1974). The existence of such features, driven either by convection or magnetic fields, have been inferred from irregular short-term variations observed by spectroscopy, photometry, and polarimetry. Starspots are also indicated in experiments of Doppler imaging of the surface of some fast rotating stars (Dupree, Baliunas & Guinan 1984).

Such small scale structures should be detectable by long baseline interferometry. The granulation and spots would appear in the spectral continuum or photospheric lines, and prominences and flares could be seen in strong chromospheric lines (H, CaII, MgI). The presence of small size

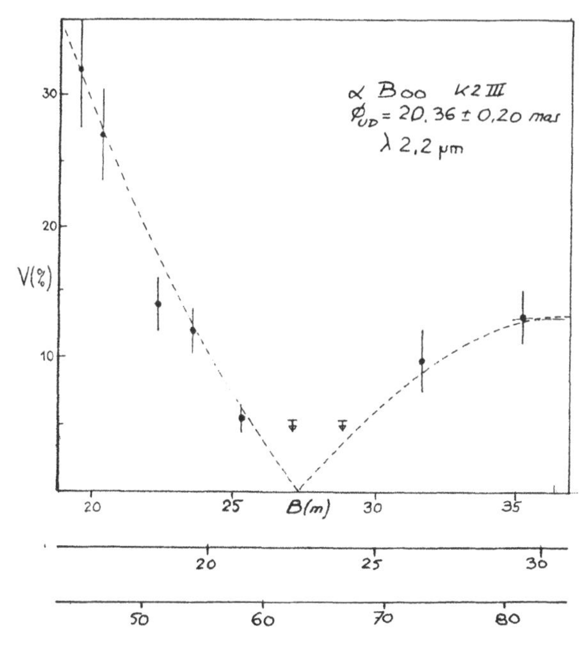

Figure 1. Limb darkening measurement for α Boo. The two baselines at the bottom are respectively for α UMa K0 III $\phi = 7.2$ mas and o UMa G5 III $\phi = 2.6$ mas at visible wavelength.

features on the stellar surface is suggested in the observations at visible wavelengths of α Aur Ab (G1III, $\phi = 6.4$mas) by the Mark III interferometer (Hummel et al. 1994) and that of β And (M0III, $\phi = 13.7$mas) from near IR observations with I2T interferometer (Di Benedetto & Bonneau 1990). Figure 2 shows the visibility curve for the star β And that does not drop to zero as expected, which suggests a detection of small structures. The solid line corresponds to a theoretical visibility. It was computed assuming a limb-darkened stellar disk with $\phi = 13.7$ mas, upon which a point source was superimposed that contributes 7 % of the total flux at 2.2 micron. Also indicated are the range of baseline needed to obtain the same measurements at visible wavelength for the stars β And and μ UMa (MO III, $\phi = 8$mas).

3. Envelopes of Early Type Stars

Interferometry can also provide strong constraints on circumstellar modeling. As a star evolves it follows a path through the HR diagram that is dependent on its mass, and at some stage will exhibit a radiative stellar wind. The mechanisms that produce a stellar wind depend on the effective temperature and spectral class of the star, and may include the following:

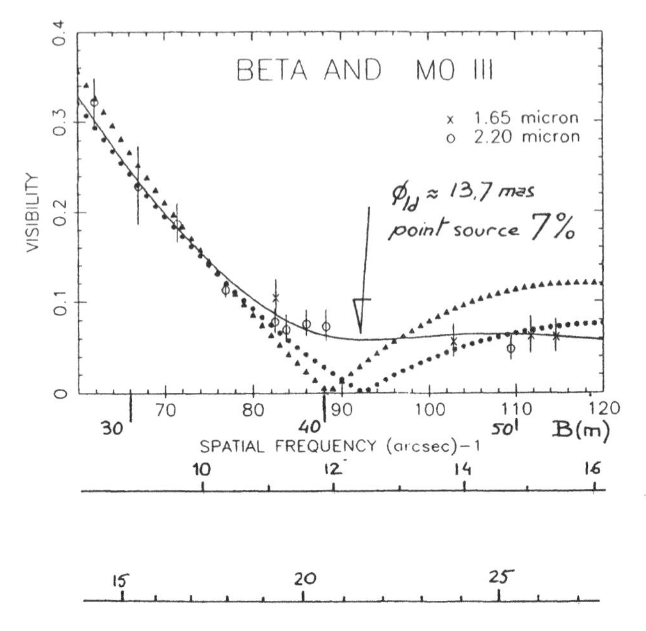

Figure 2. Detection of small structures for β And. The two baselines at the bottom are respectively for β And and μ UMa M0 III $\phi = 8$ mas at visible wavelength.

1. Thermal pressure, as in the case of the solar wind.
2. Radiative pressure on both continuum and lines, for hot stars with effective temperatures of $\sim$15000–20000 K.
3. Radiative pressure on dust for relatively cool stars.
4. Other physical phenomena such as magnetic fields or stellar rotation.

The way in which long baseline interferometry can be used to constrain different models will be illustrated using a radiative wind model we have developed at the Observatoire de la Côte d'Azur and adapted to the Be star γ Cas.

3.1. BE STARS

Be stars possess a B spectral type and at one time in their life have exhibited emission lines in their spectra, most often Balmer lines. They are not super-giants stars. They are presumed to be fast rotators spinning at 0.5 to 0.9 of their critical velocity and have a large stellar wind and high mass loss. This stellar wind seems to be at the origin of what is usually called a *two component* envelope, which is characterized by

1. An equatorial plane with high density and low expansion where the Balmer emission lines are formed.

2. A polar region with low density and high expansion where the UV absorption line profiles are Doppler shifted with velocities up to 2000 km/s.

This envelope is also responsible for a linear polarization by free electrons of about ~1.0–1.5%, and an IR excess that has been measured by IRAS. Based on this, many ad-hoc models have been computed which usually attempt to fit some Balmer line profiles and possibly the continuum emission flux. However, none of these reproduce intensity maps as a function of wavelength or incorporate high angular resolution data by computing theoretical visibilities.

3.2. A MODEL OF γ CAS

We have built a model which reproduces both spectroscopic and interferometric data that have been measured with the GI2T interferometer. It is a latitude-dependent radiative wind model for Be stars which clearly shows that the morphology of the circumstellar envelope depends strongly on the central observational wavelength and bandwidth (Araújo & Freitas Pacheco 1989; Araújo et al. 1994; Stee & Araújo 1994). This hydrodynamic code enables us to consider the effects of the viscous force in the azimuthal component of the momentum equations. The line force is the same as used by Friend & Abbott (1986), but we introduce a varying contribution of thin and thick lines from pole to equator by adopting radiative parameters which are latitude-dependent. The velocity fields and density relationships derived from the hydrodynamic equations are then used for solving the statistical equilibrium equations. By adopting the Sobolev approximation, we have calculated the electron density and hydrogen level populations throughout the envelope. We have modified this code in order to build a possible scenario for the Be star γ Cas which has been observed with GI2T during an international observational campaign in autumn 1993.

We have obtained an Hα emission profile from our model that is in good agreement with the observed spectra. Our computed intensity maps, in both the continuum and Hα, agree reasonably well with maps derived from observed visibilities. The model indicates that a radiative wind, driven mainly by optically thin lines at the equator, is a likely scenario for γ Cas. This is discussed in greater depth in Stee et al. (1994).

measurements it is The variations in visibility play an important role in understanding the data. Figure 3 illustrates two line profiles and the corresponding visibility curves, computed for two given set of parameters. Considering the spectroscopy by itself, we see that both line profiles agree roughly with the observed spectrum (dotted line). However, their corresponding visibility curves are quite different, and it is obvious that the

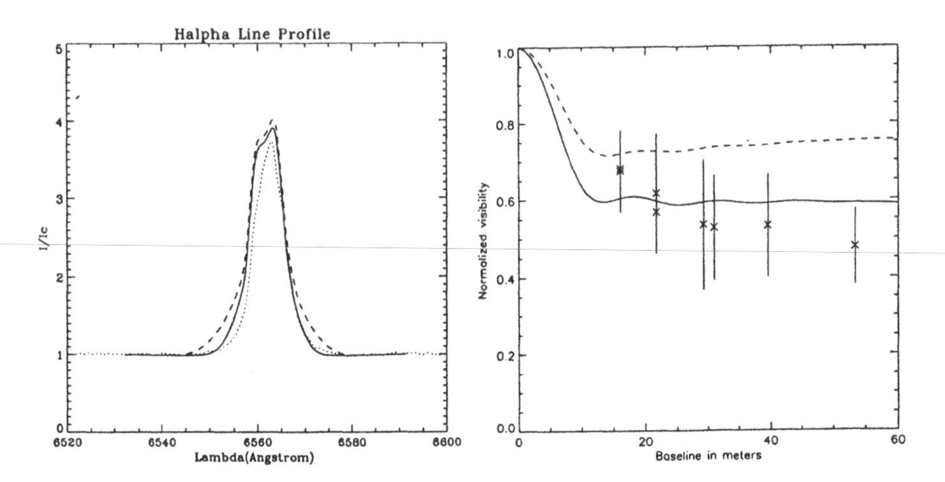

Figure 3. Two Hα line profiles and their corresponding visibility curves. Solid line and dashed line: model, dotted line: observed Hα line profile.

model with the higher visibility can be easily ruled out.

The importance of the spectral resolution, in association with a high angular resolution should also be stressed. Figure 4 shows the observed visibility along the Hα line profile and the computed visibility from our model. The error bars are large because each measurement is taken within a 4 Å bandwidth which restricts the number of detected photons and yields a low signal-to-noise ratio. Nevertheless, to a first order the model agrees well with the data, although a bump is apparent in the blue, which seems to indicate that the envelope size in the blue part of the line is systematically larger than in the red part. Such an effect could be produced if Hα radiation is being emitted by a larger or higher density region which is moving towards the observer along the line of sight.

4. Conclusion

Observations with existing and future optical interferometers will provide insight and open new horizons in problems of stellar astrophysics. The solution of these problems requires both high spectral and spatial resolution, obtainable through long baseline measurements. These must be tied into stellar models, as theoreticians have already done with spectroscopic, photometric, and polarimetric data.

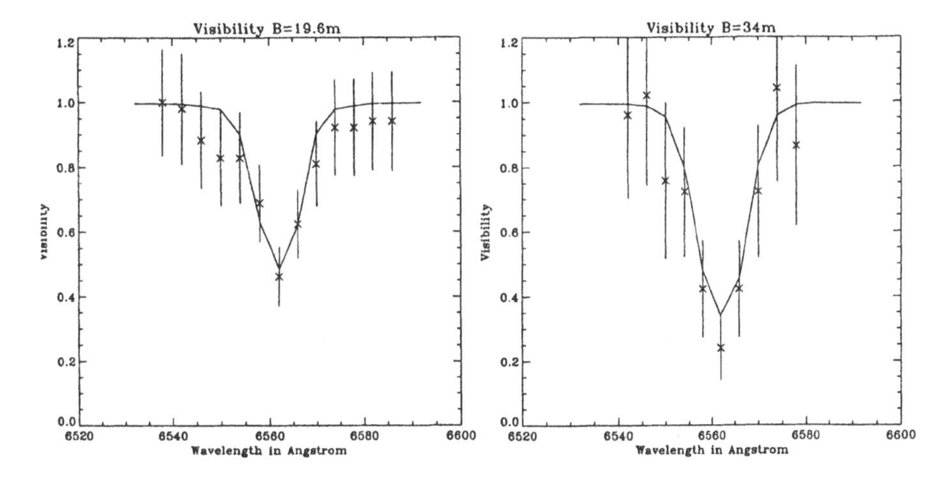

Figure 4. Visibility along Hα line profile for a 19.6 and 34 meters baseline.

References

Araújo, F.X., & Freitas Pacheco, J.A., 1989, MNRAS, **241**, 543.

Araújo, F.X., Freitas Pacheco, J.A., & Petrini, D., 1994, MNRAS, **267**, 501

Di Benedetto, P., & Foy, R., 1986, A&A, **166**, 204.

Di Benedetto, P., & Bonneau, D., 1990, ApJ, **358**, 617.

Dupree, A.K., Baliunas, S.L., & Guinan, E.F., 1984, Workshop on *High Angular Resolution Optical Interferometry*, BAAS, **16**, 797.

Friend, D.B., Abbott, D.C. 1986, ApJ, **311**,701

Hanbury Brown, R., Davis, J., Lake, R. J. W., & Thomson, R. J., 1974, MNRAS, **167**, 475.

Hummel, C.A., Armstrong, J.T., Quirrenbach, A., Buscher, D.F., Mozurkewich, D., Elias II, N.M., & Wilson, R.E. , 1994, AJ, **107**, 1859.

Schwarzchild, M., 1974, ApJ, **195**, 137.

Stee, Ph., & Araújo, F.X., 1994, A&A, in press.

Stee, Ph., Araújo, F.X., Vakili, F., Mourard, D., Arnold, L., Bonneau, D., Morand, F. and Tallon-Bosc, I. 1994, A&A, submitted.

Vakili F., Mourard D., & Stee Ph., 1994, IAU Symp. 162 on *Pulsation, Rotation and Mass Loss in Early Type Stars*, (Kluwer: Holland, 1994), pp. 435

PECULIAR STARS OF LOW LUMINOSITY

J. LIEBERT
Steward Observatory
University of Arizona Tucson, AZ 85721

AND

C.C. DAHN
U.S. Naval Observatory, Flagstaff Station
P.O. Box 1149 Flagstaff, AZ 86002-1149

Abstract. Precise trigonometric parallax measurements orders of magnitude more accurate than a milliarcsecond will contribute greatly to our understanding of peculiar, low luminosity stellar objects of several types. First, the volume of space out to which luminosities may be determined to the accuracy of the best, very-nearby stars will be increased greatly. For the relatively rare field Population II stars, this will lead to the first accurate empirical calibrations of the main sequence at the low mass end, for comparison with globular clusters of various metallicities. Parallaxes at 1 kpc or farther will be adequate to help in the discovery or confirmation of the rare carbon dwarfs – main sequence stars with carbon-rich atmospheres. For cool white dwarfs, luminosities accurate to a few per cent or better will identify unresolved binaries, and objects of unusually high and low mass. For our most numerous solar neighbors, the M dwarfs and especially those near the stellar mass limit, accurate luminosities can help in the determinations of the chemical composition and age distributions.

1. Introduction

Trigonometric parallaxes approaching or exceeding 1 mas in accuracy have been obtained for very faint stars – i.e., down nearly to the limit of the Palomar Sky Survey – in the last decade using CCD detectors. This achievement has had substantial impact on our studies of several classes of low luminosity stars, which necessarily will be faint in apparent magnitude. These

E. Høg and P. K. Seidelmann (eds.),
Astronomical and Astrophysical Objectives of Sub-Milliarcsecond Optical Astrometry, 157–162.

objects include many that are near the bottom of (1) the Population I main sequence, or dwarf M stars, (2) the Population II sequence, or subdwarf M stars, and (3) the white dwarf sequence, or degenerate dwarf stars. These parallaxes have provided the most reliable means of determining luminosity functions (LFs) for these classes of stars, or at least for calibrating the photometric systems used for this purpose.

In Fig. 1 we show an observational H–R Diagram based on U.S. Naval Observatory trigonometric parallaxes, both published and unpublished. All data points are shown with formal $\pm 1\text{-}\sigma$ error bars. Of course, the intrinsically faintest stars generally have the larger parallaxes and the smaller uncertainties. Only a few of these stars, generally in the top half of the diagram, are bright enough for HIPPARCHOS measurements. The accuracies of current parallaxes are good enough to demonstrate a dispersion among the K–M dwarfs of > 3 mag, although the faintest of these at a given color are the extreme subdwarfs of Population II. Almost invariably, the most subluminous stars exhibit high tangential velocities, and low metallicities from their spectra and colors. However, due largely to their warmer temperatures and smaller bolometric corrections near the stellar mass limit, the Population II main sequence appears to terminate at $M_V \sim +14$ or slightly fainter, compared to +18–19 for the faintest stars of Population I. If the dispersion of the observed main sequence of nearby stars in M_V were due primarily to the spread in metallicity, we would expect that below M_V of +14, the observed dispersion would grow smaller. Indeed, this is the appearance of the lower main sequence in Fig. 1. Of course, some remaining dispersion due to unresolved binaries and any pre-main sequence objects is also expected. We will comment later on the white dwarf sequence on the left side of the diagram.

In addition, parallaxes have led to the identification of unique and peculiar stars within these classes of low luminosity objects. One example of such an identification was the discovery nearly 20 years ago of the first dwarf carbon star (Dahn et al. 1977). Originally, this 13th magnitude object with a large proper motion was believed to be a low luminosity M dwarf and, hence, was placed on the U.S. Naval Observatory parallax program. A fairly accurate trigonometric parallax permitted it to be placed in the H–R diagram, but it had a peculiar location in M_V vs. B$-$V and V$-$I colors compared to other M dwarfs or subdwarfs. Subsequent spectrophotometry obtained at Lick Observatory revealed a spectrum similar to the R–type giant stars, dominated by C_2 Swan bands. Clearly the atmospheric composition of this star called G 77–61 is decisively different than observed in ordinary Population I dwarf stars! For a very long time G 77–61 remained the only known dwarf carbon star, and it was discovered because of the trigonometric parallax determintion.

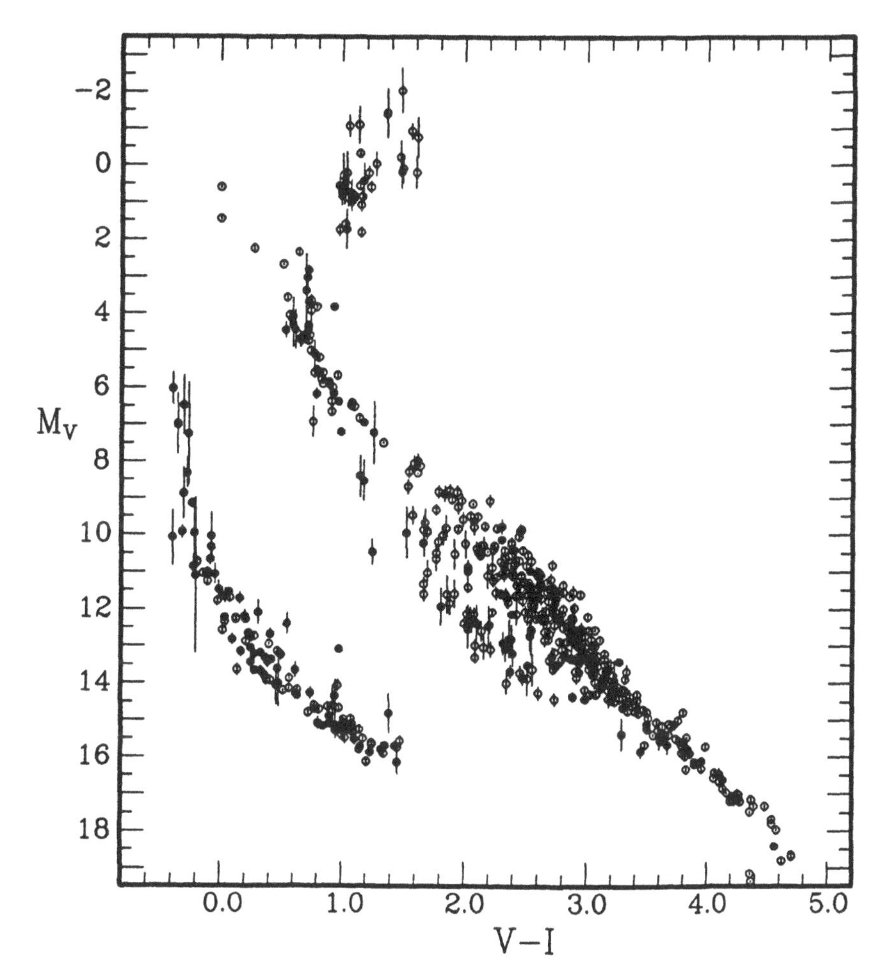

Figure 1. The M_V vs V−I diagram for a selection of 680 stars with USNO trigonometric parallaxes either published (open circles) or unpublished but nearly completed (filled circles).

When the accuracy of trigonometric parallaxes reaches an order of magnitude or two better than 1 mas, much more subtle peculiarities can be found in these classes of objects. Moreover, as we shall outline, global properties of the LFs and other parameters may be obtainable for large enough numbers of stars to be interesting.

2. Population II Stars in the Solar Neighborhood

The Population II main sequence representatives in the Galactic disk near the Sun are of great potential importance for calibrating the main sequences in globular clusters. However, the space density of Population II stars is only of the order of 1/500 that of their Population I counterparts, so one must search larger volumes of space to find rare objects. The achievement of sub-mas parallaxes would mean a valuable combination of increased accuracy and a larger sample for this class of star. In the past the F–G subdwarfs have been used primarily for this purpose, because less accurate or *no* atmospheric abundance determinations have been available for K–M subdwarfs; of course such abundances are necessary for comparison with the clusters, since the location of the main sequence is a function of metallicity.

Abundance estimates for M subdwarfs will be available soon, due largely to advances in model atmospheres analyses led by F. Allard, P. Hauschildt, K. Ruan, and their dissertation advisors R. Wehrse and M. Bessell. In a few years time, estimates of [Fe/H] and specific heavy elements may rival the determinations for globular clusters, and the stars will then be useful calibrators if their distances are known.

3. Main Sequence Carbon Stars

We have already chronicled in the Introduction some of the history of work on G 77–61, perhaps the ultimate type of rare dwarf star. It has halo-like space motions and apparently a very low abundance of heavy elements (Gass, Liebert and Wehrse 1988), though incompleteness in the opacity sets appropriate for modeling such an unusual dwarf star may result in line abundances being underestimated. At $M_{bol} \sim +9$, it has the luminosity of an early M or late K dwarf, a bit more than $10^{-2} L_{\odot}$, and T_{eff} near 4,000 K.

Its existence posed a problem for stellar evolution: Population II stars consistently show higher ratios of oxygen to carbon than stars of higher metallicity. Thus it is very unlikely that such a star formed out of material with $C/O > 1$. The normal explanation for luminous carbon stars is that these are asymptotic giant branch (AGB) objects with carbon–oxygen cores whose contents gets mixed to the surface. However, a binary star interaction is the probable solution, as it so often is for stars with peculiar abundances: G 77–61 turns out to be a 243-day spectroscopic binary, though its likely white dwarf companion was too cool for detection with the International Ultraviolet Explorer Observatory (Dearborn *et al.* 1986).

In the last few years, new discoveries have increased the number of dwarf carbon stars to about ten (cf. Green, Margon and MacConnell 1991; Warren *et al.* 1993). Their coarse spectra appear to be remarkably similar in T_{eff} to the prototype G 77–61, yet the binary hypothesis should accommodate a substantial distribution in the masses and other properties of the companions. Most, though apparently not all, of the new discoveries show kinematics of the halo population but none have abundance analyses. Two of the most recent discoveries are known to be binary, because a DA white dwarf spectrum is composite with the carbon dwarf spectrum.

Finally, the dwarf carbon stars complicate the job of those attempting to identify distant giant carbon stars in the halo, using their kinematics to study the mass distribution and structure of the extended halo. Hence, because of the evolutionary questions and galactic structure complications they pose, it is important both to find additional examples and to be able to recognize quickly this relatively rare class of star.

Sub-mas trigonometric parallaxes will permit the identification of dwarf carbon stars literally kiloparsecs away in the Galactic halo, since merely measuring a significant parallax will distinguish these from distant giants. For the modest sample now in hand, out to distances of at least 1 kpc, sub-mas parallaxes would ensure accurate measurements of the luminosities, to see how uniform this distribution is.

4. Cool and Peculiar White Dwarfs

A large increase in the number and accuracy of parallaxes for, in particular, cool white dwarfs will make possible conclusions about the mass function as well as LF of these relics of the Galaxy's early star formation history. For hotter stars of spectral type DA, simultaneous fitting of accurate CCD Balmer line profiles has yielded values of T_{eff} and *log g* of unprecedented accuracy, and a very accurate relative mass distribution (Bergeron, Saffer and Liebert 1992). These hot DA stars represent the degenerates which have formed within the last Gyr or so.

For the cooler stars, such an analysis is impossible. The hydrogen lines are weak or absent, and the physics is far less accurate. Nevertheless, a spread in the mass distribution causes a dispersion in M_V (i.e., radius) at a given color (i.e., temperature). In Fig. 1 there is some indication of the expected dispersion, but there are not many white dwarfs with well–determined parallaxes at a given color.

Should accurate absolute visual magnitudes become available for many dozens of cool white dwarfs, however, one can plot the distribution of luminosities at a given color. Since color should correlate with temperature (even if precise values are unknown), the distribution of luminosities scales closely with the radius and can therefore map the mass distribution of white dwarfs as a function of cooling age. The sample would have to be at least several times larger in order to accomplish this goal, which means reaching out to larger distances with smaller parallax errors.

White dwarfs with peculiar properties may also stand out in such diagrams based on sub-mas parallaxes. Actually, two of the cool white dwarfs in Fig. 1 that appear overluminous are the unresolved binaries L 870–2 and G 107–70, at V−I values near +0.6 and +1.0, respectively. Here the measured luminosity comes from two similar stars. A second example might be strongly magnetic white dwarfs, which can have highly unusual energy distributions (Liebert 1988).

5. Disk Dwarfs and Possible Substellar Counterparts

We conclude with the largest and most important class of low luminosity objects, the low-mass main-sequence stars of Population I. Earlier, we discussed the appearance in Fig. 1 of a decreasing dispersion in M_V fainter than +14 as the metal–poor contributors begin to disappear. There is predicted to be an opposite effect at a given metal abundance as one approaches the hydrogen-burning mass limit: the pre-main sequence time scale grows longer, approaching 1 Gyr. Then, below about 0.08 $M_{\odot}$, there are objects of "transition mass," which succeed in sustaining part of their luminosity by a limited hydrogen-burning phase, before eventually developing degenerate interiors which quenches the burning. They then cool as brown dwarfs, but not before more than 1 Gyr has elapsed, during which they persist as objects of comparable luminosity to very low mass stars. Another way of stating the problem is to say that the dispersion in luminosity as a function of age increases greatly as the mass approaches and passes below the nuclear mass limit. If one is ever to sort out the stellar, the substellar and transition mass objects of low luminosity, accurate trigonometric parallaxes are indispensible.

References

Bergeron, P., Saffer, R.A., and Liebert, J. 1992, ApJ, 394, 228
Dahn, C.C., Liebert, J., Kron, R.G., Spinrad, H., and Hintzen, P.M. 1977, ApJ, 216, 757
Dearborn, D.S.P., Liebert, J., Aaronson, M., Dahn, C.C., Harrington, R., Mould, J., and Greenstein, J.L. 1986, ApJ, 300, 314
Gass, H., Liebert, J., and Wehrse, R. 1988, A&A, 189, 194
Green, P.J., Margon, B., and MacConnell, D.J. 1991, ApJ, 380, L31
Liebert, J. 1988, PASP, 100, 1302
Warren, S.J., Irwin, M.J., Evans, D.W., Liebert, J., Osmer, P.S.,and Hewett, P.C. 1993, MNRAS, 261, 185

HIGH-PRECISION TIMING OF MILLISECOND PULSARS AND PRECISION ASTROMETRY

V. M. KASPI
IPAC/Caltech/Jet Propulsion Laboratory
770 So. Wilson Ave
Pasadena, CA 91125

Abstract.
We present the technique of long-term, high-precision timing of millisecond pulsars as applied to precision astrometry. We provide a tutorial on pulsars and pulsar timing, as well as up-to-date results of long-term timing observations of two millisecond pulsars, PSRs B1855+09 and B1937+21. We consider the feasibility of tying the extragalactic and optical reference frames to that defined by solar system objects, and we conclude that precision astrometry from millisecond pulsar timing has a bright future.

1. Introduction

Pulsars are rapidly rotating, highly magnetized neutron stars that emit a beam of light that appears as a broad-band pulse of radiation once per rotation period. What is now referred to as the "slow" pulsar population includes over 600 pulsars with rotation periods in the range 33 ms $< P <$ 5 s. What is striking about most of these sources is that their rotation is extremely stable, owing to their large rotational inertia. Rotation periods with 12 significant digits are generally straightforward to measure in just a few years of observations, and are typically limited only by measurement

E. Høg and P. K. Seidelmann (eds.),
Astronomical and Astrophysical Objectives of Sub-Milliarcsecond Optical Astrometry, 163–171.

uncertainties. The discovery of the first millisecond pulsar by Backer *et al.* (1982) made the timing of "slow" pulsars seem like child's play: the rotational period of the millisecond pulsar PSR B1937+21 has been measured with fractional uncertainty 10^{-15} (Kaspi, Taylor & Ryba 1994).

It is this rotational stability, combined with the Earth's motion around the Sun, that allows high-precision astrometry to be accomplished through pulsar timing. In this paper, we review the technique of pulsar timing with an emphasis on astrometric applications. It must be noted, however, that astrometry is only one of many byproducts of timing observations. Other equally interesting and important fields addressed in high-precision timing include general relativity, cosmology, time-keeping metrology, interstellar medium physics, orbital evolution, as well as neutron star physics.

2. Fundamentals of Pulsar Timing

2.1. TIMING OBSERVATIONS

Although a handful of pulsars are strong enough to allow the detection of individual radio pulses, for the vast majority, the pulses are buried deep within the noise of the telescope. A single timing observation therefore consists of folding the digitized telescope signal modulo the expected topocentric pulse period in a sufficiently long integration to beat down the noise. This is particularly useful because a characteristic observational property of pulsars is that the coherent summation of many individual pulses always leads to the same signature, called an "average profile," that is unique to that pulsar at that observing frequency. A fiducial point in the pulse profile, for example the pulse peak, therefore plausibly corresponds to a fixed point on the neutron star surface. Examples of two average profiles are given in Figure 1. The fiducial points chosen for the average profiles in Figure 1 are indicated by arrows. A pulse time-of-arrival (TOA) is the time at which the fiducial point of a pulse close to the mid-point of an integration arrived.

Pulsar timing consists of obtaining a sequence of TOAs, spaced typically by several weeks, over the course of many months or years. If pulsar rotation periods as observed on Earth did not change (that is, if the Earth were not accelerating with respect to the pulsar and pulsars did not lose or gain rotational energy), pulsar timing would be very simple: there would always be an integral number of neutron star rotations between any two observed TOAs, with only very small deviations resulting from measurement errors. However, life is never simple, and before the stability of pulsar rotation can be observed, one must account for a number of systematic effects.

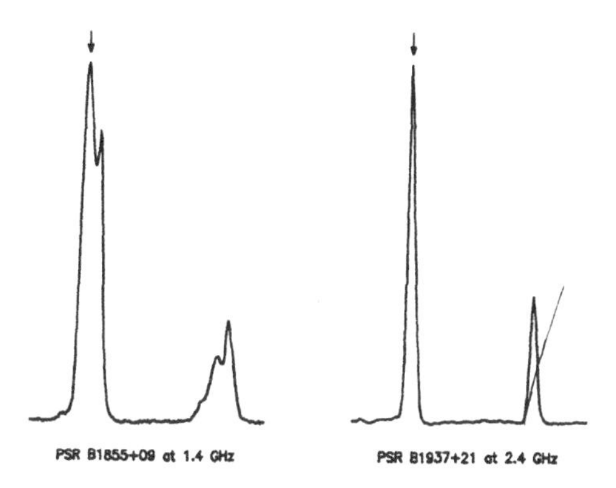

Figure 1. Average profiles for PSRs B1855+09 and B1937+21 at 1.4 and 2.4 GHz respectively. Arrows indicate the fiducial reference points for timing.

2.2. THE PULSAR REFERENCE FRAME

Even if one could observe from a reference frame not accelerating with respect to the pulsar, the observed period would change since the pulsar loses energy through magnetic dipole radiation. An approximate expression for the rotation phase is obtained by Taylor expansion:

$$\phi(t) = \phi(0) + \nu t + \frac{1}{2}\dot{\nu}t^2 + \frac{1}{6}\ddot{\nu}t^3 + ..., \tag{1}$$

where $\phi(t)$ is the rotation phase in cycles at time t, $\nu \equiv 1/P$ is the rotation frequency, and $\dot{\nu}$ and $\ddot{\nu}$ are the rotation frequency first and second derivatives. The coefficients in this expansion can be related if some physical model for the rotation is assumed. For a rotating magnetic dipole moment $\mathcal{M}$, classical electrodynamics says the total radiated power $\mathcal{P}$ is given by

$$\mathcal{P} = \frac{2(\mathcal{M}\sin\alpha)^2(2\pi\nu)^4}{3c^3}, \tag{2}$$

where α is the angle between the spin and dipole axes. Setting the rate of loss of kinetic energy equal to the radiated power yields

$$\dot{\nu} = \frac{8\pi^2(\mathcal{M}\sin\alpha)^2}{3Ic^3}\nu^3. \tag{3}$$

Pulsar timing typically allows the measurement of ν and $\dot{\nu}$. These in turn can be used to estimate various other quantities of interest. For example, the characteristic spin-down age τ of a pulsar can be estimated from $\tau = -\nu/2\dot{\nu}$, and the magnetic dipole moment $\mathcal{M}$ can be estimated via Equation 3. It can easily be shown that $\ddot{\nu}$ from dipole radiation is negligible for most pulsars, including all millisecond pulsars. However, measured values of $\ddot{\nu}$ and even higher order terms often exceed the values expected from the dipole radiation model; this is generally attributed to slight instabilities in the neutron star rotation.

2.3. THE OBSERVATORY REFERENCE FRAME

The accurate measurement of the intrinsic spin parameters ν and $\dot{\nu}$ depends crucially on the transformation of the TOAs to the neutron star's rest frame. If this is not done correctly, the parameters may be contaminated by variable relative acceleration. A transformation to the solar system barycenter is most expedient and is sufficient for all isolated disk pulsars (those in globular clusters suffer significant accelerations in the cluster potential); the approximation that the barycenter is not accelerated with respect to the pulsar rest frame is more accurate for closer pulsars. A review of the transformation is given in Backer and Hellings (1986). We discuss the main points here.

An arrival time t_b at the solar system barycenter is specified by

$$t_b = t + \frac{\mathbf{r} \cdot \hat{\mathbf{n}}}{c} + \frac{(\mathbf{r} \cdot \hat{\mathbf{n}})^2 - \mathbf{r}^2}{2cd} - \frac{D}{f^2} + \Delta_{E\odot} - \Delta_{S\odot} \tag{4}$$

where t is the observed topocentric TOA, $\mathbf{r}$ is a vector from the barycenter to the phase center of the telescope, and $\hat{\mathbf{n}}$ is a unit vector from the barycenter to the pulsar. The terms featuring the vector $\mathbf{r}$, which incorporate the Newtonian time-of-flight of the pulsed signal between the observatory and the solar system barycenter, are referred to as the solar system "Roemer delay." The observing frequency f is measured in the barycentric rest frame, and the dependence of the time delay on f^{-2} is a result of the ionized interstellar plasma. The "Einstein delay" term, $\Delta_{E\odot}$, is a combination of gravitational redshift and time dilation effects due to the motion of the Earth and other objects in the solar system. The "Shapiro delay" term, $\Delta_{S\odot}$, characterizes the general relativistic curvature of spacetime near the Sun. The correct calculation of all the above therms requires an accurate planetary ephemeris such as the Jet Propulsion Laboratory's DE200 ephemeris (Standish 1982).

2.4. BINARY PULSARS

For pulsars having binary companions, an additional transformation is required to account for the pulsar's orbital motion. Five Keplerian parameters are necessary to describe a binary orbit in which the projected motion of one object is observed: the orbital period, the orbital eccentricity, the longitude of periastron, the epoch of periastron, and the projected semi-major axis of the pulsar orbit. In some cases, due to sufficiently small orbital separations, high orbital eccentricities, or favorable alignment, the Keplerian framework does not adequately describe the orbit. For these systems, "Post-Keplerian" parameters, such as time-derivatives of the Keplerian parameters, must also be included. Measurements of such parameters can be used for precise tests of relativistic gravity theories, one of the most exciting applications of pulsar timing. Details of the parametrization using various relativistic binary models are summarized by Taylor and Weisberg (1989).

3. Pulsar Data and Precision Astrometry

It is clear from Equation 4 that the position of the pulsar plays a key role in the timing model. Imagine a pulsar in the ecliptic at some ecliptic longitude. If the TOAs from that pulsar are not corrected for Earth's motion, delays and advances will be observed such that the differences between model-predicted arrival times and observed arrival times, that is, the "residuals," will contain a systematic sine wave (ignoring, for the moment, the small eccentricity of the Earth's orbit) of period 1 yr with a phase that depends on the longitude. The sine wave amplitude will be about 500 s (corresponding to the Earth-Sun distance), much larger than all known pulse periods. By fitting out this signal, an accurate ecliptic longitude is obtained. If the pulsar is not in the ecliptic, the sine wave amplitude is reduced, and the fit also determines the ecliptic latitude. Small amplitude 1-yr sine waves in residuals can be used to fine tune the fit position.

This measurement can be done with unprecedented precision: if the input pulsar position is wrong by $\sim 1''$, the annual sinusoid has amplitude $\sim$1 ms, so for pulsars with TOA uncertainties of a few ms, the position can be known ideally to within a few arcseconds after only one year, with the uncertainty decreasing as the square root of the length of the data span. For millisecond pulsars, typical timing uncertainties are three orders of magnitude smaller, so far greater accuracy is easily achieved. For example, after 9 yr of timing PSR B1937+21 with individual TOA uncertainties of $\sim$0.2 μs, the position uncertainty is around 50 μarcsec in the reference frame defined by the DE200 planetary ephemeris.

If the pulsar has a substantial proper motion, the residuals exhibit a sine wave of linearly increasing amplitude. At a distance of 1 kpc, a pulsar moving with typical transverse velocity 200 km s^{-1} has proper motion of 40 mas yr^{-1}. Thus, it is quite difficult to detect proper motions of slow pulsars from timing measurements, however for millisecond pulsars, the measurement is straightforward. Pulsar proper motions are also measurable using interferometric techniques (e.g. Harrison, Lyne and Anderson 1993).

If the pulsar is nearby, geometric parallax, which has its greatest effect when the Earth–Sun line is perpendicular to the pulsar–Sun line, results

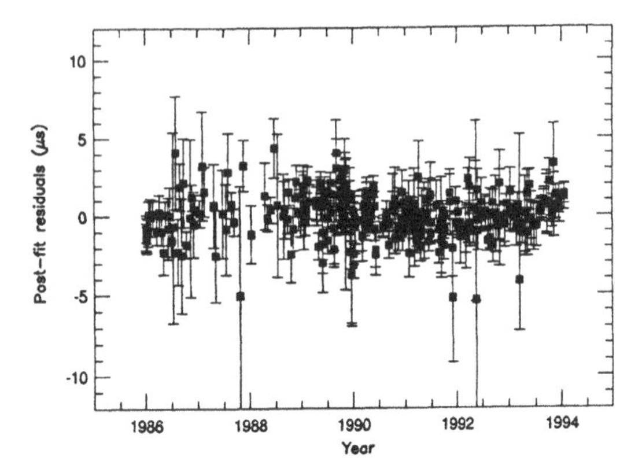

Figure 2. Post-fit residuals for PSR B1855+09 at 1.4 GHz.

in a sine wave of period 0.5 yr. (This effect is precisely described by the second Roemer term in Eq. 4). The parallax effect is tiny even by pulsar standards: for a pulsar in the plane of the ecliptic at a distance of 1 kpc, the amplitude of the parallax signal π is only 1.2 μs. A measurement of π is quite valuable however, since it allows a direct determination of the distance d to a pulsar, via $\pi = 1/d$. A timing parallax has been measured for only one pulsar, PSR B1855+09, discussed below. Pulsar parallaxes can also in principle be measured directly using VLBI, although these experiments have proven to be difficult (e.g. Gwinn *et al.* 1986).

4. Recent Results from Arecibo

We now describe recent results from a long-term, high precision millisecond pulsar timing project done at the 305-m dish at Arecibo, Puerto Rico. Bi-weekly timing observations for two millisecond pulsars, PSRs B1855+09 and B1937+21, are described in detail in Kaspi, Taylor and Ryba (1994) (hereafter KTR94). High-precision timing of PSR B1855+09 was begun in 1986 at Arecibo at 1.4 GHz, and has yielded daily-averaged arrival time uncertainties of $\sim$1 μs. Data for PSR B1937+21 goes back to 1984 at 1.4 and 2.4 GHz, with daily-averaged TOA uncertainties of $\sim$0.2 μs. The data sets analyzed by KTR94 include arrival times obtained through the end of 1992, although here we include data through the end of 1993. All TOAs in the analysis are referred to UTC.

PSR B1855+09 is a 5.4 ms pulsar in a 12 day circular orbit with a white dwarf. The pulsar's celestial coordinates and proper motion are determined with uncertainties of 0.13 mas and 0.07 mas yr^{-1} respectively, in the reference frame defined by the JPL DE200 planetary ephemeris. PSR B1855+09 is the only pulsar for which a significant measurement of timing parallax has been made. The pulsar's astrometric parameters are listed in Table 1. The residuals after removal of the best spin, astrometric and binary parameters are shown in Figure 2. The residuals are clearly dominated by random, Gaussian measurement uncertainties, which indicates that the model describes the data well. The slight hint of a cubic signal is apparent if the plot is held at a distance; this will be discussed below.

TABLE 1. Astrometric Parameters for PSRs B1855+09 and B1937+21

	PSR B1855+09	PSR B1937+21
α (J2000)	$18^h\ 57^m\ 36^s.393515(4)$	$19^h\ 39^m\ 38^s.560211(2)$
δ (J2000)	$+09^\circ 43' 17''.32370(12)$	$+21^\circ 34' 59''.14170(4)$
μ_α (mas yr^{-1})	$-2.91(4)$	$-0.134(8)$
μ_δ (mas yr^{-1})	$-5.48(6)$	$-0.452(9)$
π (mas)	1.0(3)	<0.20

PSR B1937+21 was the first discovered and is still the fastest known millisecond pulsar, having $P = 1.5$ ms. It shows no evidence for a binary companion. DE200 celestial coordinates and proper motion were measured with uncertainties of 0.05 mas and 0.01 mas yr^{-1}. The best fit astrometric parameters are shown in Table 1, and the residuals after subtraction of the best model including astrometric and spin parameters are shown in Figure 3. There is an obvious cubic trend in the residuals that indicates our model does not completely describe the rotation of the neutron star. KTR94 address this issue in detail, and demonstrate that although there are several possible origins of the "noise," including planetary ephemeris errors, a primordial background of gravitational waves, clock errors, or some poorly understood interstellar propagation phenomenon, the most likely origin is intrinsic to the pulsar itself. The lower-level cubic of opposite sign in the residuals for PSR B1855+09 support this argument.

It should be noted that we have minimized the effect these deviations from the simple spin-down model have on the astrometric parameters for PSR B1855+09 and especially PSR B1937+21 by "whitening," that is, by including sufficiently many higher order frequency derivatives to render the residuals random when determining the astrometric parameters.

5. Frame Ties and the Future

The high precision astrometry available through millisecond pulsar timing is useful for tying astronomical reference frames. An interferometric po-

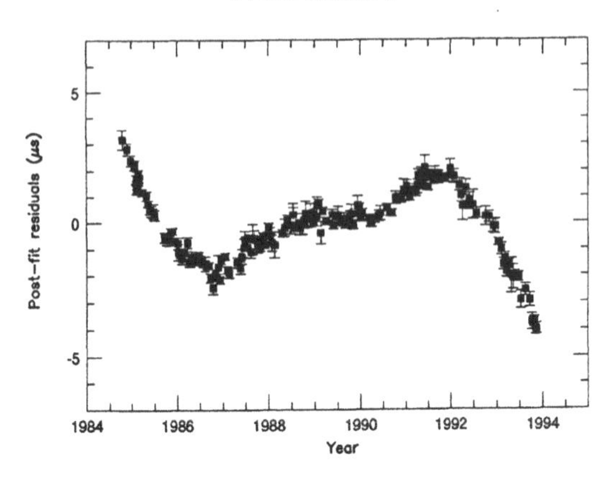

Figure 3. Post-fit residuals for PSR B1937+21 at 2.4 GHz.

sition for a pulsar together with a timing position, for example, tie the extragalactic reference frame with that defined by solar system objects. At 400 MHz, PSR B1937+21 is a 240 mJy source so is easily observable with VLBI techniques. Indeed Bartel *et al.* (1990) have observed the pulsar and obtained a VLBI position suitable for a frame tie. PSR B1855+09 by contrast is only 31 mJy at 400 MHz, and so is not well-suited to this technique.

An optical detection of a pulsar or, more likely, of the companions to binary millisecond pulsars, can in principle provide a tie with the optical reference frame. Although under 10% of the slow pulsar population are in binary systems, well over 75% of disk millisecond pulsars are in binaries, with the majority having a potentially optically observable companion. The detection of optical companions to millisecond pulsars is also interesting from an evolutionary point of view (Kulkarni *et al.* 1991). The advent of HST and other sensitive instruments such as the Keck Telescope will be a great boon to the detection of very faint companions, as well as to improved astrometry for brighter sources.

One newly discovered source deserves notice. PSR J0437−4715, discovered by Johnston *et al.* (1993), is a 5.7 ms pulsar in a 5 day circular orbit with a white dwarf. It is a 600 mJy source at 400 MHz, and is also one of the closest pulsars, with an estimated distance of only 140 pc. The source will be easily detectable by VLBI, and furthermore, the pulsar's companion has already been detected optically by Danziger *et al.* (1993). After only two years of timing, the timing position and proper motion have been measured with uncertainties of 4 mas and 3 mas yr^{-1} respectively (Bell *et al.* 1995).

The number of millisecond pulsar discoveries has sky-rocketed in recent years (see Fig. 4) owing principally to large-scale search efforts. There is every reason to believe that millisecond pulsar timing will be a booming industry in the years to come, and that precise positions and proper mo-

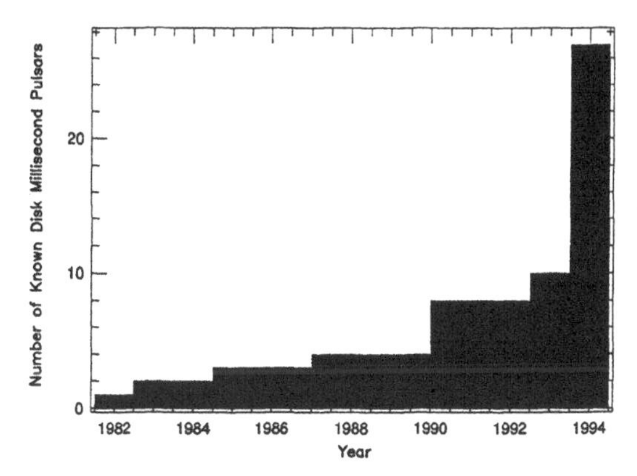

Figure 4. Number of known disk millisecond pulsars versus year.

tions will be available in the next decade for a large number of sources. Furthermore, as more millisecond pulsars are discovered, it is becoming clear that they are distributed isotropically on the sky; the possibility of a complete tie of optical, extragalactic, and dynamical reference frames from observations of millisecond pulsars may soon be within reach.

Acknowledgements

I am indebted to Joe Taylor for his foresight and advice. I also thank F. Camilo, D. Nice, and P. Ray for discussions. Financial support was provided by a Princeton University Higgins Instructorship, and by a Hubble Fellowship through NASA grant number HF-1061.01-94A from the Space Telescope Science Institute, which is operated by the Association of Universities for Research in Astronomy, Inc., under NASA contract NAS5-26555.

References

Backer, D. C. & Hellings, R. W., 1986. *Ann. Rev. Astr. Ap.*, 24, 537.

Bartel, N., Cappalo, R., Whitney, A., Chandler, J., Ratner, M., Shapiro, I. & Tang, G., 1990. In: *Impact of Pulsar Timing on Relativity and Cosmology*, X1, ed. Backer, D., Center for Particle Astrophysics, Berkeley.

Bell, J., Bailes, M., Manchester, R. N., Weisberg, J. & Lyne, A. G., 1995. in preparation.

Danziger, I. J., Baade, D. & Della Valle, M., 1993. *Astr. Astrophys.*, 276, 382.

Gwinn, C. R., Taylor, J. H., Weisberg, J. M. & Rawley, L. A., 1986. *Astron. J.*, 91, 338.

Harrison, P. A., Lyne, A. G. & Anderson, B., 1993. *Mon. Not. R. astr. Soc.*, in press.

Johnston, S., Lorimer, D. R., Harrison, P. A., Bailes, M., Lyne, A. G., Bell, J. F., Kaspi, V. M., Manchester, R. N., D'Amico, N., Nicastro, L. & Jin, S., 1993. *Nature*, 361, 613.

Kaspi, V. M., Taylor, J. H. & Ryba, M., 1994. *Astrophys. J.*, 428, 713.

Kulkarni, S. R., Djorgovski, S. & Klemola, A. R., 1991. *Astrophys. J.*, 367, 221.

Standish, E. M., 1982. *Astr. Astrophys.*, 114, 297.

Taylor, J. H. & Weisberg, J. M., 1989. *Astrophys. J.*, 345, 434.

LATE STAGES OF STELLAR EVOLUTION

P. THEJLL
Niels Bohr Institute, Blegdamsvej 17
DK-2100 Copenhagen Ø, Denmark

Abstract. A review is given of the use of high-accuracy astrometry on research on white dwarfs and the hot subdwarfs and central stars of planetary nebulae (CSPN). Predictions are made about the expected impact of HIPPARCOS, and the possible impacts of GAIA and ROEMER. Discovery of large numbers of new white dwarfs is expected, and, for the more distant hot subdwarfs and CSPN, important refinements of our current understanding of these objects. For white dwarfs independent values of mass and radius may be accurate enough to allow new understanding of the internal composition.

1. Introduction

Recently proposed astrometric space observatories (the HIPPARCOS-like but much more sensitive ROEMER, Bastian *et al.* 1993, and the astrometric interferometer GAIA, Lindegren *et al.* 1993) have the potential for drastically increasing the precision with which parallaxes and proper motions could be measured for faint stars. What are the predictable consequences of such data for research into White Dwarfs (WD), the Planetary Nebulae (PN) phase and the hot subdwarfs near the end of the horizontal branch (HB)?

White dwarf research provides testable predictions of stellar evolution as well as furnishing test-cases for many areas of physics, such as atomic physics (light-absorption under very high pressure and strong magnetic fields), quantum mechanics (the WD mass-radius relation) and general relativity (gravitational red-shift). Cosmological questions can be addressed with the WDs as they can measure the age of star-forming regions. As WD stars represent the final fate of 90% of all stars they are a key part of stud-

E. Høg and P. K. Seidelmann (eds.),
Astronomical and Astrophysical Objectives of Sub-Milliarcsecond Optical Astrometry, 173–180.

ies on stellar evolution: All the processes that stars undergo during their evolution must have the net effect of producing WDs with the observed properties.

Hot subdwarf research is of interest to investigations of stellar mass-loss, as these stars may represent the extreme case of near-complete envelope loss during single star evolution. This is the key to understanding the chemical evolution of a galaxy or cluster of stars over several stellar generations. Hot subdwarfs may furthermore explain the observed ultraviolet upturn in the flux from metal rich galaxies (Greggio&Renzini 1990) and are an important source of ionizing radiation at high galactic latitude (de Boer 1985).

2. White Dwarf research topics and the impact of astrometry

2.1. THE WD MASS

The mass-loss on the giant branches, and the Helium flash which terminates red giant branch (RGB) evolution determine the mass that WDs have. The best measurement of the mean WD mass is that by Bergeron *et al.* 1992 (BSL92 hereafter), who found $< M_{\rm WD} >= 0.53 \pm 0.13$ $M_\odot$in good agreement with a typical core mass on the RGB of close to 0.5 $M_\odot$ for stars with initial mass below about 2 $M_\odot$(Sweigart&Gross 1978). For stars of higher initial mass the evolution is much more rapid and the mass loss is also important during other evolutionary phases.

2.2. DETERMINING WD MASSES

Standard methods of mass determination for WDs have recently been discussed by BSL92. Briefly summarized the methods relevant here are: *1)* Keplers third law, *2)* Using parallaxes and a mass-radius relationship, *3)* Spectroscopic gravities and a mass-radius relationship, and *4)* Spectroscopic gravities and parallax. Precise parallaxes have an immediate impact on all these methods. Method 1 is limited by the availability of suitable systems to a very small number: 40 Eri B (Heintz 1974) and Sirius B (Gatewood& Gatewood 1978) provide the only really good data points. Other binary systems have such long periods that changes in the positions of the stars are unobservable with present technology. Stein 2051 B is an example of a star that would be analyzable if observed with μas accuracy. Method 2 uses an estimate of the effective temperature and the absolute magnitude, to give a radius which is then used with a theoretical mass-radius relationship (Hamada&Salpeter 1961) to yield the mass. Methods 3 and 4 first derive the gravity, g, from spectroscopic analysis and then (method 3) derives the mass from a mass-radius relationship or (Method 4) derives the mass from $4\pi H_\nu R^2 = f_\nu D^2$ where D is known from the parallax, H_ν is the

Eddington flux at some specific wavelength for the model spectrum that best fits the spectral features of the real star, f_ν is the observed flux at the same wavelength and R is the radius determined from $g = \frac{GM}{R^2}$.

Accurate masses for WD's would allow a reopening of the question of whether the hydrogen-rich (DA) and helium-rich (DB) white dwarfs have the same mean mass (Shipman 1979). As that argument is discussed on the basis of statistics, increasing the number of known DB would be of great help.

2.3. THE WD MASS-RADIUS RELATION

A striking property of WDs is that their internal structure is governed by quantum mechanics - therefore their bulk properties, such as the radius for a given mass, depend on quantum mechanics. The mass-radius relationship is mostly used in the predictive sense, illustrating confidence in the relation: Given a mass (or radius) for a WD, the radius (or mass) is predicted. Testing the M-R relation is limited by the availability of suitable systems. At present only the WD's Sirius B and 40 Eri B and to some extent Stein 2051B 'test' the M-R relation of Chandrasekhar. There is no real doubt that the relationship is basically valid but the issue is in fact only addressed by these few systems. If 1 in a thousand WD's is a suitable binary system then we may expect several times more test-cases if the proposed observatories are launched. The ultimate allure of the M-R relation is that it depends on chemical composition – with accurate enough independent masses and radii one could find the internal composition of WD's.

2.4. NEW WD'S WILL BE FOUND

There are several ways in which the proposed space observatories could help increase the number of analyzable binary systems with WDs, as well as WDs in general: First of all, frequent photometric monitoring of systems containing WDs would help reveal eclipsing systems - at present there are no concerted efforts to keep an eye on all known WD's (at least 2000 are known - McCook&Sion 1987) so it is not known whether some subset of the known WD's are undergoing eclipses or not. Secondly, the greatly increased accuracy of the proposed observatories will make it possible to study the motion of stars in long period binary systems that have periods that are too long for the present accuracy to give meaningful data (e.g. Stein 2051B). Common proper motion pairs will also be detected for input to detailed binary system studies.

Many WDs have been found in faint large proper motion surveys and the proposed astrometric observatories could reveal many new WDs. As about 40,000 WD's are predicted to be closer than 100 pc, the number of WD's available for study would increase by an order of magnitude or more above what is currently available. To detect previously unknown stars the mission must either operate in a sky-survey mode or operate from prepared input catalogs that contain all stars, as opposed to stars from existing catalogs. The proper motions that have been measured fall in the range from 0 to 4 arcseconds/yr with half above 300 mas/yr. Improved hardware could set lower and lower limits to the detectable proper motion (0.1 mas/yr below V=11 and 1.5 mas at V=17 for ROEMER) but procedures must be adapted to searching for fast movers in fields that will be increasingly crowded as the limiting magnitude is pushed higher and higher. Hot WDs are UV bright and any astrometric mission with UV sensitive detectors would be particularly useful for target-identification.

2.5. AGE OF THE GALAXY FROM WD COOLING

The age of the star forming region of the Galaxy can been estimated by measuring the luminosity function of cooling WDs (Liebert *et al.* 1988, and Wood 1992).

White dwarfs cool at a calculable rate that depends mainly on the present temperature of the star, but also on the internal structure and composition. The coolest WD's in the solar neighborhood set limits on the age of the disk. So far, the age measured for the disk from the intrinsically faintest WD's is rather low (9 - 10 Gyr) when compared to Globular Cluster (GC) ages (12 - 18 Gyr). As the detection and counting of WDs is fairly simple compared to GC isochrone fits, the WD age-determination has raised interesting questions regarding both the methods used in isochrone fitting, the relative timing of stellar evolution in various parts of the Galaxy and the theory of WD cooling.

The possibility for deep surveys of WD's is exciting in the context of gathering non-biased data. With the ROEMER basically all hot WDs (with cooling times of the order 1 Gyr - easily selectable from their color) could be observed out to 100 parsec with an accuracy better than 1.5 mas. This is presently possible with HIPPARCOS only for stars inside the closest 10 - 15 parsec.

3. Progenitors of WD's

By measuring distances one can calculate space densities for stars in different but connected stages of evolution and test stellar evolution theories by comparing the evolution rates. Most WDs are formed via the PN route, while a minority is formed by those stars that miss the Asymptotic Giant Branch phase and go directly from the blue HB to the WD cooling sequence. As some stars – notably the helium-rich subdwarf O stars (sdO) and the helium-rich DB WDs – are chemically peculiar, it has been suggested that such sub-classes are linked. A direct test of this suggestion could be performed if accurate rates of evolution were known for both classes of objects.

At present the space density of helium rich WDs is only indirectly known: the fraction of DB's among WD's is known and the total DA space density is known (Fleming *et al.* 1986), but no direct measurement of the DB space density exists for several reasons: DB's are rare, so large numbers are needed before a meaningful statistic can be measured, and the main method available for DB distance determination has so far been spectrophotometric: There are only parallaxes for 9% of the DB's in the McCook&Sion catalog and models for DB's have only recently become reliable (see Thejll *et al.* (1991) for a discussion of this point).

The space density of the helium rich sdO stars that sit in the high-luminosity area above the blue tip of the HB is not well known as the only method ever attempted for sdO distance determinations is unreliable: Dreizler *et al.* 1990 and Thejll *et al.* 1994 analyze the same stars with different models and find significantly different results. Trigonometric parallaxes for the sdO would give masses which, by comparison to stellar evolution predictions, could solve the problem which is of interest to the development of new radiative transfer methods for hot stars. So far, only one parallax is known for a hot sdO - the 10'th magnitude Feige 34 has a 50% accurate parallax in the revised Yale catalog (van Altena *et al.* , 1991). The stars are mostly above the galactic plane and most are far away and have not been tempting targets for astrometrists. sdO masses vary - all known values are derived from binary systems (Ritter 1990), more of which are needed.

The space density of central stars of planetary nebulae (CSPN) is not well known, but methods, based on the illuminating effect of the central star on the nebula, are available that are not applicable to the sdO case. The Zanstra method allows estimation (although somewhat uncertainly) of the luminosity of the CSPN and this gives distance estimates. Direct parallaxes are needed for determining the space density, and for testing the accuracy of the Zanstra method so that it can be used with greater confidence for far-away nebulae outside the range of parallax measurements.

4. Discussion of requirements for astrometric missions

What are the characteristics of an experiment that would accurately yield space densities and scale heights for WDs and hot subdwarfs?

For WDs the answer depends on the absolute magnitude (and hence age) one wants to study. The wide range of absolute magnitudes present strongly determines what can be done with the proposed technology. Hot WDs (say

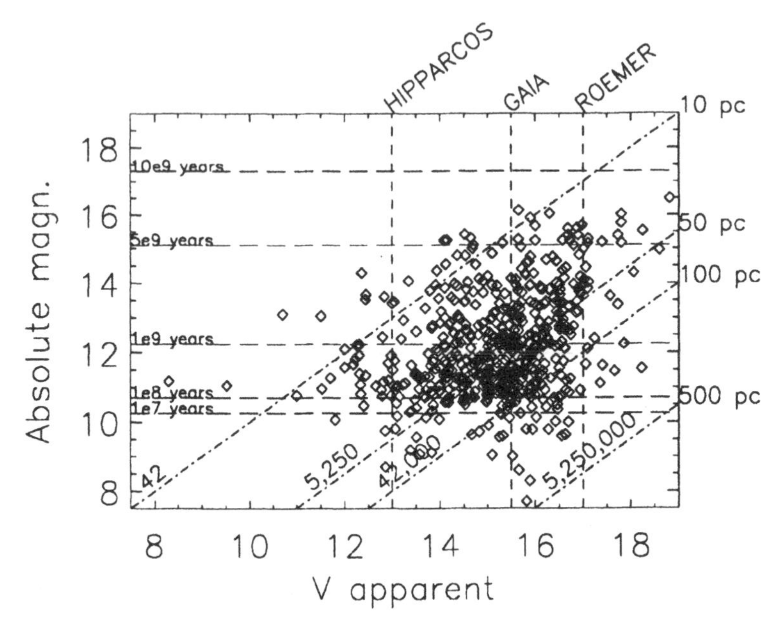

Figure 1. 575 WD's with known V magnitude and absolute magnitude plotted against each other, with horizontal lines showing the cooling times in years, diagonal lines showing the distances and vertical lines showing the limiting magnitudes for HIPPARCOS, GAIA and ROEMER. At the lower end of each distance-line is a number indicating the number of WD's that are closer than the corresponding distance, based on current measurements of WD space density. Only for the very closest WD's is the presently known sample complete.

at $M_{\rm abs} = 10$) are at 500 parsec distance (roughly 2 scale heights) at V=19 while $M_{\rm abs} = 15$ WDs require that one can observe reliably at V=23. None of the suggested missions have this capability, although the ROEMER mission is aiming for a capability of 1.5 mas accurate parallaxes at V=17, which might cover 1 scale height, or a little more, of the hot WD distribution and give 30% accurate parallaxes at the limit, which is competitive with spectrophotometric distances. GAIA would cover less than a scale height, but proposes to give parallaxes accurate to 20 μas at V=15, which in the hot WD case means distances accurate to 0.2% - virtually error free. Cool (old, Pop II) WDs would still be best dealt with spectrophotometrically unless mas accuracy can be provided at 24'th and 25'th magnitude. With scale height known, space density could be calculated and then evolution rates would follow using the known cooling rate for WDs.

Direct measurement of the space density of WDs by having complete counts of the stars in the solar vicinity - possible even for the coolest and oldest WDs - seems a preferable method, but is possible only if the large number of WDs undetected nearby the Sun are located and accounted for.

In figure 1 we see that there should be about 5000 WDs within 50 parsec of the Sun, most of which would be observable by ROEMER and to a lesser extent GAIA.

GAIA's potential importance to WD research lies in the higher accuracy attainable than with ROEMER - but as ROEMER reaches 2 magnitudes fainter the latter may well have the greater impact on WD research where accuracy in parallaxes is important but reaching further into space is the key to arguing against claims of local solar neighborhood bias.

Hot sdB and sdO and the CSPN can be seen much further away than WD's as they have luminosities from 10 to hundreds of solar luminosities. It is therefore hoped that a spectrophotometric distance determination can be performed on the stars in the complete Palomar-Green survey of high galactic latitude blue objects (Green *et al.* 1986). However, this requires advanced NLTE model atmosphere techniques that have problems in predicting the line profiles consistently in comparisons between different sets of models - particularly for the sdO stars but also, to a lesser extent, for the somewhat cooler hydrogen-rich subdwarf B (sdB) stars.

The sdB probably have the mass of a stripped He core - i.e. 0.5 $M_\odot$ - but the sdO could in principle have a range of masses from tenths of solar masses to about 2 $M_\odot$. The scale height for the sdB has been determined assuming the mass is 0.5 $M_\odot$ and it is near 500 parsec (Saffer 1994) and stars as far away as a few kiloparsec have been analyzed, so the requirement of covering a sufficient number of scale heights is met for the sdB.

Spectrophotometric mass-determinations for sdO, coupled to parallax work, relies on the basic equation in section *2.2*. Once the modelling problems were resolved one could rely on spectrophotometric values for g and then determine masses: A realistic accuracy in spectroscopically determined gravities is 0.2 or 0.1 in $log(g)$. Once π was known better than 50% to 25% one could get masses with accuracies mainly limited by $\Delta log(g)$.

HIPPARCOS can just manage to yield the required parallax accuracy for a small number of sdO bright enough and the release of data is eagerly awaited. The less problematic sdB will have their problems sorted out by more accurate missions. GAIA can, with 20 μas accurate parallaxes at V=15 give 3% accurate distances for nearly half of the presently known hot subdwarfs, and ROEMER could cover all known hot subdwarfs, putting the emphasis on improving the accuracy of spectral modelling methods.

The impact of HIPPARCOS, to some extent, and GAIA and ROEMER in particular, on resolving present problems of NLTE modelling will be great. Once the problems are sorted out, spectrophotometric methods in conjunction with parallaxes will deliver 50% accurate individual sdO masses and more accurate sdB masses. This is enough to test whether the two groups of stars have a well defined mean mass or not, and whether it is the same, which is sought-after information with impact on stellar evolution theory.

4.1. SUMMARY

The properties of the proposed observatories are such that there could be important consequences in two broad areas of research in late stages of stellar evolution: more accurate derivation of stellar properties for more stars, and detection of previously unknown stars. The discovery of many new white dwarfs, the significantly more accurate study of which will yield more and better insight into WD masses, radii, spatial distribution and hence the luminosity function, and the age-determination of the stars in the Galaxy is one of the most important potential results. For the hot subdwarfs and CSPNs reliable answers can be expected to questions about the spatial distribution and hence the role played by these stars on the road to the WD's and as tracers of stars with extreme mass-loss. Of great interest will be the many opportunities for testing various methods of measurement and analysis.

Support from the IAU, the Carlsberg foundation, and the NBI is gratefully acknowledged.

References

Bastian, U. *et al.* (1993), ROEMER, Proposal for M3
Bergeron, P., *et al.* (1992), ApJ 394, 228, **BSL92**
de Boer, K. (1985), A& 142, 321
Dreizler S. *et al.* (1990), A&A 235, 234
Fleming T. *et al.* (1986), ApJ 308, 176
Gatewood G.D.& Gatewood C.V. (1978), ApJ 225, 191.
Green R.F. *et al.* (1986), ApJSS 61, 305
Greggio&Renzini, A. (1990), ApJ 364, 35
Hamada T.&Salpeter E.E. (1961), ApJ 134, 683
Heintz W.D. (1974), AJ 79, 819
Liebert J. *et al.* (1988), ApJ 332, 891
Lindegren L., *et al.* (1993), GAIA, Proposal for an ESA Cornerstone Mission
McCook&Sion E.M. (1987) ApJSS 65, 603
Ritter H. (1990), AASS 85, 1179
Saffer R. (1994), in White Dwarfs, proceedings of the 9'th European WD Workshop.
Shipman, H.L. (1979), ApJ 228, 240
Sweigart A.V.&Gross P.G. (1978), ApJSS 36, 405
Thejll P.A. *et al.* (1991), ApJ 370, 355
Thejll P.A. *et al.* (1994), ApJ, in press
Wood M. (1992), ApJ 386, 539
van Altena W.F. *et al.* (1991), Gen. Cat. of Trig. Par.: Prel. Ver.

FINE STRUCTURE IN THE MAIN SEQUENCE: PRIMORDIAL HELIUM AND $\Delta Y/\Delta Z$

B.E.J. PAGEL
NORDITA, Blegdamsvej 17,
Dk-2100 Copenhagen Ø, Denmark

Abstract. The primordial helium abundance Y_P is important for cosmology and the ratio $\Delta Y/\Delta Z$ constrains models of stellar evolution. While the most accurate estimates of both quantities now come from emission lines in HII regions, significant information comes from effects of helium content on stellar structure including in particular the location of the main sequence as a function of metallicity and age. HIPPARCOS parallaxes with 1 or 2 mas accuracy will naturally lead to great advances in this type of study for stars with metallicities down to about 0.1 solar, but sub-mas accuracy will be needed in order to extend it to stars of still lower metallicity.

1. Introduction

One success of Big-Bang cosmology has been to predict primordial abundances of light elements (Walker *et al.* 1991; Smith *et al.* 1993); but precise determination of each primordial abundance involves difficulties, both from measurements and in extrapolating through evolution that has taken place in the meantime. While overall agreement with Big-Bang theory is not in doubt, detailed consequences such as the exact limits on baryonic and non-baryonic matter remain controversial.

Fig 1 shows the region of concordance usually accepted for primordial abundances (tall continuous vertical lines) with resulting limits on the density parameter $\Omega_{b0}h_0^2$; Ω_{b0} is the fraction of closure density from baryons now and h_0 is the Hubble constant in units of 100 km $s^{-1}Mpc^{-1}$. BDM and NDM show implied amounts of baryonic and non-baryonic dark matter. These conventional values have, however, all been challenged for various reasons (*cf.* Pagel 1994ab), resulting in the much wider limits shown by the broken lines in the figure, so any supplementary evidence can be useful.

E. Høg and P. K. Seidelmann (eds.),
Astronomical and Astrophysical Objectives of Sub-Milliarcsecond Optical Astrometry, 181–186.

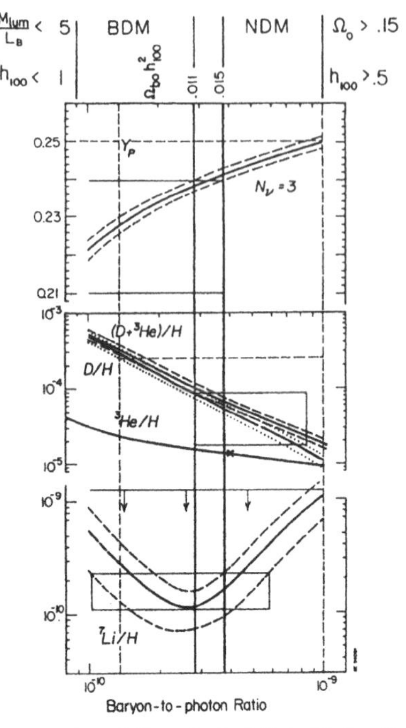

Figure 1. "Optimistic" (narrow) and "pessimistic" (wide) estimates of limitations placed on the baryonic density parameter $\Omega_{b0}h_0^2$ from adopted upper limits to primordial abundances of light elements, notably ^{4}He and deuterium.

2. Y_P and $\Delta Y/\Delta Z$

Numerous estimates of the primordial helium mass fraction Y_P (see Table 1 and detailed references in Shaver *et al.* 1983 and Pagel 1994a) agree at the 10 per cent level, but systematic errors make it difficult to achieve the precision of a few per cent needed to constrain the Big Bang. The smallest error estimates are those for the method based on extragalactic HII regions, in which one plots a regression of He against O (or N) from measurements of emission lines and extrapolates to zero O (or N). Rather little extrapolation is actually needed, but there could be additional systematic errors.

The quantity $\Delta Y/\Delta Z$ represents the amount of additional helium ejected by dying stars relative to the ejected amount of heavy elements from carbon upwards. It can be predicted from the theory of stellar evolution, assuming an initial mass function (IMF), needed because while Z comes mainly from massive stars undergoing supernova explosions, most of the additional helium comes from intermediate-mass stars (typically $5M_\odot$) which lose mass in winds and planetary nebulae. It is found observationally either from the slope of the regression of He against O in HII regions or by considering effects of helium on stellar structure and evolution.

TABLE 1. Primordial Helium (ESO 1983 *et al.*)

	Y_P	Method	First author	Problems
Sun	$< .28 \pm .02$	Interior	Turck-Chièze	κ; eq of st; ν problem
Sun	$< .28 \pm .05$	Prom. HeI	Heasley	Level pops.
B-stars	$< .30 \pm .04$	Abs. lines	Kilian	Precision
μ Cas	$.23 \pm .05$	Bin. orbit	Haywood	Precision
Subdw.	$.19 \pm .05$	Main seq.	Carney	Plx.; T_{eff}; conv.
Glob-	.23 :	RR, Δm	Caputo	Physical
ular	$.23 \pm .02$	N(HB)/N(RG)	Buzzoni	basis of
clus-	$.20 \pm .03$:	"	Cole	stellar
ters.	$.23 \pm .02$:	HB morphology	Dorman	evolution
Galactic	$.22 \pm .02$	Plan. neb.	Peimbert	Enrichment
nebulae.	.22 :	HII regions	Mezger	He^0; enr.
Extra-	$.233 \pm .005$	Irr.+BCG	Lequeux	He^0; data
galactic	$< .243 \pm .010$	BCG	Kunth	IIZw40
HII	$.228 \pm .005$	Irr.+BCG	Pagel	?
regions	$.232 \pm .003$	"	Olive	?

While conventional stellar evolution calculations give $\Delta Y/\Delta Z \simeq 2$ or less (Maeder 1992, 1993), observations give larger values, e.g. ~ 3.5 (Faulkner 1967), 5 ± 3 (Perrin *et al.* 1977) from main sequences and 3 ± 0.5 (Lequeux *et al.* 1979; Peimbert 1993) and 4 ± 1 (Pagel *et al.* 1992) from HII regions. Explanations for this discrepancy include (i) a low upper limit ($\simeq 25 M_\odot$) to the initial mass of stars undergoing supernova explosions as opposed to going into black holes; (ii) loss of Z-elements by selective galactic winds following bursts of star formation in dwarf galaxies; (iii) underestimation of oxygen abundance in HII regions by neglecting electron temperature fluctuations; and (iv) that a steep IMF (Scalo 1986) gives a value of 3 for a supernova upper mass limit of about $50 M_\odot$ (Maeder 1992). Since some of these explanations are specific to HII regions and dwarf galaxies and others more general, the importance of improving the stellar data becomes clear.

3. Effects of helium on stellar structure

Various effects of initial helium content on the evolution of stars in globular clusters are briefly summarised in Table 1. The zero-age main sequence follows quasi-homology relations; e.g. Faulkner (1967) gives:

$$L \propto (X + 1.2)^{-11.72} (Z + 0.012)^{-0.134} M^5 \quad (1)$$

for the mass-luminosity relation, and

$$L \propto (X + 0.4)^{2.67} (Z + 0.010)^{0.455} f(T_e) \tag{2}$$

for the main sequence, where X is the mass fraction of hydrogen and T_e the effective temperature. Similar relations can be derived from more modern evolutionary tracks. The mass-luminosity relation can now almost be modelled in absolute terms, given modern opacities, whereas the main-sequence relation is subject to uncertainties in convection theory and fitting model atmospheres and must in practice be calibrated assuming a certain helium abundance for the Sun, allowing for any physical and chemical changes.

Assuming these problems to be overcome, one can make an error budget for the derivation of X from either of these relations. In the first case, assuming that the mass is derived from an interferometric binary orbit as in the case of μ Cas, the major uncertainties are the distance D and the angular major axis α of the relative orbit, while the exact value of Z is not important if it is small. The resulting error budget from eq (1) is

$$\frac{\delta X}{X + 1.2} = 1.11 \frac{\delta D}{D} + 1.28 \frac{\delta \alpha}{\alpha} \tag{3}$$

or

$$(\delta X)^2 \simeq (2\, \delta D/D)^2 + (2.5\, \delta\alpha/\alpha)^2, \tag{4}$$

since $X \simeq 0.75$. Thus a parallax good to 1 per cent already leads to an error of 0.02 in X or Y and there is a still stricter requirement on α.

The error budget for the main sequence is more favourable (not counting uncertainties in the physics). Here the major uncertainties are in the distance and in the effective temperature. Taking $f(T_e) \propto T_e^7$, we have

$$\frac{\delta X}{X + 0.4} = 0.75 \frac{\delta D}{D} - 2.6 \frac{\delta T_e}{T_e} \tag{5}$$

or

$$(\delta X)^2 \simeq (0.9\, \delta D/D)^2 + (3\, \delta T_e/T_e)^2. \tag{6}$$

Models for low-metallicity stars aged 15 Gyr and with $\log T_e = 3.72$ (e.g. VandenBerg 1983) give a quasi-homology relation like eq (2) only with $(Z{+}0.003)$ instead of $(Z{+}0.010)$, but the error budget is otherwise identical. The demand on accuracy in distance is less severe by more than a factor of 2 than for the mass-luminosity relation, although there are now stringent requirements on the accuracy of T_e and reddening effects if any.

Thus most estimates of helium in metal-deficient field stars have come from main-sequence fitting. From eq (2) it follows that, at high Z, reduction of helium and Z act in opposite directions and may compensate each other,

TABLE 2. Nearby subdwarfs with $Z < 0.1Z_{\odot}$

HD	π mas	$\delta D/D$, per cent ground	LK	Hipparcos 1.5 mas	Rømer etc. 0.1 mas
103095	116 ± 5	4	1	1.3	0.1
25329	54 ± 5	9	4	2.8	0.2
201891	41 ± 6	15	12	3.7	0.25
134439-40	40 ± 5	12	6	4.0	0.25
64090	38 ± 4	11	12	4.0	0.25
193901	35 ± 6	17	17	4.3	0.3
84937	25 ± 5	20	31	6.0	0.4
108177	30 ± 7	23		5.0	0.35
94028	23 ± 7	30		5.0	0.45
19445	21 ± 6	29		7.1	0.5

whereas at low Z a decrease in helium leads to a raising or rightward shift of the main sequence independent of Z. Perrin *et al.* (1977) could find no metallicity-correlated dispersion in the main sequence of nearby disk stars and from this they deduced $\Delta Y/\Delta Z = 5 \pm 3$, a number that will certainly be greatly improved when HIPPARCOS parallax data become available and can be used in conjunction with modern, accurate opacities.

The extension of such considerations to extremely metal-deficient stars associated with the Galactic halo is considerably more challenging. Carney (1979, 1983) used ground-based parallaxes of a few extreme subdwarfs having $Z < 0.1Z_{\odot}$ with modified Yale isochrones to deduce $Y_P = 0.19 \pm 0.05$, a result that is marginally discrepant with other data in Table 1 and subject to numerous uncertainties, of which that in the distances is by no means the least significant. The problem can be seen from Table 2.

The first column gives HD numbers for the nearest stars with $Z < 0.1Z_{\odot}$, the ones above the gap being the nine considered by Carney (1983) omitting the subgiant HD 140283. The second column gives parallaxes with standard errors, from the HIPPARCOS Input Catalogue. The third column gives the corresponding percentage error in distance, which according to eq (6) translates into about the same number of units in the second decimal place of the resulting error in Y, i.e. the error is ± 0.04 for the uniquely favourable case of HD 103095 = Groombridge 1830 and unbearably large for all the others. The fourth column gives the Lutz-Kelker correction applied by Carney, which is also highly significant; full applicability of this correction to stars selected by proper motion is not completely clear (Hanson 1979; Lutz, Hanson & Van Altena 1987), so that this adds further uncertainties. Thus with only existing ground-based parallaxes available, the enterprise of trying to estimate primordial helium from the subdwarf main sequence was doomed from the start.

The fifth column gives percentage errors in distance expected from HIPPARCOS parallaxes with a standard error of ± 1.5 mas (Perryman 1994). Surprisingly, perhaps, the situation here is still quite unsatisfactory, with errors in X or Y of ± 0.03 and more, just from the distance, in all cases except Gmb 1830. The last column shows the precision obtainable from parallaxes with 0.1 mas errors, which one hopes may result from future space or interferometric projects; only in this case do the distance errors become negligible so that one can concentrate on the purely astrophysical problems.

References

Carney, B.W. 1979, *Astrophys. J.*, **233**, 877.

Carney, B.W. 1983, in Shaver, P.A. *et al.* (eds.), *Primordial Helium*, ESO, Garching, p. 179.

Dorman, B., Lee, Y.-W. & VandenBerg, D.A. 1991, *Astrophys. J.*, **366**, 115.

Faulkner, J. 1967, *Astrophys. J.*, **147**, 617.

Hanson, R.B. 1979, *Mon. Not. R. astr. Soc.*, **186**, 875.

Haywood, J.W., Hegyi, D.J. & Gudehus, D.H. 1992, *Astrophys. J.*, **392**, 172.

Lequeux, J., Peimbert, M., Rayo, J.F., Serrano, A. & Torrres-Peimbert, S. 1979, *Astr. Astrophys.*, **80**, 155.

Lutz, T.E., Hanson, R.B. & van Altena, W.F. 1987, *Bull. Amer. Astr. Soc.*, **19**, 675.

Maeder, A. 1992, *Astr. Astrophys.*, **264**, 105.

Maeder, A. 1993, *Astr. Astrophys.*, **268**, 833.

Pagel, B.E.J. 1994a, in P. Crane (ed.), ESO Workshop: *The Light Element Abundances*, Springer-Verlag.

Pagel, B.E.J. 1994b, in M. Busso, R. Gallino & C. Raiteri (eds.), *Nuclei in the Cosmos III*, Amer. Inst. Phys. publ.

Pagel, B.E.J., Simonson, E.A., Terlevich, R.J. & Edmunds, M.G. 1992, *Mon. Not. R. astr. Soc.*, **255**, 325.

Peimbert, M. 1993, *Rev. Mex. Astr. Astrofis.*, **27**, 9.

Perrin, M.-N., Hejlesen, P.M., Cayrel de Strobel, G. & Cayrel, R. 1977, *Astr. Astrophys.*, **54**, 779.

Perryman, M.A.C. 1994, *Astr. News* (ESA), Jan issue.

Scalo, J.M. 1986, *Fund. Cosmic Phys.*, **11**, 1.

Shaver, P.A., Kunth, D. & Kjär, K. (eds.), *Primordial Helium*, Garching: ESO.

Smith, M.S., Kawano, L.H. & Malaney, R. 1993, *Astrophys. J. Suppl.*, **85**, 219.

VandenBerg, D.A. 1983, *Astrophys. J. Suppl.*, **51**, 29.

Walker, T.P., Steigman, G., Schramm, D.N. & Kang, H. 1991, *Astrophys. J.*, **376**, 51.

ACCURATE BINARY MASS DETERMINATIONS: GOALS, LIMITATIONS, AND PROSPECTS

J. ANDERSEN
Niels Bohr Institute for Astronomy, Physics, and Geophysics
Astronomical Observatory
Brorfeldevej 23
DK - 4340 Tølløse, Denmark

ABSTRACT. The state of the art in accurate mass determination for binary stars is reviewed, and the angular sizes and their errors are computed for a typical system from the existing high-precision sample. It appears that sub-μas (microarcsecond) absolute astrometry will be needed in order to improve the accuracy substantially by astrometry alone. The types of system, and the kinds of data, where precision astrometry appears most promising are outlined. Finally, astrophysical applications of such accurate stellar masses, and the auxiliary data required in them, are briefly reviewed.

1. Introduction

When this contribution was first suggested, the proposed subject was the improvement in state-of-the-art mass determinations for eclipsing binary stars that could be expected to result from new parallaxes with errors in the 10-100 μas range. The ultrashort summary reply, *"Not much"*, is clearly not very useful unless backed by an explanation of the state of the art and the possibilities and limitations in improving it. Also, while determining precise stellar masses may seem to be a worthy goal in itself, a reasonably focused idea of the precise astrophysical applications of the data is a useful basis in such work.

Hence, the following will review the state of the art in mass determinations for (double-lined) eclipsing binaries, translate the results into visual binary language for comparison with the projected astrometric capabilities, and outline the practical possibilities and likely results. As will appear, very useful data can indeed result, if perhaps not always primarily on stellar masses *per se*. Finally, to place these data in a specific astrophysical context, I shall outline some of the uses of precise stellar masses that motivate their determination, and the supporting data which are also needed in such applications.

2. The State of the Art in Stellar Masses

It is elementary textbook material that in eclipsing binaries in which the lines of both stars are visible in the combined spectrum, the *individual masses* are determined from the two orbital velocities (radial-velocity amplitudes), combined with the orbital inclination as determined from the light-curve analysis. In the process, the *absolute radii* are also found, and individual *luminosities* are computed from these radii combined with effective temperatures derived from suitable calibrated (and reddening-corrected) colour index measurements.

E. Høg and P. K. Seidelmann (eds.),
Astronomical and Astrophysical Objectives of Sub-Milliarcsecond Optical Astrometry, 187–192.

These results are entirely *independent of the distance* of the system, which then follows from the apparent magnitude, reddening-corrected as appropriate.

In astrometric binaries, the astrometry yields the orbital semi-major axis a (and e, ω) and inclination of the orbit, plus the parallax, hence the *distance*. From these, the *sum of the masses* is obtained, as well as the individual *luminosities* if Δm is well determined. Determination of *individual masses* requires that either absolute astrometric orbits or a double-lined spectroscopic orbit is available. *Radii* follow from the luminosities when effective temperatures have been determined from colour indices as above.

For the basic reasons well explained by Popper (1980), the most accurate mass determinations are possible in eclipsing binaries. Published values of individual masses and radii of the components show errors down to $\sim 0.5\%$. The available sample of accurate binary mass (and radius) determinations, defined as having errors of no more than 2% in both masses and radii, comprises about 45 systems, all double-lined eclipsing binaries. These systems are listed and discussed by Andersen (1991), who also addresses in considerable detail the precautions necessary in the selection of systems as well as in the observations and analysis in order to reach this level of accuracy. Hence, that discussion will not be repeated here.

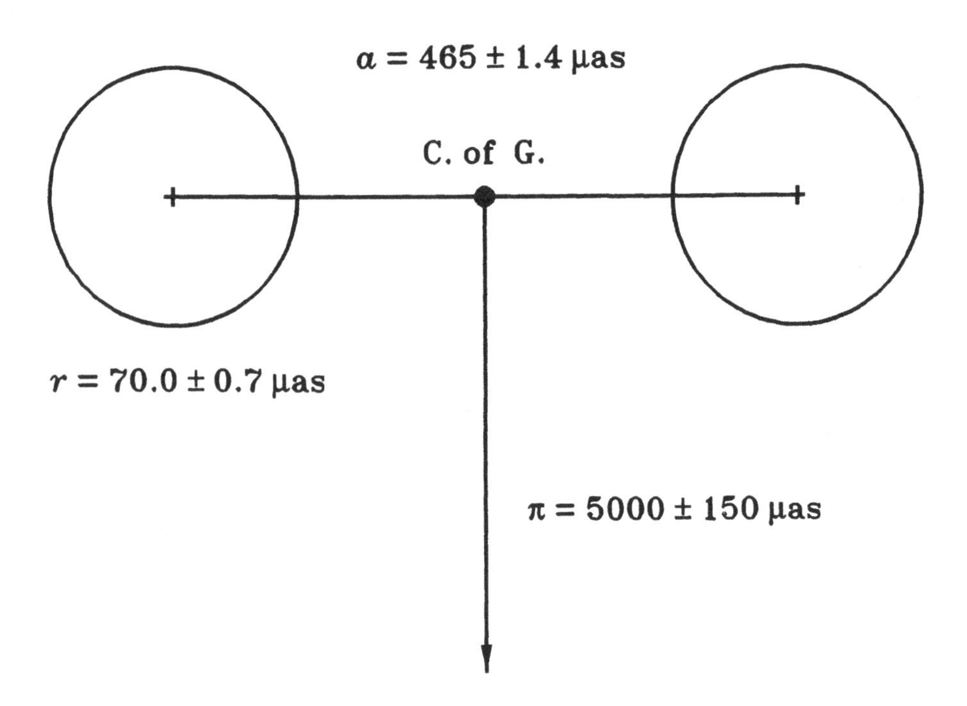

Figure 1. Angular dimensions of the typical binary system from Table 1 when placed at the typical distance of 200 pc.

From the list of stellar properties in Table 1 of Andersen (1991), the characteristics of a typical system have been extracted; they are summarised in Table 1 below:

TABLE 1. Typical properties of a well-studied binary system. The full parameter ranges observed in the sample are also given for comparison.

Parameter	Value	Range	Accuracy
Mass	2 $M_\odot$	0.6 - 23 $M_\odot$	1 %
Radius	3 $R_\odot$	0.6 - 9 $R_\odot$	1 %
Luminosity	20 $L_\odot$	0.1 - 10^5 $L_\odot$	10 %
Orbital radius	20 $R_\odot$	4 - 120 $R_\odot$	0.3 %
Distance	200 pc	15 - 2200 pc	3 %

Taking the typical system described in Table 1 and placing it at the distance indicated leads to the set of angular separations and diameters shown, with their errors, in Fig. 1.

It is obvious that sub-μas accuracy is needed to improve the determination of the intrinsic stellar parameters mass and radius substantially from astrometry alone: Assuming that the error of the parallax were only 25 μas, and the individual semi-major axes known *without error* (e.g. from ultra-precise spectroscopic orbits), the mass errors would still amount to 1.5%, not yet competitive with the best existing spectroscopic determinations.

Hence, the initial answer to the question how sub-milliarcsecond astrometry could improve the precision of state-of-the-art stellar mass determinations does seem to be: "Not much".

3. The scope for astrometric contributions

Two comments are needed to qualify the negative statement just made. First, it only applies to the types of star which are found in suitable eclipsing binary systems that have also been adequately studied, i.e., with only a couple of exceptions, main-sequence stars of Population I in the range G5 - O8 (0.6 - 23 $M_\odot$). Stars at the lowest and highest ends of the main sequence, and stars evolved to the giant stage or beyond, are either not generally found in eclipsing binary systems, or their properties (spectra, light curves, brightness) are not suitable for accurate analysis of the kind performed for the well-behaved main-sequence stars referred to above. For such stars, masses of lower accuracy may still be valuable, and/or the masses may be of interest even without the accompanying data discussed below.

It follows that astrometric studies can be profitably concentrated on systems that:

- are much closer than 200 pc, so larger errors can be tolerated, or
- have no measurable spectral lines in at least one component, or
- are not eclipsing, but have
  either: *absolute* astrometric orbits of both stars,
  or: spectroscopic orbits of both stars.

Stellar types for which such determinations would be of particular interest include, on the main sequence, O stars at the high end and K - M dwarfs and brown dwarfs at the low end; evolved stars such as giants and supergiants, hot subdwarfs, white dwarfs, or nuclei of planetary nebulae; and metal-poor Population II stars of all types.

Second, by improving the parallax error to 0.5% (Fig. 1) something *is*, of course, gained, because the error in the individual luminosities is reduced from ~10% to ~1%. It would appear to be of little specific value to do this for binary rather than single stars, though, unless some observational parameter were known particularly well for the binary stars (see below). In eclipsing binary stars, however, a further valuable result follows since the surface areas of the stars are known: The effective temperature may be determined.

From the well-known relation:

$$L = 4 \pi R^2 \cdot \sigma T_e^4$$

it is seen that if both L and R are known to 0.5-1%, we can derive T_e to an accuracy of about 0.5%, much better than with most current temperature calibrations. A recent example of this technique is given by Nordström & Johansen (1994) for the bright, nearby system β Aurigae. Again, application of this knowledge to other stars requires that very accurate observational parameters are available for their characterisation.

4. Why are accurate masses useful at all, and under what conditions?

Even the world's most accurate mass determination for a star is without interest outside the *Guinness Book of Records* unless it can be used to address some astrophysical question of wider implications. So what useful applications can one envisage for a sample of accurate empirical determinations of stellar masses (say, 1% or better)? The three main applications of which the author is aware (see further discussion in Andersen 1991) are:

- Predicting masses (and radii) for single stars from observed parameters,
- Improving "the" mass-luminosity relation, and
- Testing stellar evolution computations.

However, in all three cases, the factor limiting the accuracy with which the key relations can be defined, or theory and observation be compared, is *not* the observational error of the actual masses once it is below the 1% level. Specifically, in each of the above cases:

- As shown by the accurate data compiled by Andersen (1991), observed masses and radii for main-sequence stars with a given spectral type or colour, however accurately determined, span a range of ~30% in M and a factor 2 in R, due primarily to differences in evolution even within the main-sequence band. The best measure of post-ZAMS evolution is log g as determined fundamentally and with superior precision (0.01 dex or better, ~5% of the width of the main sequence) in eclipsing binaries, directly from M and R. Taking evolution into account allows to reproduce the observed M and R within ~5%; the excess scatter relative to the observational error of 1% is probably due to differences in chemical composition (metal abundance). Hence, knowing the mass of a star to 1% is not really useful unless log g and [Fe/H] are also known to matching accuracy.

- Similarly, the scatter of the observed points around a mean mass-luminosity relation defined from the same sample of stars cannot be explained only by observational errors in the masses (or luminosities); see Fig. 2. Most of the scatter is due to real differences in, again, evolution above the ZAMS but within the main sequence, and in metal abundance from star to star. Improving only the precision of the masses (or even M and L) would not lead to a better-defined mean relation.

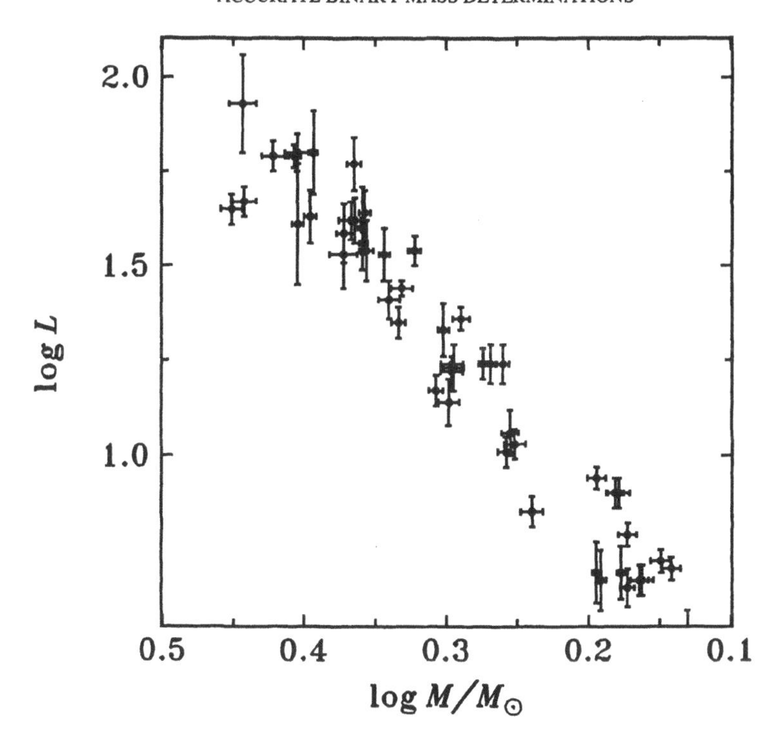

Figure 2. Zoom-in on a section of the mass-luminosity relation as defined by data from Table 1 of Andersen (1991). Note that the scatter considerably exceeds the observational errors.

- Precise observational tests of the predictions of stellar evolution theory is one of the most demanding applications of empirical data on stellar masses and radii. Basically, one evolves a model of the mass and composition of one of the binary components to match its observed radius and luminosity, fixing such model parameters as mixing length and helium abundance from a model of the present Sun. The *test* then consists of computing a model for the mass of the other star, but with the age and all other parameters unchanged, and verifying whether it fits all observed properties of that star as well, within the errors. Yet, while initial mass is the most important property of a stellar model, the metal abundance turns out in fact to be the limiting factor in fitting models to the observations as illustrated, e.g. by Andersen et al. (1989) who actually determined [Fe/H] for the system studied through high-resolution CCD spectroscopy. Again, knowing just the mass to arbitrarily high accuracy is not really useful unless the radii (or log g) and composition of the stars are also known to the highest accuracy currently possible. Attempts to draw conclusions on finer details in stellar models from broad mean relations for binaries presumed to have "Population I" metallicity and without detailed, simultaneous fits to both components in each system (e.g. Stothers & Chin 1991) are bound to remain inconclusive.

5. Conclusions

Unquestionably, many areas of astronomy will benefit from yet another quantum leap in the quality and quantity of astrometric data available for stars to ever greater distances from us. The determination and astrophysical application of accurate binary masses can be one of these areas provided the intended applications are kept clearly in mind in both selecting the systems to be studied, carrying out the observations, and securing the additional data which, as discussed above, are crucial for the masses themselves to be really useful. As always, the best results in each case will be obtained through a judicious combination of the available astrometric, spectroscopic, and photometric techniques.

References

Andersen J., 1991, Astron. Astrophys. Rev. 3, 91
Andersen J., Clausen J.V., Magain P., 1989, Astron. Astrophys. 211, 346
Nordström B., Johansen K.T., 1994, Astron. Astrophys., in press
Popper D.M., 1980, Ann. Rev. Astron. Astrophys. 18, 115
Stothers R.B., Chin C.-W., 1991, Ap.J. 381, L67

SUB-MILLIARCSECOND OPTICAL ASTROMETRY AND BINARY STARS

J.T. ARMSTRONG
Universities Space Research Association
NRL/USNO Optical Interferometer Project
U. S. Naval Observatory — AD 5
3450 Massachusetts Ave. NW, Washington, DC 20392, USA
tarmstr@atlas.usno.navy.mil

Abstract. Long-baseline optical interferometry has made it possible to measure the visual orbits of binary stars with major axes as small as 5 mas and errors of $\lesssim 100$ μas. Interferometers now nearing completion will extend these values to $a \gtrsim 500$ μas and $\sigma_a \sim 10$ μas. Observations of double-lined spectroscopic binaries with current interferometers have already yielded some mass estimates with precisions rivaling those from fitting the light curves of eclipsing double-lined systems. Luminosity estimates based on combined visual interferometric observations and velocity curves are often more precise than those from more indirect methods based on estimates of T_{eff}. New interferometers now coming into operation will make it possible to measure fundamental parameters in dozens to hundreds of binary systems.

1. Introduction

The high angular resolution of the new optical interferometers coming into operation at several sites around the world will considerably expand our knowledge of binary stars. One area that will benefit is the determination of fundamental stellar quantities—masses, radii, luminosities, and metallicities. Stellar masses can be measured only in binaries, and until recently only in eclipsing, double-lined spectroscopic binaries (SB2s) could we measure masses and radii to the 1% to 2% precision needed for comparison with stellar evolution models (Andersen 1991).

E. Høg and P. K. Seidelmann (eds.),
Astronomical and Astrophysical Objectives of Sub-Milliarcsecond Optical Astrometry, 193–202.

The results from the Mark III interferometer on Mt. Wilson, California, discussed in these proceedings by C. Hummel, demonstrate that masses and radii can be measured in non-eclipsing SB2s, and show that the distance, and thus the luminosities, can be measured as well. A few of these results already meet the demanding standards set by the results from eclipsing systems. The new interferometers will further expand the systems available for these measurements to include fainter SB2s, and to include nearby SB1s.

2. Measurement of Fundamental Stellar Parameters

Until recently, measurements of fundamental stellar parameters that are precise enough to challenge models of stellar evolution have been limited to detached, eclipsing, double-lined spectroscopic binaries (Andersen 1991). These were the only systems in which it has been possible to determine the mass ratio and the semimajor axis a (in km) of the orbit from the velocity amplitudes K_1 and K_2, and to determine the orbital inclination i from modeling the shape of the light curve when the stars eclipse one another. With the orbital eccentricity e also deduced from the velocity curves, and with the period P in hand, the masses $\mathcal{M}_{1,2}$ of the components are then calculated. An added benefit of eclipsing systems is that the light-curve modeling also produces estimates of the radii $R_{1,2}$ (also in km) of the components. However, the distance D to the system is not measured; rather, it is derived from the estimated effective temperatures $T_{\rm eff}$, the radii, and the apparent magnitude m_V. Eclipsing systems, then, yield direct determinations of stellar masses and radii, and an indirect determination of the luminosity L.

As detailed by Andersen (1991), 45 such systems have given us results that are sufficiently precise to challenge stellar evolution models. His criteria of precision are masses and radii with $< 2\%$ mean errors, although he accepts 3.5% errors for YY Gem, the only well-studied M system. But these systems occupy only part of the HR diagram, being concentrated on the main sequence; only five components have $\mathcal{M} < 1\mathcal{M}_\odot$, only eight are of spectral type later than F, and only two are red giants. Late-type dwarfs suffer from two selection effects, as pointed out by Popper (1993) among others: because they are small, they must be close together in order to be likely to eclipse; and because they are faint, they are less often discovered and harder to observe. Among binaries with giant components, the magnitude difference Δm is often too large for spectral lines of both to be measured adequately, unless they are sufficiently close in mass that they become giants at about the same time.

3. Interferometric Results

There are many more non-eclipsing binaries than eclipsing binaries, and it has always been possible to combine visual and spectroscopic orbits to obtain masses in non-eclipsing systems. But the overlap between the classes of visual and spectroscopic binaries has until recently been small. Using visual orbits, rather than eclipse light curves, to determine the orbital inclinations affords, among other things, a measurement of the distance to the system: the spectroscopic data give the physical size of the orbit to within a factor of $\sin i$, while the visual data give the angular scale and i. The disadvantage of visual data in comparison with eclipse light curves is that the stellar radii are not determined.

Speckle interferometry, and recently separate-element interferometry, have enlarged the overlap between these two classes (e.g., Griffin 1992). Several groups built two-element interferometers during the 1970s and 1980s, with typical baselines of a few tens of meters and apertures less than 10 cm. Of these, the Mark III interferometer on Mt. Wilson was the most productive, in terms of amount of data acquired. It was also unique in having a dual purpose: astrometry over the whole (available) sky, and "imaging," i.e., measurement of stellar diameters and binary orbits. The single-star astrometry of the Mark III was not precise enough to measure astrometric orbits of close binaries, but binary data from the Mark III has made contributions to the list of stars with precisely known fundamental quantities.

The Mark III, which operated on Mt. Wilson from 1986 to 1992, was a two-element Michelson interferometer with active fringe and angle tracking, automated operations, and 450 to 800 nm wavelength coverage. The 5 cm maximum aperture limited observations to $m_V \lesssim 5^{\mathrm{m}}$. Astrometric observations used one of two fixed 12 m baselines and provided stellar positions to 20 mas precision (Hummel et al. 1994b) over a $90° \times 30°$ $(\alpha \times \delta)$ region. The variable baselines of 3 to 31.5 m made it possible to measure stellar diameters larger than 2.5 mas with 0.05 to 0.1 mas precision (Mozurkewich et al. 1991, 1994), and binary orbits larger than 3 mas with 0.05 mas precision. The Mark III used a single 250 nm wide channel near λ700 nm for fringe phase tracking, and typically three channels, each 25 nm wide, at $\lambda\lambda$500, 550, and 800 nm, for scientific data. The calibration precision of the fringe visibilities ranged from 1% to a few percent in the 800 nm channel, to as much as 15% in the 450 nm channel; for the binary observations, this made it possible to measure orbits of systems with $\Delta m \lesssim 3\overset{\mathrm{m}}{.}5$, and to measure Δm in those systems to a precision of $0\overset{\mathrm{m}}{.}05$ to $0\overset{\mathrm{m}}{.}3$.

The Mark III obtained more than two dozen visual orbits with milliarcsecond or sub-milliarcsecond precision. Seventeen of these are double-lined spectroscopic binaries; of these, results have been published for five (β Ari,

α Equ, ϕ Cyg, η And, α Aur, β Per) and are in preparation for six more (Hummel et al. 1994c). It is worth remarking on some of these systems in the context of what is possible from interferometric observations. In a few of them, the precision of the masses is of the order of a few percent, or some components are in a part of the HR diagram where masses have not been well measured.

3.1. CAPELLA

The Mark III data (Hummel et al. 1994a) confirm and refine the visual orbit of the well-known giant system Capella (G1 III + G8-K0 III), one of the best-observed visual systems; they also demonstrate the accuracy of optical interferometry. The masses obtained agree well with those of Barlow et al. (1993) and have mean errors of 2.2% and 1.6% for components Aa and Ab. The Δm values from the Mark III are also consistent with other results. The limb-darkened diameters $\theta_{\rm ld} = 8.5 \pm 0.1$ mas (1.2%) and 6.4 ± 0.3 mas (4.4%), with the size of Aa reported for the first time. From the size of the orbit measured with the Mark III, the orbital parallax $\pi_{\rm orb} = 0.''0751 \pm 0.''0005$ (0.7%), giving $\log(L/L_{\odot}) = 1.895 \pm 0.007$ and 1.890 ± 0.015. (The smallest value of $\sigma_{\log L}$ in Andersen 1991 is 0.02.) If the mass ratio (which has been problematic in the past) has finally been well determined, Capella can be added to the list of stars with precisely known masses, radii, and luminosities.

3.2. ϕ CYGNI

This giant system (K0 III + K0 III) yielded 3% mass estimates in Mark III observations (Armstrong et al. 1992b); this is a slightly lower precision than Andersen (1991) recommends, but few masses of giants are known. The estimated angular diameters are 1 mas or less, so no diameter measurement with the Mark III was possible. For ϕ Cygni, $\sigma_{\mathcal{M}_1}$ and $\sigma_{\mathcal{M}_2}$ are dominated by σ_{K_1} and σ_{K_2}, while the uncertainties in the absolute magnitude M_V (0.11 and 0.13 for the more- and less-massive components, corresponding to 0.044 and 0.052 in $\sigma_{\log L}$) is dominated by $\sigma_{\Delta m}$; reducing this uncertainty depends on improved calibration of the fringe contrast, since it is difficult to measure Δm precisely when it is near zero. The diameters were not measured; Armstrong et al. (1992a) estimated uniform-disk diameters of 0.94 and 0.82 mas in the V band.

In order for ϕ Cygni to satisfy Andersen's criteria for precisely measured stars, the masses should be improved and the angular diameters measured. In this case, $\sigma_{\mathcal{M}}$ is dominated by $\sigma_{K_{1,2}}$, which are taken by Armstrong et al. (1992a) to be four times the uncertainties given by Rach & Herbig (1961) due to a discrepancy between the visual and spectroscopic values of e; thus, this system could benefit by another spectroscopic study. For measurements of the stellar diameters, there is no solution except to build a bigger interferometer.

3.3. β AURIGAE

β Aurigae (Hummel et al. 1994c) is an eclipsing system that already appears in Andersen's list, with mean errors in $R_{1,2}$ of 2.1%, in $\mathcal{M}_{1,2}$ of 1.1%, and in $L_{1,2}$ of 19%. The Mark III data improve the distance determination to 24.9 ± 0.5 pc (2%), but σ_L is dominated by $\sigma_{\Delta m}$, which remains about $0\overset{m}{.}2$. As with ϕ Cygni, improving Δm will require improved calibration.

3.4. α EQUULEI

α Equulei (Armstrong et al. 1992a) (G0 III + A5 V) fails to appear on the list of systems with precisely known parameters for two reasons: $\sigma_{\sin i}/\sin i = 3.5\%$, and $\sigma_{K_A}/K_A = 4\%$. With the improved data reduction introduced by Hummel et al. (1993), Hummel and Armstrong (1993) give $\sigma_{\sin i}/\sin i = 1.6\%$, alleviating the situation somewhat. However, the velocities of the A star in this system are difficult to measure, since the G spectrum must be subtracted before the radial velocities can be measured (Stickland 1976; Rosvick & Scarfe 1991).

4. The New Interferometers

These examples demonstrate a variety of observational challenges that must be resolved in order to obtain improved values of fundamental stellar parameters, including from resolving the disks of the components (Capella, ϕ Cygni, α Equulei), which requires longer baselines; improving the measurements of the visual orbit (α Equulei), which requires longer baselines and/or better calibration; improving the determination of Δm (β Aurigae), which also requires better calibration; and improving radial velocity measurements (ϕ Cygni, α Equulei). Except for the velocity measurements, all these challenges can be met by improved and enlarged interferometers, such as those now coming into operation.

As Table 1 shows, the typical interferometer of the 1990s has considerably larger apertures and/or longer baselines, and more collecting apertures, than the interferometers of the 1970s and 1980s. These improvements should lead to more than an order of magnitude improvement in angular resolution and to observations of $8^{\rm m}$ or $9^{\rm m}$ stars. My discussion will focus on the NPOI interferometer, which has the largest number of elements and the largest two-dimensional layout, and which is the one with which I am most familiar.

TABLE 1. New interferometers after 1990

Year	Interferometer	Apertures	Max. BL	λ (μm)	Ref.
1990	COAST (Cambridge University)	4 × 40 cm	100	0.4–0.95	Cox (1993)
1992	SUSI (Sydney University)	2 × 14 cm	640	0.4–0.85	Davis (1993)
1994	IOTA (Center for Astrophysics)	3 × 45 cm	38	0.45–0.8 2.2	Carleton et al. (1994)
1994	NPOI astrometric (USNO/NRL)	4 × 12.5 cm (35 cm in 1995)	38	0.45–0.9	Hutter (1994)
1994	NPOI imaging (USNO/NRL)	6 × 12.5 cm (35 cm planned)	437	0.45–0.9	Armstrong (1994)
1995 planned	ASEPS-0 (JPL)	2 × 40 cm	100	2.2	Colavita et al. (1994)

The NPOI consists of co-located astrometric and imaging arrays that share the feed system, optics laboratory, control system, and the designs for most of the components. The feed system supports a 12 cm beam; beam-compressing telescopes with 35 cm apertures are currently planned for installation in the astrometric array in 1995 or 1996, and in the imaging array some time after that.

The astrometric array, described in these proceedings by Hutter, Johnston, & Mozurkewich, has four siderostats as the array elements, one at the center and three at the ends of 20 m arms, for a maximum baseline of 38 m. The siderostats are permanently mounted, and their motions at the micron level due to bearing errors and to diurnal and seasonal temperature variations will be monitored by a laser metrology truss. The initial goal is a catalog of about 1000 stars with positional errors of 2 mas, on a par with the expected precision of the HIPPARCOS catalog.

The imaging array will consist of six siderostats that can be moved among a set of 30 stations placed along the arms of a Y with 250 m arms, with either two or three siderostats per arm in any given configuration. The standard configurations will have siderostats equally spaced along the arms. Tracking fringes on the short baselines between neighboring siderostats will make it possible to track fringes on the longest baselines even when the fringe contrast on those baselines is too low to detect in real time. The initial observing programs will include synthesis of images of the surfaces of nearby red giants, and, as with the Mark III, stellar diameter and binary orbit measurements.

Both arrays will detect fringes by feeding the spectrally dispersed light ($\lambda\lambda$450 to 850 nm) from each baseline into a 1×32 array of fibers, each leading to an avalanche photo-diode. This technique will give the NPOI two additional advantages over the Mark III. One is that fringe contrast calibration will be improved by the availability of data all across the visible band, making it possible to measure Δm better and detect both components in systems with larger Δm than was possible with the Mark III. The other advantage is that using multiple narrow bands, rather than a single wide band, will make it possible to track the fringe packet, not just the fringe phase modulo 2π as with the Mark III, making fringe detection and tracking more robust.

5. Prospects of New Interferometric Observations

With these improvements in interferometry in mind, we can estimate the observational capabilities of the imaging array and of the other new interferometers by extrapolation from the Mark III. With a maximum baseline of 437 m, the NPOI imaging array should be able to measure orbits as small as 250 μas with errors as small as 7 μas, and diameters as small as 180 μas with errors of 4 to 7 μas. The NPOI should perform somewhat better than this extrapolation because of the improved calibration and fringe tracking, and because up to 15 baselines will be used simultaneously; but I will ignore those sources of improvement. To be conservative, I will take the limits to be $a > 400 \pm 10$ μas and $\theta > 200 \pm 10$ μas. Even so, these extended capabilities will open up a large range of binary systems in which fundamental stellar parameters can be precisely determined.

5.1. ECLIPSING, DOUBLE-LINED DETACHED SPECTROSCOPIC BINARIES

Nineteen of the 45 systems listed by Andersen (1991) are both bright enough ($m < 9^{\mathrm{m}}$) to observe and have orbits ($a > 400$ μas) that can be resolved by the NPOI; for all of these, $\sigma_a < 10$ μas will improve the distance determinations. In systems with shallow eclipses for which the determinations of i and stellar radii from the light-curve solution have correlated errors, the measurement of the inclination from the visual orbit may improve the radius determinations as well. The new interferometers are also capable of directly measuring the angular diameters of 14 stars in ten of the Andersen (1991) systems; for both components of β Aurigae and YY Geminorum, 10 μas precision will improve the diameter determinations over the values listed in Andersen (1991), providing a further check on the light-curve models.

These measurements would have two effects. First, the determinations of the luminosities of the stars, from distances determined from the angular and physical orbit sizes, would be direct and more precise. Second, they would make possible a refinement of the calibration of the T_{eff} scale, from angular sizes determined directly from the interferometry.

5.2. NON-ECLIPSING DOUBLE-LINED BINARIES

McAlister (1985) counted the systems in the 7th Catalogue of Spectroscopic Binaries that could be reached with an interferometer with a 300 m baseline. Using the estimated separations at nodal passage of Halbwachs (1981), he counted 180 systems north of $\delta = 0°$, with 102 wider than 1 mas. I have done the same exercise with the 8th Catalogue (Batten, Fletcher, & MacCarthy 1989). I calculated expected angular separations from the apparent magnitudes, and absolute magnitudes roughly estimated from the spectral types and luminosity classes. This procedure is accurate to no better than a factor of two, but for our purposes, that is sufficient. Selecting for SB2s with , I find 201 systems. Of these, 130 systems have separations less than 3 mas, and 42 systems are fainter than 5^{m}; these 172 are thus entirely new to interferometric observation. An approximate HR diagram of the 201 systems is shown in Figure 1, in which the absolute magnitudes have been roughly estimated from the spectral types.

5.3. NON-ECLIPSING SINGLE-LINED BINARIES

If the astrometric precision is great enough, another category of spectroscopic binary becomes accessible to precise determination of stellar parameters. For SB2s, the physical scale is given by $K_{1,2}$ and P. For SB1s, we may substitute π_{trig} and a (in arc seconds) for the missing K_2. We will need 0.5% precision in π_{trig} in order to get 2% precision in $\mathcal{M}_1$ and 1% precision in $\mathcal{M}_2$. If the NPOI's astrometric precision with repeated observations and/or over small angles on the sky improves by an order of magnitude from our estimate for initial operations of 2 mas (which only experience will show), we can observe SB1s out to 25 pc with the possibility of obtaining precise results. The Gliese catalog of stars within 25 pc (Gliese & Jahreiss 1993) contains 130 SBs with $\delta > -10°$ and $m < 9^{\mathrm{m}}$, while the Batten catalog contains 32 SB1s with $\delta > -10°$ whose primaries are G dwarves brighter than $7\overset{\mathrm{m}}{.}5$ or K dwarves brighter than $8\overset{\mathrm{m}}{.}5$, and thus are within roughly 25 pc as well.

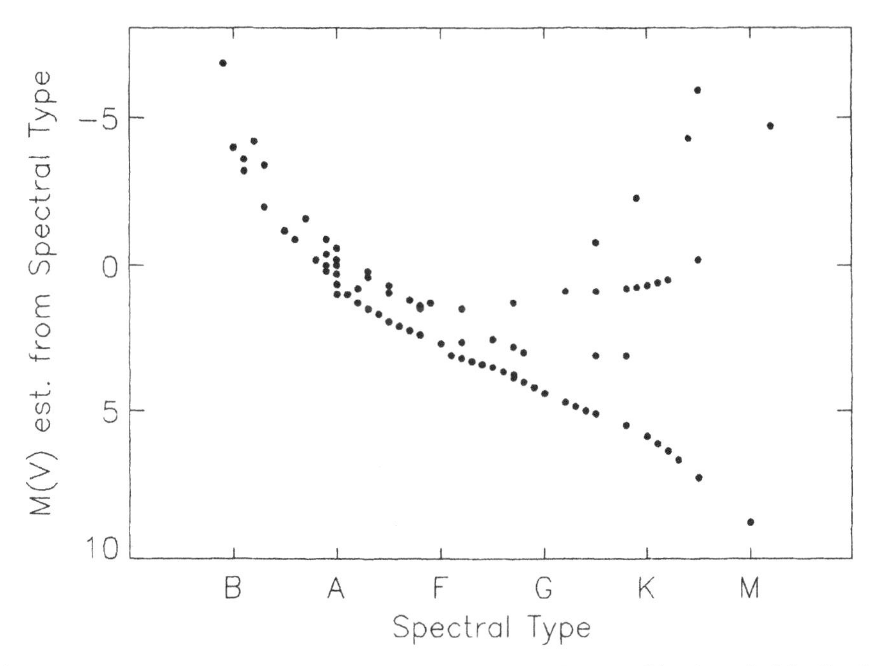

Figure 1. HR diagram of SB2s from the catalog of Batten, Fletcher, & MacCarthy (1989) resolvable with the NPOI imaging array. Selection criteria were $m < 9^{\mathrm{m}}$, orbital qualities a, b, or c, and angular separations $(a_1 + a_2) \sin i > 0.4$ mas. M_V were roughly estimated from spectral types.

6. Conclusion

I have touched on only one line of research the precise determination of stellar parameters to which high-precision astrometry in binary systems can make a contribution. Interferometry with separate apertures has started to live up to its promise as a tool for precise astrometry, having completed the list of precisely known parameters (in the sense of Andersen 1991) for one system (i.e., $R_{1,2}$ in Capella). Many more scientific contributions from interferometry are to be expected from the interferometers now coming into operation, with their longer baselines, increased collecting area, and multiple baselines.

Acknowledgements

This work was done in collaboration with D.F. Buscher, C.S. Denison, N.M. Elias II, C.A. Hummel, D.J. Hutter, K.J. Johnston, D. Mozurkewich, A. Quirrenbach, and R.S. Simon.

References

Andersen, J. 1991, A&ARev, 3, 91
Armstrong J.T. 1994, in Amplitude and Intensity Spatial Interferometry II, SPIE Proc. 2200, ed. J.B. Brekinridge, p. 62
Armstrong, J.T., Hummel, C.A., Quirrenbach, A., Buscher, D.F., Mozurkewich, D., Vivekanand, M., Simon, R.S., Denison, C.S., Johnston, K.J., Pan X.-P., Shao, M., & Colavita, M.M. 1992a, AJ, 104, 2217
Armstrong, J.T., Mozurkewich, D., Vivekanand, M., Simon, R.S., Denison, C.S., Johnston, K.J., Pan X.-P., Shao, M., & Colavita, M.M. 1992a, AJ, 104, 241
Barlow, D.J., Fekel, F.C., & Scarfe, C.D. 1993, PASP, 105, 476
Batten, A.H., Fletcher, J.M., & MacCarthy, D.G. 1989, Publ. DAO, 17, 1
Carleton, N.P., Traub, W.A., Lacasse, M.G., Nisenson, P., Pearlman, M.R., Reasenberg, R.D., Xu, X., Coldwell, C.M., Panasyuk, A., Benson, J.A., Papaliolios, C., Predmore, R., Schloerb, F.P., Dyck, H.M., & Gibson, D.M. 1994, in Amplitude and Intensity Spatial Interferometry II, SPIE Proc. 2200, ed. J.B. Brekinridge, p. 152
Colavita, M.M., Shao, M., Hines, B.E., Wallace, J.K., Gursel, Y., Malbet, F., Yu, J.W., Singh, H., Beichman, C.A., Pan X.-P., Nakajima, T., & Kulkarni, S.R. 1994, in Amplitude and Intensity Spatial Interferometry II, SPIE Proc. 2200, ed. J.B. Brekinridge, p. 89
Cox, G.C. 1993, in Very High Angular Resolution Imaging, IAU Symp. 158, ed. J.G. Robinson and W.J. Tango, p. 163
Davis, J. 1993, in Very High Angular Resolution Imaging, IAU Symp. 158, ed. J.G. Robinson and W.J. Tango, p. 163
Gliese, W., & Jahreiss, H. 1993, Third Catalogue of Nearby Stars (preliminary version), available on the Astronomical Data Center CD ROM
Griffin, R.F. 1992, in Complementary Approaches to Double and Multiple Star Research, IAU Colloquium 135, ed. H.A. McAlister and W.I. Hartkopf (ASP Conf. Ser., San Francisco), p. 98
Halbwachs, J.L. 1981, A&ASupp, 44, 47
Hummel, C.A., Armstrong, J.T., Quirrenbach, A., Buscher, D.F., Mozurkewich, D., Simon, R.S., & Johnston, K.J. 1993, AJ, 106, 2486
Hummel, C.A., Armstrong, J.T., Quirrenbach, A., Buscher, D.F., Mozurkewich, D., Elias, N.M. II, & Wilson, R.E. 1994a, AJ, 107, 1859
Hummel, C.A., Mozurkewich, D., Elias, N.M. II, Quirrenbach, A., Buscher, D.F., Armstrong, J.T., Johnston, K.J., Simon, R.S., & Hutter, D.J. 1994b, AJ, 108, 326
Hummel, C.A., et al. 1994c, in preparation
Hutter, D.J. 1994, in Amplitude and Intensity Spatial Interferometry II, SPIE Proc. 2200, ed. J.B. Brekinridge, p. 71
McAlister, H.A. 1985, in Calibration of Fundamental Stellar Quantities, IAU Symp. 111, ed. D.S. Hayes, L.E. Pasinetti, and A.G. Davis Phillip (Reidel, Dordrecht), p. 97
Mozurkewich, D., et al. 1994, in preparation
Mozurkewich, D., Johnston, K.J., Simon, R.S., Bowers, P.F., Gaume, R.A., Hutter, D.J., Colavita, M.M., Shao, M., & Pan X.-P. 1991, AJ, 101, 2207
Popper, D.M. 1993, ApJLett, 404, L67
Rach, R.A., & Herbig, G.H. 1961, ApJ, 133, 143
Rosvick, J.M., & Scarfe, C.D. 1991, MNRAS, 252, 68
Stickland, D.J. 1976, MNRAS, 175, 473

HST IN SEARCH OF BINARIES AMONG FAINT MEMBERS OF THE HYADES CLUSTER

O. FRANZ, K.J. KREIDL, L.H. WASSERMAN,
A.J. BRADLEY, G.F. BENEDICT, R.L. DUNCOMBE,
P.D. HEMENWAY, W.H. JEFFERYS, B. McARTHUR,
E. NELAN, P.J. SHELUS, D. STORY,
A.L. WHIPPLE, L.W. FREDRICK and W.F. van ALTENA

Abstract. The HST Astrometry Science Team is using the Fine Guidance Sensors (FGS) in the Transfer Function (TF) Scan mode to search for binaries among the faint members of the Hyades cluster. To date (March 1994), nine binaries have been discovered among 24 stars examined. The closest pair (total V=13.5) has a separation of 0.051 arcsec; the faintest (sep=0.287 arcsec) has magnitudes V=15.0 and 16.5; neither object posed a challenge to the capabilities of FGS. For another pair, two observations 152 days apart show a 13 deg change in position angle, indicating rapid orbital motion. One decade should suffice to define the orbit with angular dimensions of sub-millisecond of arc accuracy.

Clearly, this work will soon permit mass determinations for low-luminosity members of the Hyades cluster. Moreover, information on the frequency of binaries will provide insight into the role of duplicity in star formation and in the dynamic evolution of the cluster. To be truly useful, a census of binaries in the Hyades (and other clusters) must ultimately reach cluster members fainter than those currently under investigation, requiring astrometry with sub-millisecond of arc accuracy at near-infrared wavelengths.

E. Høg and P. K. Seidelmann (eds.),
Astronomical and Astrophysical Objectives of Sub-Milliarcsecond Optical Astrometry, 203.

PROSPECTS FOR DETECTING UNSEEN COMPANIONS OF PERIODIC VARIABLES

A. PIGULSKI
Wrocław University Observatory,
Kopernika 11, 51-622 Wrocław, Poland

Abstract. All periodic variables with stable periods such as pulsating stars, eclipsing binaries and pulsars offer the possibility of detecting unseen companion(s) by means of the light-time effect. We discuss the limitations of the method (the visibility of the effect) for different types of periodic variables. Special attention is paid to the ranges of mass ratios and orbital periods in which unseen companions can be found. We also indicate several systems with light-time effect in which hypothetical companions can be detected by speckle interferometry or precise astrometric observations. In these cases, the detection of the companions may lead to the determination of the components' masses.

1. Introduction

Any object emitting periodic signal is a candidate for the detection of the light-time effect (hereafter LTE) provided that it belongs to a binary or multiple system. This effect causes apparent period changes in accordance to the motion along the binary orbit. By the detection of LTE in period changes of a periodic variable, the presence of its unseen companion can be uncovered. Independently of the fact whether the companion is observed directly or not, the study of LTE may give an important contribution to our knowledge of the binary.

2. Limitations of the method

A periodic variable used to the study the LTE should have a stable period. ¿From all known periodic variables, some types of pulsating stars, eclipsing

E. Høg and P. K. Seidelmann (eds.),
Astronomical and Astrophysical Objectives of Sub-Milliarcsecond Optical Astrometry, 205–208.

binaries, and pulsars have enough stable periods to be promising candidates for the study of LTE.

In order to estimate the visibility of the effect in the O-C diagram, let us consider—for simplicity—LTE in a double system with a circular orbit. In the O-C diagram, the changes of the period P of the signal emitted by the visible component due to LTE will have the form of a sine curve with semi-amplitude equal to $P_{\rm orb}K_{\rm vis}/(2\pi c)$, where $P_{\rm orb}$ is the orbital period, and $2K_{\rm vis}$ is the range of the primary's radial-velocity curve.

The amplitude of LTE depents proportionally on $P_{\rm orb}$ and $K_{\rm vis}$ and is equal to the time which light needs for passing the projected orbit. The LTE can be detected only if this amplitude is larger than the typical error of the determination of the phase of periodic signal. Since this error determines the scatter of points in the O-C diagram, the visibility of LTE will, if fact, depend on the $P_{\rm orb}/P$ ratio. Consequently, for a given orbital period, short-period variables are more suitable for the study of LTE than the long-period ones. Furthermore, the amplitude of the effect is proportional to the mass ratio q (defined as $M_{\rm unseen}/M_{\rm visible}$). Assuming that the range of the variation due to LTE is larger than $0.02\,P$, we estimated the minimum mass ratios for different types of periodic variables (see Table 1). In the $P_{\rm orb}$—q plane, shown in Fig. 1, we present the lower limits for the detection of LTE for all periodic variables listed in Table 1.

TABLE 1. Minimum mass ratios q and minimum masses of unseen companions which can be found by means of LTE in a system with $P_{\rm orb}$ = 100 yr. Minimum $P_{\rm orb}$ needed for the detection of a 1 $M_\odot$ companion is given in the last column.

Type of visible object	Typical mass	Typical period	Minimum q	Minimum $M_{\rm unseen}$	Minimum $P_{\rm orb}$
RR Lyrae star	0.6 $M_\odot$	0.5 d	0.05	0.03 $M_\odot$	1.3 yr
Classical Cepheid	8 $M_\odot$	8 d	0.4	3 $M_\odot$	500 yr
β Cephei star	12 $M_\odot$	0.2 d	0.007	0.08 $M_\odot$	2.5 yr
δ Scuti star	2 $M_\odot$	2 h	0.005	0.01 $M_\odot$	60 d
pulsating WD	1 $M_\odot$	500 s	0.0005	0.5 $M_{\rm Jupiter}$	1 d
eclipsing binary	3 $M_\odot$	2 d	0.06	0.2 $M_\odot$	9 yr
typical pulsar	1.5 $M_\odot$	1 s	10^{-6}	0.5 $M_{\rm Earth}$	no limit
millisecond pulsar	1.5 $M_\odot$	5 ms	4×10^{-9}	0.2 $M_{\rm Moon}$	no limit

As one can see from Fig. 1, the shorter is the period of the signal the less massive companions can be discovered by means of LTE. In the systems containing pulsars or pulsating white dwarfs, there is a possibility that planetary companion(s) will be discovered.

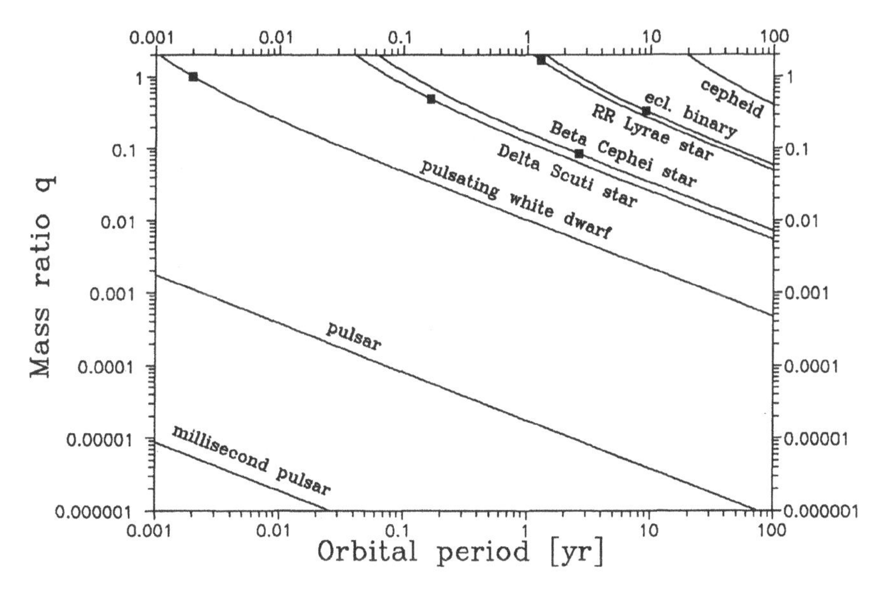

Figure 1. The lower limits for the detection of LTE for several types of periodic variables. A circular orbit and the most favourable inclination ($i = 90°$) were assumed. The filled squares indicate the case of a $1M_\odot$ companion.

3. Conclusions and examples

Apart from the discovery of the presence of an unseen companion, the detection of LTE can yield another result: a spectroscopic orbit can be found from the O-C diagram if the observations cover more than one orbital cycle. This is very important in the case of pulsating variables, the radial velocities of which are strongly affected by the pulsation.

If the presence of a LTE companion (i.e., companion discovered by means of LTE) can be confirmed by other independent observations this can yield additional information about the system. The high-quality spectrographic observations can reveal the secondary spectrum or provide radial-velocity curve of the primary (it should be consistent with the observed LTE). On the other hand, some LTE companions (if not very faint in comparison with the primary) can be resolved by interferometric methods yielding a relative visual orbit. Next, the absolute visual orbit can be obtained with the precise astrometric observations. This is very important because a combination of spectroscopic and visual (astrometric/speckle) orbits can lead to the determination of the masses of the components as well as other astrophysical parameters of the system.

With this in mind, we prepared a somewhat arbitrary list of objects (Table 2) which probably have companions because quite convincing periodic changes of their periods—presumably due to LTE—are observed. Independent confirmation of the presence of a companion for any of these objects would constitute an interesting and important contribution to their study.

This work was supported by the IAU grant and the reasearch KBN grant Nr 2 P304 001 04.

TABLE 2. Some periodic variables for which period changes attributable to LTE are observed.

Variable name	Type	P [d]	$P_{\rm orb}$ [yr]	Reference
RX Aur	cepheid	11.624	53.8	Szabados (1988)
IU Aur	EB	1.812	0.81	Mayer & Drechsel (1987)
AH Cep	EB	1.775	62.9	Drechsel et al. (1989)
FZ CMa	EB	1.273	1.47	Moffat et al. (1983)
IM Aur	EB	1.247	3.78	Bartolini & Zoffali (1986)
TU UMa	RR Lyr	0.558	23	Szeidl et al. (1986)
BW Vul	β Cep	0.201	33.5	Pigulski (1993)
SZ Lyn	δ Sct	0.121	3.22	Paparó et al. (1988)
KZ Hya	δ Sct	0.060	9.3	Liu et al. (1991)
PSR 1820−11	pulsar	0.28 s	0.98	Phinney & Verbunt (1991)

References

Bartolini C. and Zoffali M. (1986) Light-time effect in IM Aurigae, *Astron. & Astroph.*, **168**, pp. 377-379

Drechsel H., Lorenz R. and Mayer P. (1989) Solution of the light curves with third light contribution: the eclipsing binaries LY Aurigae and AH Cephei reconsidered, *Astron. & Astroph.*, **221**, pp. 49–61

Liu Y., Jiang S. and Cao M. (1991) A further proof on the binary model of HD 94033, *IBVS No.* **3606**

Mayer P. and Drechsel H. (1987) Up-to-date parameters of the eclipsing triple system IU Aur, *Astron. & Astroph.*, **183**, pp. 61–65

Moffat A.F.J., Vogt N., Vaz L.P.R. and Grønbech B. (1983) The early B-type eclipsing binary FZ CMa (HD 52942): a masive triple system, *Astron. & Astroph.*, **120**, pp. 278–286

Paparó M., Szeidl B. and Mahdy H.A. (1988) The high amplitude Delta Scuti star SZ Lyncis revisited, *Space Science Review*, **149**, pp. 73-82

Phinney E.S. and Verbunt F. (1991) Binary pulsars before spin-up and PSR 1820−11, *Monthly Not. Royal Astron. Soc.*, **248**, pp. 21P–23P

Pigulski A. (1993) The light-time effect as the cause of period changes in β Cephei stars. III. BW Vulpeculae, *Astron. & Astroph.*, **274**, pp. 269–273

Szabados L. (1988) Possible light-time effect in the cepheids FN Aquilae and RX Aurigae, *Publ. Astron. Soc. Pacific*, **100**, pp. 589–593

Szeidl B., Oláh K. and Mizser A. (1986) Period changes of RR Lyrae stars II, TW Her, VZ Her, AV Peg and TU UMa, *Comm. Konkoly Obs. No.* **89**, pp. 57–110

2.2 GALACTIC POPULATIONS, KINEMATICS AND DYNAMICS

A GALACTIC POPULATION CENSUS IN SUPPORT OF ASTROMETRIC MEASUREMENTS

M.A.C. PERRYMAN
Astrophysics Division, European Space Agency
ESTEC, 2200AG Noordwijk, The Netherlands

Abstract. A galactic counterpart of the Sloan Digital Sky Survey, consisting of two dedicated 2–3-m class telescopes (one in each hemisphere), could provide precisely those stellar data necessary to complement the information furnished by a future astrometric space mission: in particular, an automated acquisition and analysis system could provide spectral types, metallicities, and radial velocities of several tens of millions of stars down to about 15–16 mag. Such a dedicated spectroscopic stellar survey would provide considerably more information about the details of the stellar distribution within our Galaxy than is known at present. It is suggested that a galactic version of the Sloan Digital Sky Survey, utilising the GSC-II as an observing list, and capitalising on recent developments in multi-fibre spectroscopic survey capabilities, would be a timely initiative.

1. Requirements derived from a future astrometric mission

"*Optical astronomy has long lagged behind other astronomical disciplines of more recent origin in the production of survey data of well-known and characterised completeness and in digital form, both for imaging and spectroscopy. This has had an adverse impact in many areas, since it is still true that for the most part an object is not 'understood' until it has been identified in the optical and its nature as revealed by optical observations is understood*". I quote the introduction to a description of the Sloan Digital Sky Survey (Gunn & Knapp 1993), which employs a dedicated 2.5-m telescope, for an observing programme of five years duration commencing in late 1995. The goals include, for half the northern sky, four-band imaging with a limiting magnitude of ~ 23 mag, combined with a spectroscopic survey capable of yielding 10^6 galaxy redshifts complete to ~ 19 mag, and 10^5 quasar redshifts complete to ~ 20 mag.

E. Høg and P. K. Seidelmann (eds.),
Astronomical and Astrophysical Objectives of Sub-Milliarcsecond Optical Astrometry, 211–215.

Following the success of the Hipparcos space astrometry mission, concepts for a follow-up astrometry mission—GAIA (Lindegren & Perryman, this volume) and ROEMER (Høg, this volume)—have been proposed for ESA's long-term scientific programme. Preliminary studies indicate that a future mission could provide positions, parallaxes and proper motions at the 20 μas level, for $\sim$ 50 million stars brighter than $V = 15-16$ mag, along with multi-epoch multi-colour photometry. The scientific potential in areas such as stellar evolution and galactic dynamics is immense (Lindegren & Perryman, this volume). Such results would underline the absence of complementary stellar information, which could be acquired from ground, and which would be required for full exploitation of the astrometric data.

The global observing programme for Hipparcos was based on an 'Input Catalogue' of 120000 stars, which was laboriously assembled from a variety of existing ground-based observational catalogues, complemented by auxiliary data acquired specifically for the programme (Turon *et al.* 1992). The Input Catalogue provides an important database for an astrophysical interpretation of the Hipparcos astrometric data, although it will be inhomogeneous and incomplete in particular in terms of spectral types and, especially, radial velocities. The importance and feasibility of acquiring bulk radial velocity data through a concerted observational initiative was recognised by several individuals, and a specific proposal to fund and carry out such a programme was proposed, before the launch of Hipparcos, by Dr R. Griffin in Cambridge. Despite its uncontested scientific appeal, and its small relative cost (say, 0.5 per cent) compared with that of Hipparcos, necessary support was not forthcoming, and it was left to individuals to acquire these data for subsets of the Hipparcos programme through separate initiatives. The proposals were well-received, by ESO amongst others, and this programme now looks set to provide very valuable information for exploitation of the Hipparcos astrometric data (Mayor *et al.* 1991).

Radial velocities are important not only for completing the kinematic information—they represent the third component of the space velocity in addition to the two proper motion components (with 20 μas/yr corresponding to 1 km/s at 10 kpc). Repeated measurements provide a powerful way of identifying and characterising binary systems; this information again complements the Hipparcos astrometric data, where multiple systems prove a considerable complication to the main reduction process, but are scientifically important because of the possibility of mass determinations on the basis of well-determined distances and orbits. And separating the effects of time-dependent photocentric motions—and perspective acceleration—will demand such supporting observations.

A dedicated radial velocity programme for Hipparcos would not have ceased with multiple measurements of the 120000 programme stars; the

Tycho Catalogue, derived from the star mapper observations, will provide proper motions at the 10–20 mas/yr level (and perhaps at the 2 mas/yr level in combination with Astrographic Catalogue positions) for slightly more than one million stars! For the majority of these stars, radial velocities and spectral types will, unfortunately, be unknown.

Determination of elemental abundances and ratios are used to provide information on the star formation history of the Galaxy, information on the overall chemical enrichment (which is sensitive to its dynamical evolution), and the timescales of formation of different components of the Galaxy. Metallicity determinations would obviously provide much additional information and could, in principle, be acquired at the same time.

2. A Dedicated Spectroscopic Survey

If a deep, microarcsec class, astrometric mission is adopted by ESA, it will provide a formidable quantity of highly-accurate stellar distances and space velocities. One of the lessons that must be learned from the Hipparcos programme is the importance, and difficulty, of acquiring the data necessary for (i) efficient operation of the satellite; (ii) an optimum conduct of the scientific observing programme; (iii) a rigorous reduction of the satellite data; and (iv) a proper exploitation of the scientific results. These are not satisfied simply with the positions used to identify and observe the target objects. Photoelectric standards, accurate colour indices, variability data, and information on double and multiple systems, were needed. For the number of stars involved this was just achievable by *ad hoc* methods. Even so, interpretation of the astrometric results will highlight the incompleteness and inhomogeneity of certain associated data.

An astrometric catalogue with tens of millions of objects will require dedicated resources to provide the data necessary for the observations and data reductions, and the complementary astrophysical data necessary for a scientific proper exploitation. Spectroscopic follow-up on a vast scale will be required, and the only realistic way of achieving this would seem to be through a substantial multi-fibre spectroscopic survey—I will follow Longair (1993) and classify a set of reduced observations as a survey if it satisfies the following criteria: (a) completeness, (b) systematic observation and reduction of the data, (c) reliability, and (d) accessibility.

An indication of the capabilities of the present generation of fibre-fed spectrographs is the AAT's two-degree field, multi-fibre system (2dF). This employs a 4-component atmospheric dispersion compensated corrector, coupled with a double-buffered robotically driven 400-fibre spectroscopic facility, in turn feeding two separate spectrographs employing 1024×1024 CCDs; the resulting minimum object separation is 25 arcsec (Taylor

1994). (The Sloan survey uses a focal plane 2048 × 2048 CCD array with a pair of fibre-fed spectrographs, allowing the measurement of 660 spectra per field, using pre-drilled, interchangeable plug-plates, at a resolving power $\lambda/\Delta\lambda = 2000$, and with a minimum object separation of about 53 arcsec, and spectroscopic exposure times of one hour.)

Da Costa (1994) has already noted that "*there can be little question that stellar science with [the AAO 2dF] has virtually unlimited potential... However, it should be kept in mind that the key to successful observing with 2dF is positions, positions, positions*". This problem is also, of course, faced by the Sloan Survey, which will generate positions for the spectroscopic observations during a first-pass photometric survey. Fortunately, for both an astrometric space mission and this proposed ground-based complementary observing programme, positions of the target objects down to 15–16 mag, with an accuracy well matched to the requirements, should be available from the ambitious GSC-II digitising programme. GSC-II (Jenkner, private communication) is an all-sky survey planned by the STScI, following on from their highly successful Guide Star Catalog, intended to provide the positions of 2×10^9 objects down to $V = 18$ mag, with an absolute positional accuracy of 0.59 arcsec (worst case) by 2025, proper motions to better than 0.008 arcsec/yr, and two-colour photometry to better than 0.2 mag. If funded, this should provide the main elements of the *a priori* catalogue required for both the space and ground-based observations.

Concerning telescope automation, the 106th Annual Meeting of the Astronomical Society of the Pacific was held in June of this year in Flagstaff, the subject of the meeting being 'Robotic Telescopes', dealing with the construction, scheduling and planning of automatic telescopes. AutoScope Corporation, for example, is reported to be developing a fully automated, fibre-fed spectrograph and matching two-metre class automatic telescope, with other groups considering how such a system could be managed.

Achievable signal-to-noise figures can be roughly estimated from the values quoted for the AAT 2dF performances. Scaling the figures given by Taylor (1994) to a 3-m telescope, would yield a S/N of roughly 20–25 at $\Delta\lambda = 2$ Å, for $V = 16$ mag, seeing in the range 1.5 to 0.8 arcsec, and an integration time of 10 minutes. This should yield a radial velocities to 5–10 km s^{-1}, depending on spectral type and, for late-type stars, an abundance estimate accurate to 0.2–0.3 dex, with dwarfs and giants being readily distinguishable. Kurtz (1991; see also the related articles from IAU Commission 45 on progress in stellar classification) refers to plans for the upgraded MMT with a 300-fibre spectrograph to provide 1000 classification-quality spectra per hour for $V < 14$ mag.

Is an extended, space-related project feasible? Assuming 600 active fibres, and scaling from the AAT/MMT experiences, a project lasting for

10 years and using two telescopes, one in each hemisphere, could indeed generate spectral information on 50 million objects. Evidently, the precise performances of such a system, required S/N ratio and spectral resolution over given wavelength ranges necessary to permit automated MK-type spectral classification, and to establish useful estimates of metallicities and radial velocities, would require much more detailed studies and optimisation, with considerable innovative software development and ingenuity necessary for the automatic collection, reduction, classification and interpretation of the many terabytes of data generated.

3. Conclusions

Optically, our Galaxy looks not so very different, qualitatively and quantitatively, to what it did fifty or more years ago, the biggest advances having been made at other wavelengths (e.g., at 21 cm), and through studies of the interstellar medium. At the 1981 Vatican meeting on *Astrophysical Cosmology* M.S. Longair made a plea for the central importance of large systematic surveys in providing the fundamental data needed for cosmology. Subsequently (Longair 1993), he added: "*I would extend my statement about the significance of surveys for astrophysical cosmology to the whole of astronomy and astrophysics. The unwritten hope is that by making the right types of systematic observation, the answer to astrophysical and cosmological problems can simply be read directly from the data—there would be a minimum need for interpretation or theory to understand the important answers*".

It seems an appropriate time to consider the non-trivial question of precisely what complementary data are needed, if any, and how they might be obtained.

References

da Costa, G. (1994) in AAO Newsletter, No. 69.

Gunn, J.E., Knapp, G.R. (1993) The Sloan Digital Sky Survey, *ASP Conference Series 'Sky Surveys: Protostars to Protogalaxies'*, Vol. 43, Soifer, B.T. (ed.), p267.

Kurtz, M.J. (1991) The Need for Automation in the Reduction and Analysis of Stellar Spectra. *IAU Transactions XXIB*, 331.

Longair, M.S. (1993) A Survey of Surveys, *ASP Conference Series 'Sky Surveys: Protostars to Protogalaxies'*, Vol. 43, Soifer, B.T. (ed.), p313.

Mayor, M. *et al.* (1991) Radial Velocities for Stars of the Hipparcos Mission. *IAU Highlights of Astronomy*, **9**, 433

Taylor, K. (1994) in AAO Newsletter, No. 69.

Turon, C. *et al.* (1992) The Hipparcos Input Catalogue, ESA SP-1136.

GALACTIC KINEMATICS ON THE BASIS OF MODERN PROPER MOTION DATA

M. MIYAMOTO
National Astronomical Observatory
Mitaka, Tokyo 181, Japan

Abstract. An accumulation of high precision astrometric data in conjunction with high-precision monitoring of the Earth's orientation, motivates "Galactic Astronomy". As regards local kinematics, all of the three components of both the vorticity and the shear of stars can be completely determined, in addition to the velocity ellipsoid. We can now be released from the constraint of the "axisymmetric" galaxy. The determination of the proper motion of the LMC will be crucial to understanding the global structure and dynamics of the Galaxy with the dark halo and MACHO's motions.

1. Introduction

With an accumulation of accurate astrometric data for a huge number of stars in a large galactic volume, provided by astrographs (NPM and SPM projects), photoelectric meridian circles equipped with the 2-D CCD micrometer (La Palma, Bordeaux, USNO, and Tokyo Meridian Circles), and HIPPARCOS, we are now entering the second brilliant era of the "Stellar Astronomy" (cf. Trumpler and Weaver 1953) since the time of Oort and Lindblad. During the first era, attention focused on probing Galactic Rotation and Stellar Populations, while in the current era we shall take a great interest in clarifying the Galactic Dark Halo.

Since the availability of proper motion data provides a primary impetus to Stellar Astronomy, the present review talk concentrates on the proper-motion analysis based on a large astrometric catalogue (ACRS) available today. Two topics are discussed: the local kinematics of the Galaxy and the detection of the LMC proper motion.

E. Høg and P. K. Seidelmann (eds.),
Astronomical and Astrophysical Objectives of Sub-Milliarcsecond Optical Astrometry, 217–226.

2. Local Kinematic Parameters of the Galaxy

The precessional constant plays a primary role in the reference frame for accurate proper motions. In order to obtain absolute proper motions, we need precise knowledge of the motion of the observationally accessible origin (the equator and equinox) of the reference frame, with reference to a background inertial frame. Planetary precession is determined to a sufficient accuracy (Lieske *et al.* 1977). A long-standing question is the accurate determination of the luni-solar precession.

In the first place, the change in the orientation of the observer's platform – the earth – should be monitored with reference to fixed points (QSO's) observable on the celestial sphere. Alternatively, the proper motions of stars determined directly with reference to background fixed points (galaxies) would be absolute, independent of the Earth's orientation. Such an idea was proposed in the 1930's by W.H.Wright, and is now applied successfully to the geodesic VLBI observations for monitoring the Earth's orientation, and to the NPM and SPM projects for the absolute determination of proper motions. However, we make the analysis of the proper motion data in rather classical fashion, for the moment, since the ideal observations mentioned above are not yet definitive. This situation will not change in the analysis of the proper motion data to be provided by the HIPPARCOS Catalogue.

Nowadays, we have the nominal luni-solar precessional constant to an accuracy of eight figures: $\psi = 5038''.7784/\text{cy}$ at J2000.0 (IAU (1976)). The present value of the constant results from Fricke's correction Δp to Newcomb's value at the fourth and fifth digits (Fricke 1977). Fricke also determined the so-called fictitious equinoctial motion Δe, which is inherent in the optical determination of the equinox. The current fundamental reference system FK5 was constructed on the basis of Fricke's corrections (Fricke *et al.* 1988).

The importance for galactic astronomy of settling the fifth and sixth digits of the precessional constant should be stressed here. The fifth digit corresponds to the order of magnitude of $0''.1/\text{cy}$ (1 mas/yr) and the galactic rotation indicated by the combination of the Oort constants B–A is about $0''.5/\text{cy}$ (5 mas/yr). On the other hand, the rotation of the galactic warp around the axis joining the sun and the galactic center is about $0''.05/\text{cy}$ (0.5 mas/yr), as mentioned later. Therefore, at these digits the proper motion analyses for determining the precessional constant is strongly coupled to the general galactic rotation.

Now, in the galactic rectangular coordinate system (x_1, x_2, x_3) with the origin at the sun and the three axes pointing to the galactic center, the direction of galactic rotation, and the north galactic pole, respectively, the three-dimensional Ogorodnikov-Milne model for the systematic stellar velocity field at the solar neighborhood $\boldsymbol{r}$ $(= x_1\boldsymbol{i} + x_2\boldsymbol{j} + x_3\boldsymbol{k})$ is expressed as

$$\boldsymbol{V} = \boldsymbol{S} + \boldsymbol{\nabla} Q + \boldsymbol{\omega} \times \boldsymbol{r} \ , \tag{1}$$

where

$$Q = \frac{1}{2}\sum_{i,j} D_{ij}^{+} x_i x_j \ , \tag{2}$$

$$\boldsymbol{\omega} = D_{32}^{-}\boldsymbol{i} + D_{13}^{-}\boldsymbol{j} + D_{21}^{-}\boldsymbol{k} = \frac{1}{2}\mathrm{rot}\boldsymbol{V} \ , \tag{3}$$

$$D_{ij}^{\pm} = \frac{1}{2}\left(\frac{\partial V_i}{\partial x_j} \pm \frac{\partial V_j}{\partial x_i}\right) \ \text{ for } i,j = 1,2,3 \ , \tag{4}$$

and $\boldsymbol{S}$ (S_1, S_2, S_3) is the mean flow (centroid) at the sun, so that the solar motion is given by $-\boldsymbol{S}$. D_{ij}^{+} denotes the shear of stars in the (x_i, x_j)-plane, D_{ii}^{+} the dilatation, and D_{32}^{-}, D_{13}^{-}, and D_{21}^{-} the rotations of stars around the x_1, x_2, and x_3 axes, respectively. All of these kinematic parameters are evaluated at the position of the sun. If the stellar velocity field is axisymmetric or $V_R = 0$, D_{12}^{+} and D_{21}^{-} are reduced to the familiar Oort constants A and B, respectively.

The equations of condition for least squares fitting are derived from eqs. (1)–(4). Given radial velocity data together with proper motion data, then we can solve these equations for all twelve unknowns (S_1, S_2, S_3, ω_1, ω_2, ω_3, D_{12}^{+}, D_{13}^{+}, D_{23}^{+}, D_{11}^{+}, D_{22}^{+}, D_{33}^{+}). It is noticed that nine parameters of the twelve – all except D_{ii}^{+}'s – can be completely determined on the basis of proper motion data only. We profit here from the two-dimensionality, the proper motions. The proper motions in stellar kinematics are twice as informative as radial velocities, if combined with a suitable distance estimate.

Hereafter, we concentrate on the determination of the nine parameters $S_1, \ldots, D_{23}^{+}$. The equations of condition are reduced to

$$\begin{bmatrix} \mu_\alpha \cos\delta \\ \mu_\delta \end{bmatrix} = \boldsymbol{M}\boldsymbol{X} \quad \text{with } \boldsymbol{X}^T = (S_1\ S_2\ S_3\ \omega_1\ \omega_2\ \omega_3\ D_{12}^{+}\ D_{13}^{+}\ D_{23}^{+}) \ , \tag{5}$$

where the 2×9 elements of the matrix $\boldsymbol{M}$ are given by Miyamoto and Sôma (1993) (referred to as Paper I), where in eq.(30), "α–α_{GP}" is substituted for "α".

The rotation vector $\boldsymbol{\omega}$ deserves comment. If we have proper motions (μ_α, μ_δ) given in a rigorously non-rotating reference frame, then, we can derive from eq.(5) a solution for the general galactic rotation $\boldsymbol{\omega}$ exclusively. However, the conventional reference frame, FK5, is still considered to be imprecise at the order of magnitude of $0''.1$/cy (1 mas/yr). Therefore, the proper motions (μ_α, μ_δ) given in the FK5 system contain a contribution from the rotation of the frame itself. This situation will remain in the HIPPARCOS Catalogue to be released, since HIPPARCOS could not observe QSO's as fixed points on the celestial sphere.

Thus, the equatorial rectangular components of $\boldsymbol{\omega}$ in eq.(5) should be written as

$$\begin{bmatrix} \omega_1 \\ \omega_2 \\ \omega_3 \end{bmatrix} = \boldsymbol{N} \begin{bmatrix} D_{32}^- \\ D_{13}^- \\ D_{21}^- \\ \Delta p \\ \Delta e + \Delta\lambda \end{bmatrix} , \qquad (6)$$

where the 3×5 elements of the matrix $\boldsymbol{N}$ are trigonometric functions of the directions of the x_1, x_2, and x_3 axes and the ecliptic pole. Δp, Δe, and $\Delta\lambda$ denote the luni-solar precessional correction, the fictitious equinoctial motion correction, and the planetary precessional correction to the FK5 system, respectively.

It is noticed here that eq.(6) gives only three conditions for the five unknowns D_{32}^-, D_{13}^-, D_{21}^-, Δp, and $\Delta e + \Delta\lambda$. Therefore, applying least squares to eqs.(5)–(6), we can determine at most three designated unknowns of the five. This situation is a fundamental limitation on the analysis of proper motions. Thus, the important point to note is that the determination of the rotation left in the reference frame depends on a model of the galactic velocity field, while the determination of the galactic velocity field depends on the rotation left in the reference frame. In order to deal with this shortcoming, what we can do is to select sample stars which seem to fit a simple velocity-field model of the Galaxy, and to determine first the rotation of the reference frame.

It is known that K-M giants are an old and well-relaxed population of stars and are expected to have already reached a steady-state in the galactic potential. Such a state of a stellar system means that the system should exhibit only a simple plane-parallel galactic rotation described by the Oort constants $A = D_{12}^+$ and $B = D_{21}^-$. In Paper I, we analyzed the proper motions of about 30000 K-M giants chosen from the astrometric catalogue ACRS Part 1 (Corbin and Urban 1991) on the FK5 system. Starting from an initial trial of the correction $\Delta p \sim -0''.3$/cy suggested independently of the stellar kinematics by the VLBI and LLR observations (McCarthy and Luzum 1991, Williams *et al.* 1991), we have proven that the K-M giants are, indeed, in a steady state. Then, applying the plane-parallel galactic rotation model to these stars, we have determined a rational set of the corrections Δp and $\Delta e + \Delta\lambda$ to the FK5 system together with the Oort constants and the solar motion (see Table 1).

TABLE 1. Kinematic Parameters Derived from Proper Motions Given by ACRS Part 1

Kinematic Parameters	Least Squares Method K-M Giants $\|z\| \leq 0.5$ kpc	Maximum Likelihood Method K-M Giants $\|z\| < 1.0$ kpc	Least Squares Method Young Stars 0.5 kpc $\leq r \leq$ 3.0 kpc
Δp ($''$/cy)	-0.267 ± 0.028	-0.214 ± 0.022	-0.27 (given)
$\Delta e + \Delta\lambda$ ($''$/cy)	-0.116 ± 0.026	-0.075 ± 0.037	-0.12 (given)
S_1 (km/s)	$+13.6 \pm 0.3$	$+13.4 \pm 0.31$	$+8.7 \pm 0.8$
S_2 (km/s)	$+23.3 \pm 0.3$	$+20.3 \pm 0.38$	$+15.9 \pm 0.8$
S_3 (km/s)	$+11.9 \pm 0.3$	$+12.2 \pm 0.22$	$+9.1 \pm 0.7$
S_{total} (km/s)	29.5	26.7	20.3
$A = D_{12}^{+}$ ($''$/cy)	$+0.263 \pm 0.012$	$+0.243 \pm 0.011$	$+0.285 \pm 0.019$
$B = D_{21}^{-}$ ($''$/cy)	-0.176 ± 0.010	-0.193 ± 0.010	-0.260 ± 0.015
V_θ (km/s)	-177.1 ± 6.2	-175.7 ± 6.0	-219.9 ± 9.8
D_{13}^{+}	0	0	-0.059 ± 0.011
D_{13}^{-}	0	0	$+0.059 \pm 0.011$
D_{23}^{+}	0	0	$+0.039 \pm 0.010$
D_{32}^{-}	0	0	$+0.039 \pm 0.010$
D_{zz} (km/s/kpc^2)	-	15.6 ± 2.2	-
σ_R (km/s)	-	$+31.3 \pm 0.4$	-
σ_θ (km/s)	-	$+25.2 \pm 0.5$	-
σ_z (km/s)	-	$+21.2 \pm 0.5$	-
$\epsilon_{\mu\alpha}$ ($''$/cy)	-	0.56 ± 0.02	-
$\epsilon_{\mu\delta}$ ($''$/cy)	-	0.52 ± 0.02	-
total number adopted	20292	22629	1892

A reason why the FK5 system is still rotating should be discussed here: In his determination of the rotation left in the FK4 system, Fricke used 512 FK4/FK4 Sup stars, which are biased to young ages (about 60 % of these stars are O-B stars, Supergiants, and Bright Giants). Fricke adopted the plane-parallel galactic rotation model, so that the stars chosen are implicitly assumed to be in a steady-state. If the stars perform rotations other than the familiar rotation given by the Oort constant B, the model attributes all the rotation components in the galactic plane to the precessional and equinoctial motions of the reference frame. However, since Fricke's sample of the 512 stars is biased to young ages, the sample is not guaranteed to be in a steady-state. Furthermore, the nearest distance limit of 70 pc is still not sufficient to avoid localized velocity fields (cf. Gould-belt), and the motion of the more distant young stars is, on the other hand, liable to be disturbed by the galactic warp (Miyamoto *et al.* 1993).

So long as the analysis is based upon least squares, we miss the important information, the velocity dispersion of stars. In order to determine the dispersion as well and to examine how the previous result (Paper I) is modified by inclusion of dispersion, Tsujimoto and Miyamoto (1994) have examined again the local kinematic parameters and the rotation of the FK5 system by the Maximum Likelihood Method (Murray 1983), on the basis of the same sample of K-M giants as before.

We have obtained a similar result (Table 1) to our previous one within the formal standard error. However, the present result gives smaller corrections to the FK5 system than before. The velocity dispersions (σ_R, σ_θ, σ_z) thus determined for the K-M giants agree with the generally accepted values. We also estimated the error of the proper motion ($\epsilon_{\mu\alpha}$, $\epsilon_{\mu\delta}$). The estimation agrees quite well with the error cited in the ACRS Part 1. Moreover, in the present case, we tried to estimate the second derivative, D_{zz} in Table 1, of the galactic rotation V_θ with respect to the height z kpc from the galactic plane. The result gives the relation $V_\theta(z) = V_\theta(0) - 16z^2$ km/s, which implies a shearing rotation of the K-M giants.

Having determined the rotation remaining in the FK5 system, we can go one step further. On the basis of the proper motion data given by the ACRS Part 1 for about 2000 O-B stars, Supergiants, and Bright Giants (referred to as young stars), we have examined the kinematic behavior of the HI galactic warp, whose optical counterpart is considered to be the young stars.

Applying the corrections given in Paper I to the proper motions, we can now determine all three components of the general galactic rotation of the young stars (cf. eq.(6)). The result for the young stars is given in the fourth column of Table 1. The rotation of the young stars is indicated by D_{21}^-, D_{13}^-, and D_{32}^-. As described in Miyamoto *et al.* (1993), the young stars, as the optical counterpart of the HI warp, are now streaming around the galactic center in an inclined plane of the kinematic warp with the velocity −225 km/s, and simultaneously the plane is rotating around the galactic center – sun – anticenter line in the increasing sense of the present warp with an angular velocity of 4 km/s/kpc.

3. The Galactic Dark Halo

On the basis of the radial velocities of HII regions/reflection nebulae, Blitz (1979) and Brand and Blitz (1993) have demonstrated that the galactic rotation curve is not declining in the region exterior to the solar circle.

Moreover, Hartwick and Sargent (1978) have used the radial velocities of outlying globular clusters and dwarf spheroidals to show that the velocity dispersion of the galactic satellites remains nearly constant out to about 60 kpc from the galactic center. These results are considered to provide strong evidence that our Galaxy has a massive dark halo. However, different analyses made by Miyamoto *et al.* (1980) and Lynden-Bell *et al.* (1983) for the radial velocity data of the galactic satellites have not always resulted in evidence for a massive dark halo, and Lynden-Bell *et al.* have described the current situation as "Slippery Evidence on the Galaxy's Invisible Heavy Halo".

Now, if we could indeed determine the spatial motion (especially the proper motions) of the Magellanic System, its galactic orbital motion would provide an important constraint for the mass distribution of the dark halo out to 50 kpc from the galactic center. Furthermore, the determination of the absolute proper motions of the Magellanic Clouds (member stars) is of great importance for estimating the spatial motion and distance of MACHO's, which are considerd to be a physical constituent of the halo.

The theoretical orbital motion of the system (Murai and Fujimoto 1980, Lin and Lynden-Bell 1982) predicts that the LMC should have a heliocentric proper motion of about 2 mas/yr pointing to the direction of trailing the Magellanic Stream. Recently, Kroupa *et al.* (1994) and Jones *et al.* (1994) have detected the proper motions of the LMC, resulting in the total amount of $\mu = 1 - 2$ mas/yr pointing to the direction of trailing the Magellanic Stream. The results are fortunately compatible with the theoretical prediction.

Kroupa *et al.* (1994) used proper motions taken from a modern astrometric catalogue, the PPM Catalogue, for 35 stars which are proven by Sanduleak to be LMC members (hereafter, referred to as Sanduleak stars). The PPM catalogue is compiled on the FK5 system. However, as we have shown recently (Paper I), the FK5 system still needs modification as regards its corrections for the precession and equinoctial motion (the second column of Table 1). Moreover, the proper motion systems of different astrometric catalogues usually differ systematically. Therefore, a different astrometric catalogue with improved precessional and equinoctial motions gives a different result for LMC proper motions. We have recently examined the LMC proper motions (Miyamoto and Sôma 1994), based on the other modern catalogue, the ACRS Part 1 (Corbin and Urban 1991).

In order to separate the differences between the PPM and ACRS Part 1 catalogues into systematic and random parts, we use the catalogue comparison method developed by Brosche (1966), which has been later modified into a more practical form (Bien *et al.* 1978) and applied in the construction of the FK5. The catalogue difference, ACRS Part 1 − PPM, is composed

TABLE 2. Proper Motions of the Large Magellanic Cloud

Data Source	Proper Motion System	μ-total ($''$/cy)	PA from NGP (degree)
Photographic Plates (Jones *et al.* 1994)	galaxies	0.12 ± 0.04	282
PPM (Kroupa *et al.* 1994)	FK5	$0.16 \pm 0.09^*$	306
PPM (Miyamoto, Sôma 1994)	improved FK5	0.15 ± 0.09	287
ACRS (Miyamoto, Sôma 1994)	FK5	0.05 ± 0.09	122
ACRS (Miyamoto, Sôma 1994)	improved FK5	0.09 ± 0.09	156
ACRS 6 stars (Miyamoto, Sôma 1994)	improved FK5	0.55 ± 0.33	190

* Kroupa *et al.* gave 0.17. But, the proper motions of PPM No. 354939 in their Table 1 is erroneous.

of four types of differences in α, δ, μ_α, and μ_δ for about 250000 common stars all over the celestial sphere. These differences are expanded into triple orthogonal series, respectively, of products of Fourier, Legendre, and Hermite polynomials, and are tested for statistical significance, term by term, according to Brosche (1966).

We have examined about 300 terms of the series and adopted only statistically significant terms for the systematic difference in α, δ, μ_α, and μ_δ under the significant level of 0.1 % in the F-test. Details of the comparison will appear elsewhere (Miyamoto and Sôma 1994). Figures 1 and 2 show the contours of the systematic difference of $\mu_\alpha \cos\delta$ and μ_δ, respectively, between ACRS Part 1 and PPM in the LMC region. The asterisks in the figures indicate the 35 Sanduleak stars. It is found that the systematic difference of the proper motions between ACRS Part 1 and PPM amounts to the value to be determined for the LMC proper motions.

Adding the systematic difference $\Delta\mu_\alpha \cos\delta$ and $\Delta\mu_\delta$ thus obtained to the proper motions given by the PPM for the Sanduleak stars, we have found the mean proper motions of the LMC on the ACRS system. Table 2 compares the LMC proper motions reported hitherto by various authors. In the table, the first column gives data sources and authors, and the second column the proper motion system used, where "galaxies" indicates the

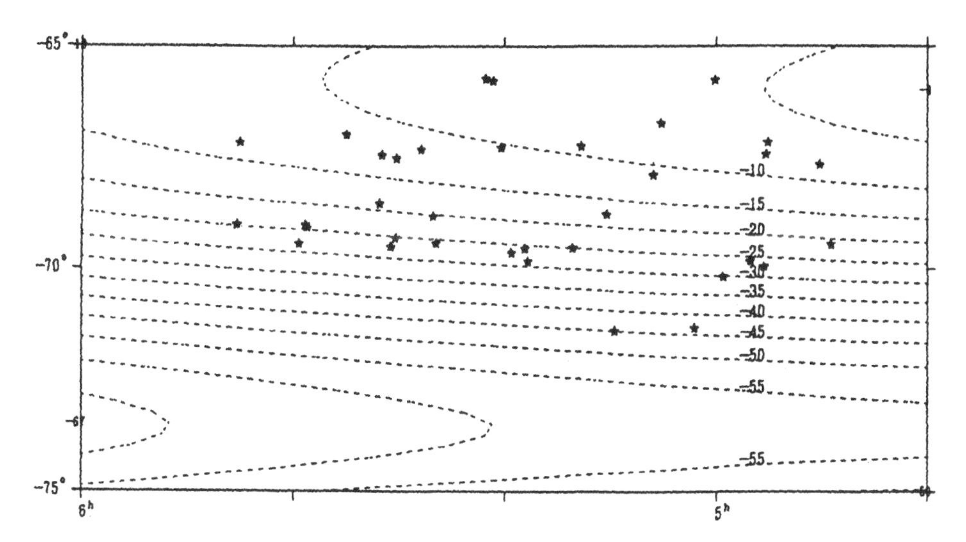

Figure 1. $\Delta\mu_\alpha \cos\alpha$ (ACRS–PPM): Tested are 293 orthogonal functions, of which 163 functions are selected. (unit:0''.01/cy)

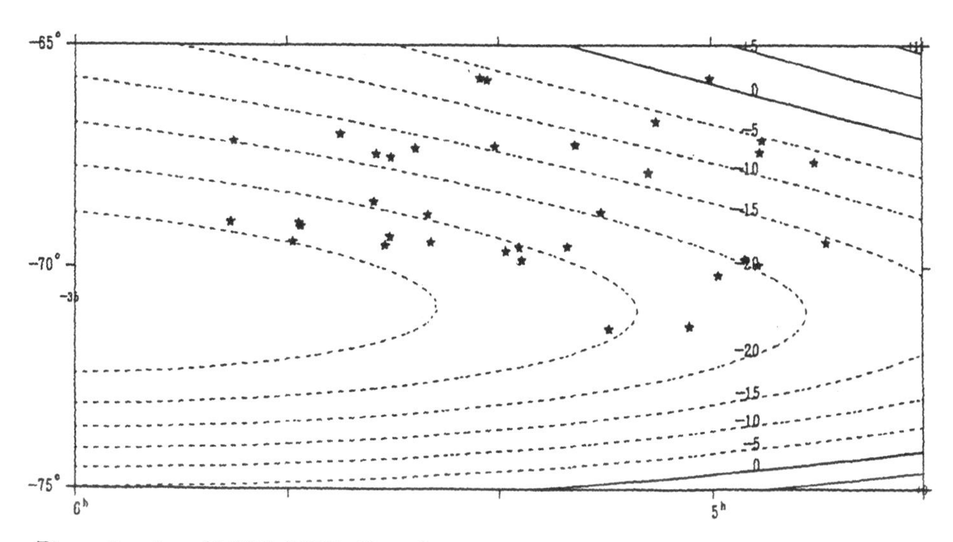

Figure 2. $\Delta\mu_\delta$ (ACRS–PPM): Tested are 293 orthogonal functions, of which 143 functions are selected. (unit:0''.01/cy)

proper motions with reference to galaxies, and "improved FK5" means that the FK5 proper motion system was improved by introducing the corrections determined in Paper I.

Comparing the second and third lines in Table 2, we find that the corrections Δp and $\Delta e + \Delta\lambda$ to the FK5 system do not damage the result given by Kroupa *et al.* (1994), but the corrections considerably affect the μ_α-component. The same is true for other cases. The fourth and fifth lines in Table 2 give the results based on the proper motion system of the ACRS Part 1. The remarkable point here is that the direction of the proper motion vector is almost reversed in comparison with the results given by the first to the third line. The last line gives, only for reference, the mean proper motions of 6 Sanduleak stars found in the ACRS Part1.

Though the present results are marginal in view of the standard error, the largest contrast with the previous results is that the ACRS proper motion system makes the LMC move toward the Magellanic Stream, contrary to the theoretical prediction. But, we remember that Lin and Lynden-Bell (1977) have once simulated the Magellanic Stream advancing ahead of the LMC. The LMC proper motions detected so far are not yet definitive.

References

Bien, R., Fricke, W. and Schwan, H. (1978) *Veröff.Astron.Rechen-Institut Heidelberg*, **No. 29**.
Blitz, L. (1979) ApJ, **231**, L115.
Brand, J. and Blitz, L. (1993) A&A, **275**, 67.
Brosche, P. (1966) *Veröff.Astron.Rechen-Institut Heidelberg*, **No. 17**.
Corbin, T.E. and Urban, S.E. (1991) *Astrographic Catalogue Reference Stars (ACRS)*, U.S. Naval Observatory.
Fricke, W. (1977) *Veröff.Astron.Rechen-Institut Heidelberg*, **No. 28**.
Fricke, W., Schwan, H. and Lederle, T. (1988) *Veröff.Astron.Rechen-Institut Heidelberg*, **No. 32**.
Hartwick, F.D.A. and Sargent, W.L.W. (1978) ApJ, **221**, 512.
Jones, B.F., Klemola, A.R. and Lin, D.N.C. (1994) AJ, **107**, 1333.
Kroupa, P., Röser, S. and Bastian, U. (1994) MNRAS, **266**, 412.
Lieske, J.H., Lederle, T., Fricke, W. and Morando, B. (1977) A&A, **58**, 1.
Lin, D.N.C. and Lynden-Bell, D. (1977) MNRAS, **181**, 59.
Lin, D.N.C. and Lynden-Bell, D. (1982) MNRAS, **198**, 707.
Lynden-Bell, D., Cannon, R.D. and Godwin, P.J. (1983) MNRAS, **204**, 87p.
McCarthy, D.D. and Luzum, B.J. (1991), AJ, **102**, 1889.
Miyamoto, M., Satoh, C. and Ohashi, M. (1980) A&A, **90**, 215.
Miyamoto, M. and Sôma, M. (1993) AJ, **105**, 691 (Paper I).
Miyamoto, M., Sôma, M. and Yoshizawa, M. (1993), AJ, **105**, 2138.
Miyamoto, M. and Sôma, M. (1994) *to be published.*
Murai, T. and Fujimoto, M. (1980) PASJ, **32**, 581.
Murray, C.A. (1983) *Vectorial Astrometry*, Adam Hilger Ltd, Bristol.
Trumpler, R.J. and Weaver, H.F. (1953) *Statistical Astronomy*, Univ. California Press.
Tsujimoto, T. and Miyamoto, M. (1994) *to be published.*
Williams, J.G., Newhall, X.X. and Dickey, J.O. (1991) A&A, **241**, L9.

SUB–MILLIARCSECOND ASTROMETRY OF STAR CLUSTERS AND ASSOCIATIONS

FLOOR VAN LEEUWEN
Royal Greenwich Observatory
Madingley Road, Cambridge, UK

Abstract. Some of the possibilities created by sub–milliarcsecond astrometry in the study of both nearby and distant star clusters are presented.

1. Introduction

Sub–milliarcsecond astrometry can bring a much larger number of clusters and associations within the range of direct distance determinations, thus contributing very significantly to the study of star formation (mass-functions) and stellar evolution. The spatial resolution of some nearby clusters and associations will create the possibility of observing the 3–dimensional distributions of mass, radial kinetic energy and angular momentum. For those clusters it will remove the noise on the HR–diagram due to spread in distance moduli and make it possible to detect escaped former cluster and association members on the basis of 3-d positions and velocities. In the more distant future, sub–milliarcsecond astrometry opens the way to measuring accelerations in nearby star clusters, thus making it possible to measure their masses.

2. Distances for clusters and associations

Measuring parallaxes to an accuracy level of 1 to 10 microarcsec brings the order of 1000 open clusters, several associations and a number of globular clusters within range of a direct distance determination. Currently, differences in metal abundances, reddening corrections and age produce serious uncertainties in cluster distances (Lyngå, 1980). Figure 1 shows the impact of improvements in accuracy on absolute magnitudes at various distances

E. Høg and P. K. Seidelmann (eds.),
Astronomical and Astrophysical Objectives of Sub-Milliarcsecond Optical Astrometry, 227–232.

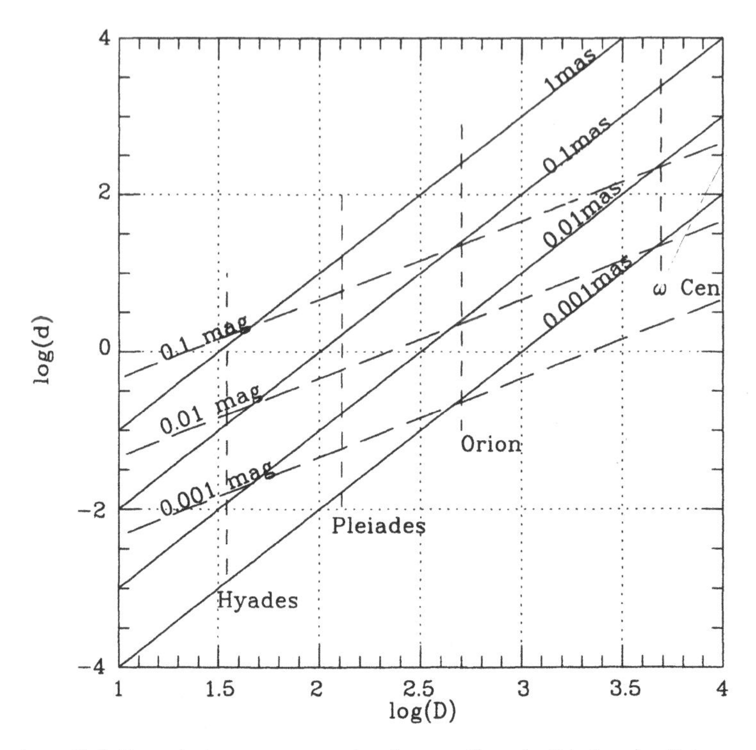

Figure 1. Relations between accuracies in parallax (solid lines), distance moduli (long–dash lines), the spatial resolution (d) and the distances (D) of clusters and associations. The 1 mas parallax accuracy is obtained with Hipparcos, 0.01 mas could be obtained with the revised ROEMER and GAIA concepts.

from the sun. The positions of some clusters and associations are shown in comparison. It is clear from Fig.1 that even though the impact of Hipparcos has been enormous, there is still a considerable improvement needed to make it possible to use the clusters and associations in the detailed calibrations of the HR–diagram: the HR-diagram constructed from preliminary Hipparcos results indicates a considerable intrinsic spread in absolute magnitudes for very nearby stars with the same $(B-V)$ colour index (Perryman and Heger, 1993).

In a concept like the ROEMER or the GAIA satellite (Lindegren et al, 1993, Høg, 1994), where high accuracy astrometry is combined with intermediate band-width photometry, it will become possible to calibrate the variations in luminosity as a function of age and metallicity using interstellar reddening corrections obtained from two-colour diagrams, provided the photometric passbands are carefully chosen and can be reproduced on the ground. A detailed HR-diagram thus obtained can then be used as the observational basis for the study of stellar evolution over ages ranging from less than a million years in some associations to 15 billion years in some globular clusters, taking into account fully differences in metallicity.

The nearby younger star clusters and associations will provide data on the mass spectrum of star formation processes using observed luminosity functions that include the detection of members in the cluster halo (Kholopov and Artyukhina, 1972, Van Leeuwen, 1980). Due to mass segregation, even by an age of 10^8 years a strong gradient in the mass spectrum may be observed between the cluster centre and the halo. Observations of the mass spectrum based only on data from the projected cluster centres can thus lead to wrong conclusions (Taff, 1974). This is further complicated by the escape of cluster members, for which the probability depends on the mass of the star and its chance of passing through the cluster centre.

Combined high precision astrometric and photometric studies of both young and old star clusters and associations are essential for filling in the details in the stellar evolution theories. The same applies for clusters with different metal abundances. The comparisons between clusters of different ages can show how their mass functions change due to the dynamical evolution of these systems, and thus also how the clusters and associations populate the galaxy with new stars.

3. Spatially resolved systems

The nearest clusters are well studied down to magnitude 14 to 16. The limit is set by the availability of early photographic plates. Accuracies in (relative) proper motions have reached levels of 0.1 to 0.2 mas/year (Vasilevskis et al, 1979), allowing the study of internal motions. Two main problems remain: the relative positions of cluster stars are only known projected along the line of sight, and for the fainter stars the proper motions are no longer sufficient to clearly distinguish members from non-members. The effect of the first problem is that a considerable uncertainty remains in the interpretation of the kinematics: it is necessary to deconvolve the observed projected proper motion dispersions with the density distribution of the cluster, which itself has been derived from differentiation of the projected density distribution. This would not be too bad if the number of stars involved were large, but in open clusters these numbers are small, leaving considerable uncertainties in matters like the actual presence of low mass stars in the cluster centre (Van Leeuwen, 1983). A lively discussion on this problem in the Pleiades can be found in Kholopov, 1971a, Mirzoyan and Mnatsakanian, 1971, Kholopov, 1971b and Mirzoyan et al, 1980.

The uncertainty in the number of faint members and their masses causes problems in the determination of the mass-function and therefore also in the potential energy function of the cluster. The difficulties currently existing for fainter cluster members are illustrated by the discrepant results between proper motion surveys of the Pleiades halo and flare star studies (Jones, 1980).

With the improvement of parallax accuracies to a level of 1 to 10 μas the positions of individual stars in a nearby cluster or association can be mapped in 3-dimensions, eliminating partly the need for a selection of members on the basis of proper motions only. Within the area of an open cluster there will always be a few tens of stars (comparable to the number of stars within 10 pc from the sun) that are not cluster members, but simply passers by. These can be distinguished on the basis of their relative space velocity. Assuming radial velocities can be obtained at accuracies of $0.1km/s$, the three positional and three velocity coordinates of each cluster star can be obtained. This also applies of course to all other stars in and around the cluster. Members which are in the process of escaping fast as the result of three body interactions or which are slowly drifting away as the result of interactions with much heavier members can also be located, thus making the reconstruction of the mass function more complete. The improvement of parallaxes will also remove the intrinsic spread in distance moduli which at the moment still confuses the Pleiades HR-diagram with a noise of 0.015 magnitudes.

The space density distribution of the cluster stars can be translated into a potential energy distribution, provided all the mass in the cluster is known. It is thus important that membership probabilities are obtained as far as possible down the mass spectrum. An additional problem, as well as a unique source of information on pre-main sequence stellar evolution, is that the faintest stars of a cluster have often not yet evolved into main sequence objects, causing uncertainties in the mass–luminosity relations. The observed velocities and the potential energy distribution provide an estimate of the total kinematic energy of the star, and thus of the distribution of energy over stars of various masses. Comparisons between different mass-groups should reveal the extent to which exchange of energy has taken place in older systems, or for very young systems, whether there are differences in the energy distribution for stars of different masses that would reflect differences in primordial conditions. This information is supplemented by the energies from the escaping stars, where some of the fast escaping stars might possibly pinpoint the binary system they left behind.

The angular momentum of a cluster star determines the chance of stars paying a visit to the cluster centre, and exchange a little energy. The radial energy determines how frequently this will happen. As was shown by Van Leeuwen (1983) and Terlevich (1984), the angular momentum dispersion for stars in the cluster halo increases with time due to the galactic gravity

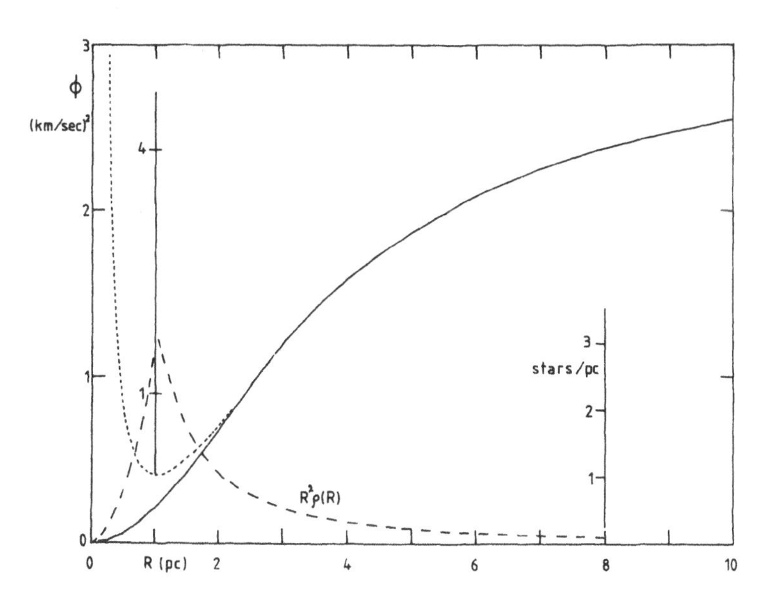

Figure 2. The radial potential energy distribution (dotted line) for a star with a transverse velocity of $0.44 km/s$ at $1pc$ from the centre in an open cluster of 1000 solar mass, compared with the radial distribution of B and A type stars in the Pleiades (dashed line). The vertical line starting from the minimum of the effective potential energy distribution compares the radial kinetic energy per unit mass with the observed squared velocity dispersion in the cluster centre, $0.49(km/s)^2$ (Van Leeuwen, 1983)

gradient over the cluster. This leads, together with mass segregation in the cluster centre, to under–population of the cluster centre in low mass stars and makes the distribution of these stars, after a while, almost unaffected by short distance interactions with other members in the cluster centre. In this (temporary) state of quasi–equilibrium the observed density distribution of the low mass stars and their observed angular momentum and kinetic energy distribution are related through the potential energy distribution function. The agreement between the potential energy distribution, as derived from the total density distribution and as derived from the faint stars, should indicate whether the total mass distribution in the cluster is determined entirely by the known cluster stars, or whether there is the need for additional (dark) matter.

4. Accelerations

If it were possible to measure the radial accelerations in a star cluster, then the distribution of mass in the cluster could be measured. Figure 2 shows that accelerations (the derivative of the solid line) in the halo of the Pleiades are expected to range up to $10^{-11} m/s^2$ (equivalent to $6 \times 10^{-8} AU/yr^2$). Much higher accelerations will be found close to the centre of the cluster, but these do not reflect the cluster's mass distribution but rather are the result of pericentre accelerations in the stellar orbits.

It can be shown that given observations with positional accuracy σ_p, spanning an interval T and evenly spaced at intervals of Δt, the accuracy of the acceleration measurements approximates:

$$\sigma_{acc} = 13.4 \sigma_p T^{-2.5} \Delta t^{0.5}$$

For a 5–year GAIA mission, this could lead to accuracies of $0.6 \mu as/yr^2$. However, the main gain will be later in the next century, when the full effect of the factor $T^{-2.5}$ is felt. This is similar to the situation that started 100 years ago: the positions in star clusters and associations recorded from that time onwards on photographic plates enable us today to study the kinematics of these systems. A mission such as GAIA or ROEMER can act as a first epoch for the determination of accelerations in the next century and thus lay the basis for direct measurements of masses of stellar systems, in a way comparable to the current use of the earliest photographic plates in studies of cluster dynamics. Until such a project is properly finished, however, the study of the dynamics of star clusters, and to some extent associations also, will depend on the availability of (old) photographic material.

References

Høg, E. (1994) present volume
Jones, B.F. (1980) **Astron.Journ.** 86, 2, p290
Kholopov, P.N. (1971a) **Sov. Astr.–AJ**, Vol 15, No.3, 415
Kholopov, P.N. (1971b) **IBVS** 566
Kholopov, P.N. and Artyukhina, N.M. (1972) **Sov. Astr.–AJ**, Vol 15, No.5, 760
Van Leeuwen, F. (1980) in **Star Clusters**, p157, ed. J.E.Hesser, Reidel, Dordrecht.
Van Leeuwen, F. (1983) PhD Thesis, Leiden University
Lindegren, L. (1993) GAIA proposal for ESA post Horizon 2000 program
Lyngå, G. (1980) in **Star Clusters**, p13, ed. J.E.Hesser, Reidel, Dordrecht.
Mirzoyan, L.V. and Mnatsakanian, M.A. (1971) **IBVS** 528
Mirzoyan, L.V., Mnatsakanian, M.A. and Oganyan, G.B. (1980) in **Stellar physics and Evolution**, ed. L.V.Mirzoyan, Yerevan
Perryman, M.A.C. and Heger, D. (1993) in **ESA Bulletin** No 75, p7
Taff, L.G. (1974) in **Astron.Journ.** 79, 11, p1280
Terlevich, E. (1984) PhD Thesis, Cambridge University
Vasilevskis, S., Van Leeuwen, F., Nicholson, W., and Murray, C.A. (1979) **Astron.Astroph.Suppl.** 37, 333

DETECTABILITY OF GALACTIC KINEMATICAL PARAMETERS FROM PROPER MOTIONS

R.L. SMART
Space Telescope Science Institute

AND

M.G. LATTANZI
Space Telescope Science Institute,
Affiliated with the Astrophysics Division,
SSD, ESA; on leave from Oss. Astr. di Torino

Abstract. We discuss the inclusion of a warp structure (as discussed by Miyamoto et al. 1993) to a realistic galactic model. We investigate how the proper motions are effected by changes in the inclination, phase angle and spin velocity of the warp. It is shown that accurate proper motions of disk stars do reflect the variation of warp parameters. A simulation of OB stars with an apparent magnitude limit of 7.0, mimicing the early type content of the HIPPARCOS catalogue, indicates that variations of phase are probably not observable while detection of velocity variation is within the limit of the current precision.

1. Introduction

Highly accurate and precise proper motions are essential for a 3-dimensional detailed analysis of galactic structure, for example, the presence and role of dark matter as an essential constituent of the Milky Way (and of the local Galaxy in particular); the independent determination of K_z and, ultimately, of the local mass density; the warped structure of the Galaxy; and the dynamics of the bulge. This contribution is an attempt to estimate the possibilities of present and future high accuracy proper motion compilations like HIPPARCOS and GAIA (Lindegren and Perryman, this conference) in addressing those outstanding questions of galactic structure and evolution.

E. Høg and P. K. Seidelmann (eds.),
Astronomical and Astrophysical Objectives of Sub-Milliarcsecond Optical Astrometry, 233–237.

In our model special parameters are chosen to simulate the particular problem for which we seek to characterize its influence on the proper motions of a tracer population or all of the stars. We do not yet attempt to make our model match the observed proper motions; instead we only measure the changes in a given distribution as a function of the simulation parameters, hence providing limits for existing data, or, future observations. For this discussion we have limited our sample to OB stars with apparent magnitude brighter than 7th. This is for two reasons, to replicate the completeness limit of the HIPPARCOS catalogue and to pick stars that should follow the warp that is seen in HI observations.

The next section deals with the particulars of our basic model. In section 3 we discuss the inclusion of warp into our Galaxy and in section 4 we analyze the effect of this warp on the proper motions of typical young disk tracers. We change three parameters of the warp; inclination to the galactic plane, longitude of the ascending node, or "phase angle", and the modulus of the circular velocity in the warp plane. Special emphasis is given to the condition for the "observability" of the warp in the proper motions.

2. Parameters of the Galactic Model

We have constructed this simulation using the IDL computer language, which proved powerful for operating directly on whole vectors and matrices. The distribution of a given parameter is decided from a-priori assumptions, the parameter space is then "populated" using a Monte Carlo type process.

For example, we use an exponential density distribution of the form $\rho = \rho_o \exp(-z/z_o)$, with the parameters taken from Gilmore and Wyse (1985, hereafter GW) to populate the Z coordinate space. We assume that the stars are uniformly distributed in X and Y space which provides a simple geometric distribution function. The luminosity function is taken from Bachall and Soneria (1980, hereafter BS) with a modification that steepens the function when assigning a magnitude to halo stars, producing fainter stars for this population, as suggested by Gilmore (1983).

The simulation requires a user supplied direction, angular width and distance limit. In the following investigation, we chose 1° slices at 45° intervals out to 3000 pc. We then remove those stars that are below some given magnitude cutoff. The absorption is calculated using the procedure described in BS, reaching a maximum at b=1°. In the discussion by BS they warn the reader against using this procedure below b=$20°$ and further work on the model will investigate alternatives to this extrapolation. The generation of stars is continued until the number counts in eight 1° areas of the slice match the number densities in BS to within 10%.

We now assign a velocity to the star. The initial galactocentric (U, V, W)

velocities are drawn at random from the peculiar velocity ellipsoids with dispersions taken from Mihalas and Binney (1981, hereafter MB). We add to these peculiar velocities the galactic rotation for the thin and thick disk V=(220,140) respectively and the solar motion [(+9, +12, +7) from MB], all in km/s.

Therefore our model of the Galaxy has 5 basic input parameters for each of the three populations, these are shown in Table 1.

TABLE 1. Galactic Simulation Parameters

Component	Thin disk	Thick disk	Halo
ρ_o [GW]	1.0	0.02	0.00125
z_o pc [GW]	300.	1000.	4500.
σ_u km/s [MB]	24	80	130
σ_v km/s [MB]	15	60	87
σ_w km/s [MB]	12	60	86

3. Inclusion of a Galactic Warp

The original impetus for undertaking this project was to understand how accurate proper motions might enable us to understand more completely the galactic warp. Our current understanding is based on HI, HII and CO radio observations, IRAS point source number counts and proper motions of OB type stars (Sparke 1993 for review). The generally accepted view is that the sun is almost directly on the line of nodes of this warp and that the warp starts just outside the solar orbit. Figure 1 shows the overall orientation where the XYZ frame is centered at the sun ($\odot$), i is the angle of inclination of the warp, ϕ is the phase angle.

To include a warp we have to consider the velocities in the correct order, before the application of the thin disk rotation and the local solar motion. The warp velocity is added directly in the warp frame using the longitude in this frame (l_w). Then we rotate the velocity and position vectors to the galactic frame. This change is shown in equation 1 (Note: The $\mathbf{R}_1$, $\mathbf{R}_2$, $\mathbf{R}_3$ are rotations matrices about the x, y and z axis respectively).

$$\begin{pmatrix} U' \\ V' \\ W' \end{pmatrix} = \mathbf{R_3}\,(-\theta)\;\; \mathbf{R_1}\,(-i) \left(\mathrm{V_{warp}} \begin{pmatrix} \sin l_w \\ \cos l_w \\ 0 \end{pmatrix} \right) + \begin{pmatrix} U \\ V \\ W \end{pmatrix} \qquad (1)$$

The U', V', W' are then corrected for the motion of the LSR and the solar motion to give U'', V'', W'' which in turn provide proper motions in

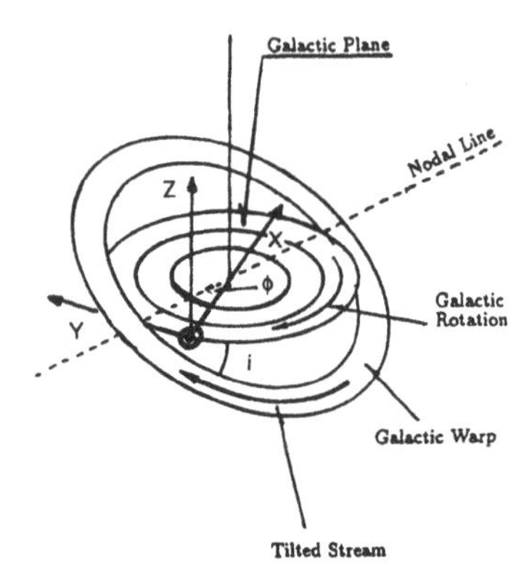

Figure 1. Geometry of the Simulated Galactic Warp (modified from Miyamoto et al. 1993). The sun is indicated with the symbol ⊙.

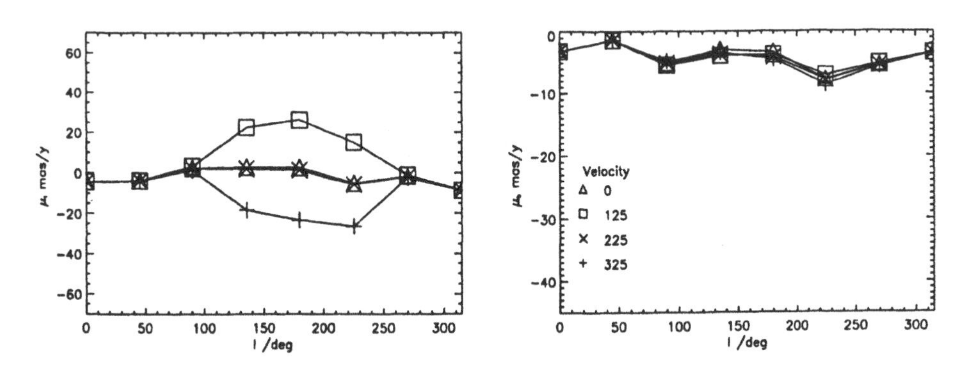

Figure 2. The effect of varying warp stream velocity on the proper motion in galactic coordinates (μ_l on the left and μ_b on the right) vs galactic longitude for OB stars brighter than 7th apparent magnitude.

galactic coordinates and radial velocity viz:

$$\begin{pmatrix} \mu_l \\ \mu_b \\ V_r \left(\frac{1}{4.74\,r}\right) \end{pmatrix} = \frac{1}{4.74\,r} \mathbf{R_3}\,(b)\ \mathbf{R_2}\,(l) \begin{pmatrix} U'' \\ V'' \\ W'' \end{pmatrix} \tag{2}$$

where r is the sun-star distance, l and b are the galactic longitude and latitude of the star.

4. Discussion

We have examined the effect of varying the inclination, velocity within the 'tilted stream' (see figure 1) of the warp, and the phase angle on the proper motions. Figure 2 shows the effect of varying the velocity. The main discriminant is the longitude proper motion where for no warp and a warp with velocity 225km/s we have a maximum difference of about 4mas/yr. Examination of similar graphs for phase angle shows that for small changes of 5° we see no difference and for 30° we see changes of only 3mas/yr. From this simulation we conclude that velocity differences in the galactic warp might be observable by HIPPARCOS, while phase angle differences probably will require a factor of ten improvement in precision.

5. Acknowledgments

We wish to thank Stefano Casertano, Larry Taff and Kavan Ratnatunga for many useful and stimulating discussions. RLS expresses his thanks to the IOC of symposium 166 and the American Astronomical Society for providing support to attend this meeting. We would also like to thank the referee for valuable comments that increased the clarity of this paper.

References

Mihalas, D. and Binney, J., 1981, *Galactic Astronomy*, W.H.Freeman, San Francisco.
Gilmore, G., 1983. *Mon. Not. R. astr. Soc.*,**207**,223.
Gilmore, G., and Wyse, R. F. G., 1985. *AJ*,**90**,2015.
Bahcall, J. N., and Soneira, R. M. 1980. *ApJS*,**44**,73.
Miyamoto, M., Soma, M., and Yoshizawa, M. 1993. *AJ*,**105**,2138.
Sparke, L.S., 1993. *Back to the Galaxy*, Aip Conf. **278**, eds S. Holt and F. Verter.

PROBING NON-AXISYMMETRY WITH PROPER MOTIONS

JAMES BINNEY
Department of Physics
1 Keble Road, Oxford, OX1 3NP, England

Abstract. The potential of the Milky Way is almost certainly not axisymmetric: the centre is believed to be dominated by a bar, and beyond the solar radius some non-axisymmetric feature of the potential appears to warp the disk. There are grounds for believing that the outer potential is mildly elliptical. Sub $\mathrm{mas\,yr^{-1}}$ proper motions of objects that lie near the plane several kiloparsecs from the Sun would play a crucial role in refining our understanding of these non-axisymmetries.

1. Introduction

Until recently it has usually been assumed that the Milky Way is fundamentally an axisymmetric system – the only non-axisymmetric features in its gravitational potential were assumed to arise from spiral structure. In the last few years two potential sources of more significant non-axisymmetries have attracted attention: (i) triaxiality of the Milky Way's dark halo would be expected to distort the outer galaxy from axisymmetry, and (ii) several independent arguments point towards the inner $\sim 2\,\mathrm{kpc}$ of the Milky Way being dominated by a rapidly rotating bar. My purpose here is to ask what high-precision astrometry can contribute to our knowledge of the shape of the Milky Way.

At optical wavelengths obscuration severely limits our ability to probe the large-scale structure of the Milky Way. Since the severity of the problem posed by obscuration has steadily lessened as technological advances have increased the sensitivity of detectors and opened up new wavebands, I shall ignore obscuration in this exploratory survey. Thus I ask 'what could be learned about non-axisymmetry in the Milky from proper motions?' without regard to the feasibility of obtaining the necessary data.

E. Høg and P. K. Seidelmann (eds.),
Astronomical and Astrophysical Objectives of Sub-Milliarcsecond Optical Astrometry, 239–246.

My treatment will be superficial in that I shall largely ignore the interconnectedness of most measurements of galactic structure. The great variety of the observational material available to us—the position, parallax, proper motion, radial velocity, spectral type and metallicity of a star are in principle all measurable—can be adequately exploited only by comparing them with pseudo-observations of dynamical models of the Milky Way (e.g., Binney 1994). Since such models are not yet available, we cannot yet fully exploit kinematical observations.

2. NON-AXISYMMETRY OF THE OUTER GALAXY

As is now well known, the circular-speed curve of the Milky Way is flat or continues to rise at the solar radius R_0 and beyond (e.g., Fich & Tremaine 1991), whereas the prediction of models in which the disk, bulge and metal-poor halo all have position-independent mass-to-light ratios Υ, is that the circular speed v_c should decline from about R_0 outwards. This situation could simply indicate that $\Upsilon_{\rm disk}$ increases outwards. Indeed, the detection of gravitational micro-lensing events along lines of sight to the Magellanic Clouds and the bulge offers some support to this conjecture by suggesting the existence of dynamically important numbers of sub-stellar objects. If the flatness of the rotation curve is caused by an increase with radius of $\Upsilon_{\rm disk}$, it would be natural for the outer galaxy to be axisymmetric. However, it is widely suspected that v_c is held constant by a dark halo of non-baryonic matter. A non-baryonic dark halo would be dynamically distinct from the disk, and there are reasons for believing that it would be non-axisymmetric (Binney 1978, Frenk *et al.* 1988, Warren *et al.* 1992), and/or that its principal axes would not coincide with those of the Milky Way's visible components (Dekel & Shlosman 1983, Toomre 1983).

2.1. KINEMATIC EFFECTS OF TRIAXIALITY

Kuijken & Tremaine (1994) have presented a detailed investigation of the kinematic effects of a triaxial halo, albeit with emphasis on the role of radial velocities rather than proper motions. They suppose that the intersection of the Milky Way's equi-potential surfaces with the plane have ellipticity ϵ_Φ, and that the short axes of these ellipses make angle ϕ_b with the Sun-centre line. Then they seek constraints on the quantities $c_\Phi \equiv \epsilon_\Phi \cos 2\phi_b$ and $s_\Phi \equiv \epsilon_\Phi \sin 2\phi_b$.

Three local measures lead to an interesting upper limit, $|s_\Phi| \lesssim 0.05$, on s_Φ. These are (i) the value of the vertex deviation l_v (the angle between the long axis of the local velocity ellipsoid and the direction to the galactic centre) for stars of velocity dispersion $\sigma_r \gtrsim 30\,{\rm km\,s^{-1}}$ (which may be assumed to be phase-mixed and little affected by very local features such as

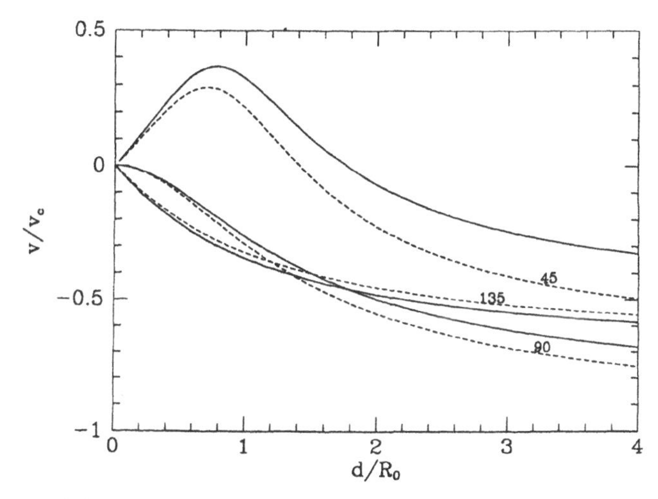

Figure 1. Line-of-sight velocities for material seen at distance d down three lines of sight ($l = 45°, 90°, 135°$) in a galaxy with flat circular-speed curve and elogated potential ($s_\Phi = 0, c_\Phi = 0.1$ and in the corresponding axisymmetric galaxy (full and dashed curves).

spiral arms), and (ii) the values of the constants K and C that are defined by the generalized Oort equations for the line-of-sight velocity and proper motion of a disk object seen at distance d at galactic longitude l,

$$v_{\rm los}(d,l) = \Big(K + A\sin 2l + C\cos 2l\Big)d + O(d^2) \quad (1)$$

$$\mu(d,l) = B + A\cos 2l - C\sin 2l + O(d). \quad (2)$$

The values of K, C and l_v are all consistent with zero. This conclusion is reinforced by the limit $|v_R(\mathrm{LSR})| < 9\,\mathrm{km\,s^{-1}}$ placed on the radial component of velocity of the LSR by observations of objects such as globular clusters and OH/IR stars that should have no net motion with respect to the centre. For $\epsilon_\Phi = 0.1$, the cited upper limit on s_Φ implies that the Sun lies within $\sim 15°$ of a principal axis.

Given that we lie close to a principal axis, it is difficult measure ϵ_Φ from radial velocities alone. In Figure 1 radial velocity v is plotted as a function of distance d along lines of fixed l for the cases $c_\Phi = 0.1$ (full curves) and $c_\Phi = 0$ (dashed curves). Although c_Φ significantly affects the radial velocities observed at given (d,l), the shapes of the full and dashed curves are similar, so the value of c_Φ cannot be determined unless one somehow knows the underlying circular-speed curve, or knows the distances to observed objects in terms of R_0. Figure 2, which plots the ratio $d_v \equiv v_{\rm los}/\mu_l$ for the models and lines of sight of Figure 1, suggests how accurate proper motions could be used to resolve this ambiguity. This ratio is clearly independent of the velocity normalization of the model, and one cannot simultaneously

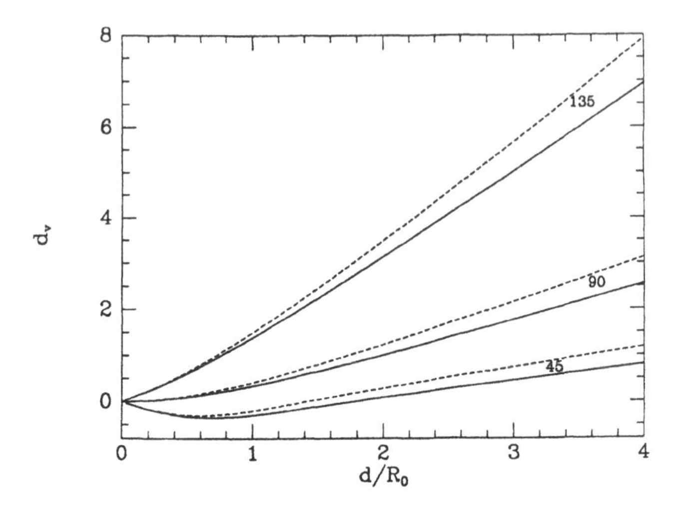

Figure 2. The ratio $d_v \equiv v_{los}/\mu_l$ for objects seen at distance d down three lines of sight ($l = 45°, 90°, 135°$) in a galaxy with elogated potential ($s_\Phi = 0, c_\Phi = 0.1$; full curves) and in the corresponding axisymmetric galaxy (dashed curves).

superimpose all three full curves on the corresponding dashed curves by a single revision of R_0. For example, the curves for $l = 135°$ require that R_0 be revised by a smaller factor than do the curves for $l = 45°$. Hence if one could make this plot from accurate observational material, one could determine whether the outer Milky Way is axisymmetric or not without knowledge of either v_c or R_0.

Although Kuijken & Tremaine conclude that c_Φ cannot be determined by radial velocities alone, they obtain an estimate, $c_\Phi \simeq 0.08$ from Merrifield's (1992) discussion of the flaring of the HI disk. Kuijken & Tremaine point out that this value is supported by the local value of the Oort ratio $X^2 \equiv \sigma_\phi^2/\sigma_R^2 = 0.42 \pm 0.06$. This value is considerably smaller than that expected for a flat or rising circular-speed curve (e.g. Cuddeford & Binney 1994), but is compatible with v_c constant at $v_c = 197 \pm 9\,\mathrm{km\,s^{-1}}$ and $c_\Phi = 0.08 \pm 0.014$. In the model favoured by Kuijken & Tremaine, the Sun's instantaneous angular velocity is $\Omega_0 = 5.1 \pm 0.6\,\mathrm{mas\,yr^{-1}}$. If one makes the plausible assumption that the peculiar motion of the massive black hole candidate Sgr A* is negligible, then the VLBI proper-motion study of Backer (1994) amounts to a direct measurement of $\Omega_0 = 6.55 \pm 0.34\,\mathrm{mas\,yr^{-1}}$ in mild conflict with the prediction of Kuijken & Tremaine.

2.2. KINEMATIC SIGNATURE OF A WARP

Whether or not the dark halo is triaxial, its principal axes are likely to be misaligned with those of the inner galaxy. Such misalignment would provide a natural explanation of the fact that the disk of the Milky Way, like many other galactic disks is warped (e.g. Binney 1992). The shape of the disk, which is flat inside $r \simeq 10\,\mathrm{kpc}$ and then bends to the north at $l > 0$ and southwards at $l < 0$ is most easily traced in the distribution of the Milky Way's gas (Henderson, Jackson & Kerr 1982). Recently the DIRBE/*COBE* team has shown that the stellar disk is also warped (Freudenreich *et al.* 1994) as would be expected if the warp in the gas layer arises from misalignment of the potentials of the dark halo and inner galaxy.

In a popular theory of how this misalignment results in a warp (Sparke & Casertano 1988), the entire warped disk is precessing about the symmetry axis of the dark halo as if it were a rigid body. Let ω_p be the angular velocity of the precession. Then the velocity of a star at position $\mathbf{r}$ is $\mathbf{v} = (\boldsymbol{\Omega}(r) + \boldsymbol{\omega}_p) \times \mathbf{r}$, where Ω is the local circular frequency. Extracting the perpendicular component of this by dotting through with the unit vector $-\widehat{\boldsymbol{\Omega}}(r_\odot)$ [$\boldsymbol{\Omega}(r_\odot)$ points to the South], we find

$$\mu_b(\mathbf{r}) = \frac{(\mathbf{r} - \mathbf{r}_\odot) \cdot \boldsymbol{\omega}_p \times \widehat{\boldsymbol{\Omega}}(r_\odot) + \boldsymbol{\Omega}(r) \times \widehat{\boldsymbol{\Omega}}(r_\odot) \cdot \mathbf{r}}{|\mathbf{r} - \mathbf{r}_\odot|}. \tag{3}$$

$\boldsymbol{\omega}_p$ is perpendicular to the halo's equatorial plane and thus perpendicular to the warp's line of nodes (which is the intersection of the equatorial planes of halo and inner disk). At large radii the disk is expected to align with the halo's plane. So a lower limit on the angle ψ between $-\boldsymbol{\Omega}(r_\odot)$ and $\boldsymbol{\omega}_p$ is given by the tilts of the largest reliably observed rings; since the ring at $r = 20\,\mathrm{kpc}$ rises $\sim 3\,\mathrm{kpc}$ out of the plane of the inner disk, we have $\sin\psi \simeq \tan\psi \gtrsim \frac{3}{20}$. Finally exploiting the fact that the Sun lies near the line of nodes, we have for objects in the flat inner disk at $l = 0$ or $l = 180°$

$$\mu_b \simeq \pm\tfrac{3}{20}\omega_p, \tag{4}$$

where the plus sign applies at $l = 0$ and minus sign at $l = 180°$. The second, less informative, term in (3) is non-zero for objects at $r \gtrsim 10\,\mathrm{kpc}$. From Figs 1 & 2 of Sparke & Casertano (1988) one finds $\omega_p \simeq 0.014(v_c/r_d)(\epsilon_\Phi/0.07)$. So adopting $r_\odot/r_d = 2$ and $\Omega(r_\odot) = 6.5\,\mathrm{mas\,yr^{-1}}$, we have finally

$$\mu_b(l = 0, r \lesssim 10\,\mathrm{kpc}) \simeq 0.027(\epsilon_\Phi/0.07)\,\mathrm{mas\,yr^{-1}}. \tag{5}$$

If Sgr A* is indeed a massive black hole, its peculiar motion might be small enough for its proper motion to be given by (5). The uncertainty in μ_b(Sgr A*) is currently $\sim 0.23\,\mathrm{mas\,yr^{-1}}$ (Backer 1994). Alternatively, equation (5) could be checked by averaging the values of μ_b for all stars in the centre and anticentre directions.

3. Triaxiality of the bulge

Several independent lines of evidence now point to the inner $\sim 2\,\text{kpc}$ of the Milky Way being dominated by a bar, which is probably rapidly rotating (e.g. Weiland *et al.* 1994; Stanek *et al.* 1994). The nearer end of the bar is thought to lie at positive longitudes, such that the line from the centre to the Sun makes an angle of 15° to 45° with the long axis of the bar. It has been argued that the distribution and motions of interstellar gas clouds suggest that corotation lies near $R_{\text{corot}} = 2.5\,\text{kpc}$ (Binney *et al.* 1991; hereafter BGSBU) although larger values of R_{corot} are also popular (e.g. Combes 1994).

Spaenhauer, Jones & Whitford (1992) have determined the proper motions of 429 K and M giants in Baade's window ($l = 1°, b = -4°$). Unfortunately, these are relative proper motions and it remains to determine the mean motion of the sample. However, Zhao Rich & Spergel (1994) find evidence for the bar in the kinematics of the 62 stars of the Spaenhauer *et al.* sample for which Rich (1988, 1990) has obtained spectra. Zhao *et al.* divide these stars into metal-poor ([Fe/H] < -0.2) and metal-rich ([Fe/H] > 0) groups, containing 15 and 39 stars, respectively. By assuming that all the stars lie 8 kpc from the Sun and using Rich's radial velocities, Zhao *et al.* plot the space velocities of the two groups. If the galactic potential were axisymmetric, one would expect the projections onto the (v_r, v_l) plane of the groups' velocity ellipsoids to align with the (v_r, v_l) directions. Actually, neither ellipse aligns with these directions, and the major axes of the two ellipses are nearly at right angles to one another.

Zhao *et al.* analyse this phenomenon in terms of orbits in a three-dimensional generalization of the BGSBU bar model. They find that most of the orbits that carry stars through Baade's window are stochastic rather than regular, and can be broadly classified into prograde and retrograde. The ellipse in the (v_r, v_l) plane formed by the velocities of a suitable selection of prograde orbits has an orientation similar to that of the metal-rich stars, while the velocity ellipse of a selection of retrograde orbits resembles the ellipse of the metal-poor stars. Unfortunately, it is not possible to determine the sense of rotation of observed stars without very much more accurate distances to these stars than are currently available.

The n-body model of the galactic bulge presented by Sellwood (1993) suggest that the Zhao *et al.* result has a deep evolutionary significance. Since the earliest days of n-body simulations of galactic disks, it has been known that bars form naturally in such systems. More recently it has become clear that bars formed in thin disks can have appreciable vertical extension (Combes *et al.* 1990, Raha *et al.* 1991). Moreover, when seen in projection from a point in the original disk plane, these thickened bars

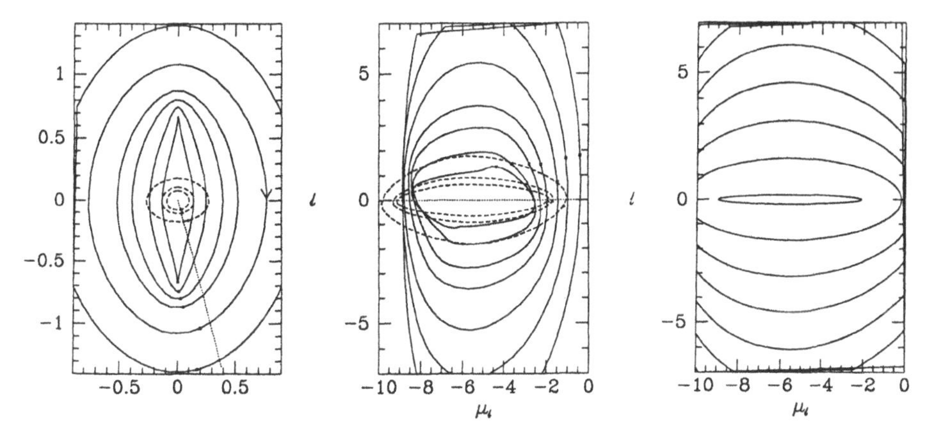

Figure 3. Orbits in the BGSBU model of the bulge/bar. The left-hand panel shows the orbits in real space with the axes marked in kpc; the diagonal straight line points towards the Sun's position. The centre panel shows as a function of l the proper motions μ_l of particles on these orbits in $\mathrm{mas\,yr^{-1}}$. The right-hand panel shows how the centre panel would look if the potential had the same circular speed curve $v_c(r)$ but were axisymmetric. Stars circulate clockwise in the left-hand panel and anti-clockwise in the other two panels. In the left panel x_1 orbits are shown by full curves and x_2 orbits by dashed curves.

have the peanut-like shape characteristic of the bulges of some edge-on disk galaxies, including our own. Zhao *et al.* find that the ellipsoid in the (v_r, v_l) plane formed by the velocities of stars at $(l, b) = (1°, -4°)$ in Sellwood's model, is similar to the ellipse of the metal-rich stars. This suggests that the metal-rich stars formed in the thin disk, and that only the metal-poor stars are part of the old, slowly rotating, bulge.

Observations in the infrared and radio bands make it possible to do astrometry of the bulge outside Baade's window (e.g. Blum, Sellgren & Terndrup 1994), which, from the dynamical point of view, is inconveniently close to the galactic rotation axis. Figure 3 shows where objects such as gas clouds and young stars, that move on nearly closed orbits, would be expected to lie in the (l, μ_l) plane. The left-hand panel shows several closed orbits in real space. The centre panel shows the orbits' traces in the (l, μ_l) plane when viewed along the dotted line of sight of the left panel. Since the assumed velocity of the Sun ($220\,\mathrm{km\,s^{-1}}$) is larger than the peak tangential velocity of any of these orbits, the (l, μ_l) traces are dominated by the reflex of the Sun's motion, and are confined by it to $\mu_l < 0$. For comparison, the right-hand panel shows the (l, μ_l) traces of circular orbits in a potential with the same underlying circular-speed curve as the BGSBU model. The non-axisymmetry of the BGSBU potential has the effect of restricting the range in μ_l associated with the x_1 orbits. This effect is particularly pronounced at negative μ_l, where the model predicts that there should be a pronounced accumulation of objects around $\mu = -8.5\,\mathrm{mas\,yr^{-1}}$ for $|l| \lesssim 4°$.

4. Conclusions

If proper motions of sub mas yr^{-1} accuracy can be secured for objects that are widely distributed through the Milky Way, we should be able to constrain strongly suspected asymmetries due both to the dark halo and the central bar. Clearly the requirement for a wide distribution of objects with proper motions will be hard to meet at optical wavebands. Moreover, since the predicted motions are small and common to nearly all objects in a given field, it is essential that proper motions be relative to the frame established by extragalactic objects.

References

Backer, D.C., 1994. In *Unsolved Problems of the Milky Way*, ed. Blitz, L., Kluwer, Dordrecht

Binney, J.J., 1978. MNRAS, **183**, 779

Binney, J.J., 1992. ARA&A, **30**, 51

Binney, J.J., 1994. In *Galactic and solar system optical astrometry*, ed. L. Morrison, p. 141, Cambridge University Press, Cambridge

Binney, J.J., Gerhard, O.E., Stark, A.A, Bally, J, Uchida, K.I., 1991. MNRAS, **252**, 210

Blum, R.D., Sellgren, K, Terndrup, D.M., 1994. In *Unsolved Problems of the Milky Way*, ed. Blitz, L., Kluwer, Dordrecht

Combes, F., 1994. In *Unsolved Problems of the Milky Way*, ed. Blitz, L., Kluwer, Dordrecht

Combes, F., Debbasch, F., Friedli, D., Pfenniger, D., 1990. A&A, **233**, 82

Cuddeford, P., Binney, J.J., 1994. MNRAS, **266**, 273

Dekel, A., Shlosman, I., 1983. In *IAU Symposium 100, Internal kinematics & dynamics of galaxies* ed. Athanassoula, E., pp. 187–188, Reidel, Dordrecht

Fich, M., Tremaine, S., 1991. ARA&A, **29**, 409

Frenk, C.S., White, S.D.M., Davis, M., Efstathiou, G., 1988. ApJ, **327**, 507

Freudenreich, H.T., *et al.*, 1994. ApJ, **429**, L69

Kuijken, K., Tremaine, S.D., 1994. ApJ, **421**, 178

Henderson, A.P., Jackson, P.D., Kerr, F.J., 1982. ApJ, **263**, 116

Merrifield, M.R., 1992. AJ, **103**, 1552

Raha, N., Sellwood, J.A., James, R.A., Kahn, F.D., 1991. Nat, **352**, 411

Rich, R.M., 1988. AJ, **95**, 828

Rich, R.M., 1990. ApJ, **362**, 604

Sellwood, J.A., 1993. In *Back to the Galaxy*, eds Holt, S.S., Verter, F., p. 133, AIP, New York

Spaenhauer, A., Jones, B.F., Whitford, A.E., 1992. AJ, **103**, 297

Sparke, L., Casertano, S., 1988. MNRAS, **234**, 873

Stanek, K.Z., Mateo, M., Udalski, A., Szymanski, M., Kaluzny, J., Kubiak, M., 1994. ApJ, **429**, L73

Toomre, A., 1983. In *IAU Symposium 100, Internal kinematics & dynamics of galaxies* ed. Athanassoula, E., pp. 177–186, Reidel, Dordrecht

Warren, M.S., Quinn, P.J., Salmon, J.K., Zurek, W.H., 1992. ApJ, **399**, 405

Weiland at al, 1994. ApJ, **425**, L81

Zhao, H.S., Rich, R.M., Spergel, D.N., 1994. ApJ, **000**, 000

GALACTIC DYNAMICS AND DARK MATTER WITH SUB-MILLIARCSECOND ASTROMETRY

D. PFENNIGER
Geneva Observatory
University of Geneva
CH-1290 Sauverny
Switzerland

ABSTRACT. Improving the astrometric accuracy by one or two orders of magnitude over ground-based techniques will not only change our raw knowledge about the Galaxy, but it will also modify 1) the fundamental questions that can be addressed, and 2) the stellar dynamical concepts used so far. More detail in Galactic structure, such as the shape and flow in its putative bar, will be accessible. Also, with the instruments of the next generation the large scale dark matter distribution in the Galaxy, whether distributed in a spheroidal smooth halo or a massive outer disc made of cold clumpy gas, will be measurable. Techniques used for mapping the cosmic flow and mass distribution at Mpc scales and more might be applied to the solar neighbourhood to find the degree of clumpiness of the local matter distribution.

1. Galaxies in an Astrometric Context

The improvement by a factor ~ 10 on the accuracy of stellar astrometric measurements by the Hipparcos satellite over ground-based observations, and by a possible further factor ~ 50 down to perhaps $20\,\mu$as by future astrometric satellites[1] will have profound consequences on the way we describe the Galaxy, and consequently other galaxies. Indeed many theoretical concepts or models presently operative in stellar dynamics, such as the concept of isolating integrals, need a certain degree of symmetry in the stellar distribution (such as time-invariance or axisymmetry) in order to be useful in practice.

The age-velocity dispersion relation and the vertex deviation were already features incompatible with a strictly time-independent or axisymmetric Galaxy. We can expect that such discrepancies with first order Galactic models will turn up more often with the publication of the Hipparcos measurements, and all the more so with the results of even more accurate instruments, since the accuracy level will increasingly exceed the level at which symmetric galaxy models are relevant.

While better data is coming, in galactic studies we can notice a slow but radical theoretical shift in the way galaxies are perceived, as sketched below. The observational objectives that have been considered as important in the past, such as the precise determination of the scale-lengths, or Oort's constants of axisymmetric and stationary Galaxy *models*, will become irrelevant for more precise models.

E. Høg and P. K. Seidelmann (eds.),
Astronomical and Astrophysical Objectives of Sub-Milliarcsecond Optical Astrometry, 247–250.

2. Galactic Dynamics at Large and Small Scales, Spiral Evolution

In fact the detailed dynamical modeling of galaxies, which is increasingly performed by N-body simulations, shows a consistent but quite opposite picture than theoreticians had in mind for years. A disc galaxy can no longer be viewed as a stationary object over several Gyr[2], that would form rapidly in an early phase and then would remain frozen for tens of rotational periods. On the contrary, disc galaxies seem to form progressively over Gyrs by a continuous accretion of gas at the disc periphery (e.g. [3]), and are sometimes subject to faster episodes of tidal or merger interactions with dwarfs or other galaxies. All the evidences we have by probing the Universe at cosmological distances is that galaxies did significantly evolve during the last 10 Gyr (see e.g. [4]). The morphology-radius relationship[5] in galaxy clusters is particularly suggestive that spirals in such dense environments are destroyed by crossing a cluster only once. Contrary to elliptical galaxies, spiral galaxies should be viewed as relatively flexible, easily perturbed objects.

To understand the structure and evolution of an isolated disc galaxy one needs to include the gas dissipative component, that has to be efficient anyway to produce initially a condensed and flat disc. An energy dissipative disc tends to increase the fraction of rotational energy to pressure energy, because, contrary to energy, angular momentum is hardly dissipated. Ineluctably the disc reaches the Safronov-Toomre gravitational instability threshold well before every bit of matter is settled into circular motion. A striking property of the observed disc galaxies is that when they are close to this instability threshold, they are only marginally stable[6]. Marginal stability means that the dissipative factors are still acting now, and that their effect is counterbalanced by the global heating of the gravitational instability.

The main visible effect of such a marginal stability is to make galactic discs highly responsive to dynamical perturbations: such discs are strong *amplifiers* of the effect of a perturbation. In fact, most disc galaxies are developing instabilities producing spiral arms or stellar bars *most of the time*, hence the "spiral" name. The most massive and persistent structures resulting from large scale instabilities are the bars, which redistribute angular momentum and mass efficiently by gravitational torque in a few rotational periods only. As shown by N-body simulations[7],[8], the final mass distribution of barred galaxy models adopts an exponential profile, precisely observed in the optical part of disc galaxies. Also, it is now clear that most disc galaxies display a substantial degree of non-axisymmetry such as bars and ovals in the central parts, and asymmetric HI discs and warps in the outer parts.

So, the old question to understand the persistence of spiral arms becomes irrelevant if spiral arms in fact grow, wind up, and fade away in a recurrent way. The typical time-scales of these fluctuations is the dynamical time, $10^8 - 10^9$ yr.

A central feature of spirals is the bulge, that for a long time was seen as an old stellar population component. But the large dispersion in abundances, metal-rich stars with large motions out of the plane[9], and bluish bulges make it difficult to maintain the scenario that bulges formed rapidly in an early phase. Also, the absence of bulges in late-type galaxies should be understood. As an alternative, dynamical studies show that bulges may grow progressively as a result of the above mentioned instability episodes. Bars may dissolve into bulges by various mechanisms, such as gas accretion or galaxy satellite merging inside the bar[10],[2].

Another predicted feature associated with bars is their tendency to take a peanut or box shape when the bar is seen edge-on[11]. So the distinction between a bar and a bulge is artificial from the point of view of the dynamics. This property of stellar bars will be testable in the Galaxy, because its bulge does display a boxy shape, particularly in the near-infrared.

The global picture that emerges is that spirals may evolve along the Hubble sequence over Gyrs, conformably to the existing irreversible processes: general energy dissipation, bulge formation, star formation, and nucleo-synthesis. Spirals begin as late-type, asymmetric, small, bulgeless, metal-poor but gas-rich galaxies. During this secular evolution, the disc is symmetrized by more rotations, mass accretes at the warped disc periphery, the density and rotation velocity increase by the general dissipation, the bulge grows by internal instabilities, star formation increases the stellar mass and consumes the gas, and stellar evolution enriches the ISM gas. Faster phases of evolution with possible starbursts occur when galaxy satellites merge. Mergers between equal-sized galaxies produce elliptical galaxies by a rapid dynamical mixing of discs. Smaller mergers (satellite mass $< 10\%$ primary galaxy mass) produce big S0 bulges by heating the disc to large heights[2].

3. Dark Matter in Disc Galaxies: Cold Molecular Hydrogen?

Then, in this evolution scenario one should also understand why the M/L ratio decreases systematically along the Hubble sequence: most ($\sim 99\%$) of the mass in Sd galaxies is "dark", while most ($\gtrsim 50\%$) of the mass in Sa galaxies shines as a "normal" stellar population. A consistent proposition[12] is that during the process of galaxy evolution (which is dynamically unavoidable whatever dark matter is made of) most of dark matter has been transformed into stars.

Therefore, not only dark matter should be mostly in some form of hydrogen, but furthermore this hydrogen should be in a sufficiently diluted phase in order to make stars later on. This excludes Jupiters and brown dwarfs. In this picture dark compact objects can form at most the fraction of dark matter that might exist in late type spirals, a minor part of the initial dark matter in early type galaxies.

In a critical investigation on the way gas mass is detected in the ISM (mainly by HI and CO), our conclusion[13] is that most gas may be invisible with present techniques if it is molecular hydrogen (the lowest energy state of hydrogen) at a temperature near the coldest cosmic temperature of 3 K. This gas would be metal poor, in a highly clumpy state, along a fractal mass distribution that would extend at small scale, down to a few tens of AU, the already observed fractal cold (5–100 K) gas already observed in molecular clouds (in HI, CO and FIR) at scales in the range of $\sim 0.01-100$ pc. Most of this gas would be located in the HI outer galactic discs, where neutral hydrogen is known to be proportional (by a factor $10-30$) to dark matter anyway. One can see that in such a fractal structure, with a fractal dimension $D \sim 1.7$, clumps collide frequently at every scale. Yet the estimated energy dissipation time-scale is very long, several Gyr[13]. Also with $D < 2$ (less than a surface!) the sky covering factor of the smallest clumps is very small, so the fractal looks *in average* much more transparent to the background sources than it would if the gas would have a larger D, like the one of common gases.

4. Objectives for Future Uses of Astrometric Data

So, theory offers plenty of justifications to look at the Galaxy with a much finer spatial resolution than available before. At the expected level of accuracy planned for future astrometric satellites, $\ll 0.1$ mas, the huge amount of data will permit to map directly the star distribution and the first velocity moments over most of the optical Galaxy. The problems will be then to describe all the asymmetries, beginning by its bar and its warp, and then smaller scale bumps, such as spiral arms, or even smaller mass irregularities. According to self-consistent bar models[14],[8] the streamings in the stellar velocity field due to a Galactic bar are expected to reach tens of km/s within the bar and still several km/s beyond the solar radius.

Combining these kinematical data with dynamical assumptions will allow to map the total local matter distribution around the Sun by injecting the first velocity moments of stellar populations into the Jeans equations. Contrary to earlier models, no symmetries should be assumed. If the velocity tensor is only partly known, in principle a method similar to the POTENT method[15] might be used. Currently this method yields the total mass distribution at 100 Mpc scale, knowing the radial velocity and distances of thousand of galaxies. This will be even more useful at the $100-1000$ pc scale because the transverse velocities will be known. Then the local distribution of dark matter near the Sun could be determined. We expect that dark matter, if associated with cold gas, might be more abundant outside the several ~ 100 pc wide hot gas bubble we live in.

With a parallax accuracy $\lesssim 0.1$ mas it will be possible to map the matter distribution in the Galactic nearby halo. Then from the kinematics it should be possible to discriminate whether dark matter is distributed smoothly in a round or fat spheroid as classically assumed, or, as we propose, in an outwards thickening outer disc made of cold clumpy gas.

REFERENCES

[1] Lindegren L., et al.: 1993, in *European Space Agency Mission Concepts*, Tome 1
[2] Pfenniger D.: 1992, in *Physics of Nearby Galaxies*, XIIth Moriond Astrophysics Meeting, T.X. Thuan et al. (eds.), Editions Frontières, Gif-sur-Yvette, p. 519
[3] Evrard A.E.: 1992, ibid. [2], p. 375
[4] Durret F. (ed.): 1994, *Clusters of Galaxies*, XIVth Moriond Astrophysics Meeting, Editions Frontières, Gif-sur-Yvette, in press
[5] Whitmore B.C.:1992, ibid. [2], p. 425; and 1994, in [4]
[6] Kennicutt R.C.: 1989, ApJ **344**, 685
[7] Hohl F.: 1971, ApJ **168**, 343
[8] Pfenniger D., Friedli D.: 1991, A&A **252**, 75
[9] Rich R.M.: 1992, ibid. [2], p. 153
[10] Pfenniger D., Norman C.: 1990, ApJ **363**, 391
[11] Combes F., Debbasch F., Friedli D., Pfenniger D.: 1990, A&A **233**, 82
[12] Pfenniger D., Combes F., Martinet L.: 1994, A&A **285**, 79
[13] Pfenniger D., Combes F.: 1994, A&A **285**, 94
[14] Pfenniger D.: 1984, A&A **141**, 171
[15] Deckel A., Bertschinger E., Faber S.M.: 1990, ApJ **364**, 349

ASTROMETRIC TESTS OF GALACTIC EVOLUTION

GERARD GILMORE
Institute of Astronomy, Madingley Rd, Cambridge, UK
gil@mail.ast.cam.ac.uk

1. Introduction

There are many fundamental aspects of Galactic structure and evolution which can be studied best or exclusively with high quality three dimensional kinematics. Amongst these we note as examples determination of the orientation of the stellar velocity ellipsoid, and the detection of structure in velocity-position phase space. The first of these is the primary limitation at present to reliable and accurate measurement of the Galactic gravitational potential. The second is a critical test of current standard models of Galactic formation and evolution.

2. Measuring Gravitational Potentials

The classical method of measuring mass utilises the motions of stellar tracers in the Galaxy. This essentially measures the gravity-pressure-angular momentum balance of a suitable dynamically-relaxed tracer population. Appropriate analysis allows the gravitational force to be derived, and from this, *via* Poisson's equation, the mass density that generates that potential. All such analyses are applications of the collisionless Boltzmann equation. All are in practice approximate. The primary limitation in present analyses is the absence of accurate 3-dimensional kinematics, of the type which would be provided by a high precision astrometric project.

2.1. THE COLLISIONLESS BOLTZMANN EQUATION

The dynamics of any large stellar system are governed by the collisionless Bolzmann equation

$$\frac{Df}{Dt} \equiv \frac{\partial f}{\partial t} + \frac{\partial \vec{x}}{\partial t} \cdot \frac{\partial f}{\partial \vec{x}} + \frac{\partial \vec{v}}{\partial t} \cdot \frac{\partial f}{\partial \vec{v}} = 0, \tag{1}$$

E. Høg and P. K. Seidelmann (eds.),
Astronomical and Astrophysical Objectives of Sub-Milliarcsecond Optical Astrometry, 251–258.

where f is the phase space density at the point $(\vec{x}, \vec{v})$ in phase space (*ie* there are $f(\vec{x}, \vec{v})d^3\vec{x}d^3\vec{v}$ stars in a volume of size $d^3\vec{x}$ centered on $\vec{x}$ with velocity in the volume of size $d^3\vec{v}$ about $\vec{v}$).

The collisionless Boltzmann equation is satisfied by *any* stellar population, whether other stars are present or not. This arises because stars do not interact except through long-range gravity forces, and those are being described through a smooth background potential. Consequently, f does not have to describe the entire Galaxy; one can concentrate on any subsample of stars, and apply the collisionless Boltzmann equation to it. Such subsamples are *tracer populations*, since one may use their kinematics to trace the potential of the Galaxy, irrespective of what generates this potential.

If one has a steady-state tracer population, and a time-independent potential, as the large-scale field in the Milky Way apparently is to an adequate approximation for the present purpose, then

$$\frac{\partial f}{\partial t} = 0. \tag{2}$$

While the Galaxy is not rotationally symmetric, the essential features of the present analysis do not depend on this asymmetry, so it may be suppressed for now. It is convenient to write out the collisionless Boltzmann equation in cylindrical polar coordinates (r, ϕ, z) in which $z = 0$ is the disk plane of symmetry, with corresponding velocity components (v_r, v_ϕ, v_z):

$$v_r \frac{\partial f}{\partial r} + v_z \frac{\partial f}{\partial z} + \left(\mathcal{K}_r + \frac{v_\phi{}^2}{r} \right) \frac{\partial f}{\partial v_r} - \frac{v_r v_\phi}{r} \frac{\partial f}{\partial v_\phi} + \mathcal{K}_z \frac{\partial f}{\partial v_z} = 0 \tag{3}$$

where the accelerations $\dot{v}_r, \dot{v}_\phi, \dot{v}_z$ explicit in the Boltzmann equation have been equated to the forces (real and fictitious) that cause them, and ϕ-gradients in f and in the potential set to zero. The vector $\vec{\mathcal{K}}(r, z)$ is the gravity force. Then clearly knowledge of $f(\vec{x}, \vec{v})$ allows the force components $\mathcal{K}_r$ and $\mathcal{K}_z$ to be derived. Note, though, that a general function f will not allow a unique solution for $\mathcal{K}_r$ and $\mathcal{K}_z$: f has five independent variables (we suppressed any ϕ-dependence) and so cannot in general be made to satisfy the equation above, which contains only two functions of two variables. Since the equation cannot easily be solved in general with real data, one simplifies the analysis and proceeds by taking velocity moments. Multiplying through by v_z and by v_r and integrating over all velocity space produces Jeans' equations:

$$\nu \mathcal{K}_z = \frac{\partial}{\partial z}(\nu \sigma_{zz}^2) + \frac{1}{r}\frac{\partial}{\partial r}(R \nu \sigma_{rz}^2) \tag{4}$$

$$\nu \mathcal{K}_r = \frac{1}{r}\frac{\partial}{\partial r}(r \nu \sigma_{rr}^2) + \frac{\partial}{\partial z}(\nu \sigma_{rz}^2) - \frac{\nu \sigma_{\phi\phi}^2}{r}, \tag{5}$$

where $\nu(r,z)$ is the space density of the stars, and $\vec{\vec{\sigma}}(r,z)$ their velocity dispersion tensor (*ie* $\sigma_{ij}^2 = \langle v_i v_j \rangle$). In this way we have separated the two force components (radial and vertical in this case), and can in principle derive them both from measurements of the moments of the velocity distributions and density of a stellar tracer population.

2.2. THE TILT TERM

The term involving σ_{rz} in the Jeans equation is related to the tilt of the velocity ellipsoid in the (r,z) plane. It is our lack of knowledge of this term which is the greatest limitation in determination of mass distributions in the Galaxy. Two specific examples will suffice to illustrate this. This term has been treated explicitely in derivation of the surface mass density and scale length of the Galactic disk (Kuijken & Gilmore 1989abc,1991; Fux & Martinet 1994; Cuddeford & Amendt 1992), which utilize eqn (4). It has also been discussed in terms of eqn (5), for the disk asymmetric drift by Gilmore, Wyse & Kuijken (1989), and when relating the shape of the stellar distribution in the galactic halo to the shape of the distribution of dark matter by (for example) van der Marel (1991).

The two limiting cases may be discussed analytically, and obviously include orientation of the stellar velocity ellipsoid in cylindrical or in spherical polar coordinates. As an example, Kuijken & Gilmore assumed the long axis of the velocity ellipsoid always points towards the centre of the Galaxy, as is appropriate for a round potential. In this case, they could treat the σ_{rz} term as an extra force term. If the Galaxy's disk, like those of other spirals, has a vertical scale height which is constant with radius and a radially exponential surface density profile $\mu \propto e^{-r/\mathrm{h_r}}$, (see Fux & Martinet 1994) then $\sigma_{zz} \propto \mu \propto e^{-r/\mathrm{h_r}}$, and $\nu \propto \mu \propto e^{-r/\mathrm{h_r}}$. from this Kuijken & Gilmore were able to derive a relationship between σ_{rz} and σ_{zz}. Since the details of this derivation have not previously been published, we present them here.

2.3. DERIVATION OF THE TILT TERM σ_{RZ}

Define the coordinate frames as:
Cylindrical polar (r, z, ϕ), with ϕ suppressed by symmetry, and with velocity components $\dot{z}$ and $\dot{r}$;
Spherical polar, (R,θ, ϕ), where we make θ the angle subtended at the Galactic centre between the radius vector R and the radial planar coordinate r, so that $tan\theta = z/r$, and suppress ϕ by symmetry. The relevant velocity components, to avoid too many levels of subscripts, are (A, B, C), with B suppressed by symmetry. Thus, the A component of velocity is the radial velocity away from the Galactic centre, the C component is perpendicular to it, and at $z = 0$ one has $\dot{r} \equiv A$, and $\dot{z} \equiv C$.

The coordinate transformations by simple trigonometry are:

$$\dot{z} = A sin\theta + C cos\theta, \quad (6)$$
$$\dot{r} = A cos\theta - C sin\theta \quad (7)$$

By definition of the orientation of the velocity ellipsoid, $\mathrm{Cov}(A, C) = 0$, and

$$\sigma_{rz} = \mathrm{Cov}(\dot{r}, \dot{z}) \quad (8)$$
$$= (\dot{z} - \langle\dot{z}\rangle)(\dot{r} - \langle\dot{r}\rangle) \quad (9)$$
$$= \dot{z}\dot{r} \quad (10)$$
$$\dot{z}\dot{r} = (A sin\theta + C cos\theta)(A cos\theta - C sin\theta) \quad (11)$$
$$= A^2 sin\theta cos\theta - C^2 cos\theta sin\theta \ (+AC \ terms) \quad (12)$$
$$\Longrightarrow \sigma_{rz} = sin\theta cos\theta(\sigma_{AA} - \sigma_{CC}) \quad (13)$$
$$\dot{z}\dot{z} = (A sin\theta + C cos\theta)(A sin\theta + C cos\theta) \quad (14)$$
$$= A^2 sin^2\theta + C^2 cos^2\theta \ (+AC \ terms) \quad (15)$$
$$\Longrightarrow \sigma_{zz} = \sigma_{AA} sin^2\theta + \sigma_{CC} cos^2\theta \quad (16)$$

Defining $\sigma_{AA} = \alpha^2\sigma_{CC}$ gives

$$\sigma_{zz} = (\alpha^2 sin^2\theta + cos^2\theta)\sigma_{CC} \quad (17)$$

and

$$\sigma_{rz} = sin\theta cos\theta(\alpha^2 - 1)\sigma_{CC} \quad (18)$$
$$= \frac{sin\theta cos\theta(\alpha^2 - 1)}{\alpha^2 sin^2\theta + cos^2\theta}\sigma_{zz} \quad (19)$$

(Note: At $z = 0$ of course $A \equiv \dot{r}$, and $C \equiv \dot{z}$, so $\sigma_{rr} = \alpha^2\sigma_{zz}$, but this cannot be true in general.)

Since $\tan\theta = z/r, \sin\theta \propto z, \cos\theta \propto r$, and hence

$$\sigma_{rz} = \frac{rz(\alpha^2 - 1)}{(\alpha^2 z^2 + r^2)}\sigma_{zz} \quad (20)$$

This then leads directly to:

$$\frac{1}{\nu r}\frac{\partial}{\partial r}(r\nu\sigma_{rz}) = 6\sigma_{zz}\left\{\frac{4z^3}{(4z^2 + r^2)^2} - \frac{rz}{\mathrm{h_r}(4z^2 + r^2)}\right\} \quad (21)$$
$$= \mathrm{T}(r, z)\sigma_{zz} \quad (22)$$

The vertical Jeans' equation now becomes

$$\mathcal{K}_{z,\mathrm{eff}} = \mathcal{K}_z - \mathrm{T}(r, z)\sigma_{zz} = \frac{1}{\nu}\frac{\partial}{\partial z}(\nu\sigma_{zz}). \quad (23)$$

Given a 'true' $\mathcal{K}_z$, this linear equation can be solved for σ_{zz} (especially since T has an analytic z-integral), and hence the effective force $\mathcal{K}_{z,\mathrm{eff}}$ calculated, with its corresponding potential. Kuijken & Gilmore were then able to proceed by assuming, as a first approximation, that the tracer population moved under the action of this potential.

Given the quality of observational data currently available relevant to the determination of the Galactic $\mathcal{K}_z$force law, the validity or otherwise of this assumption is the primary uncertainty in present determinations of the mass distribution in the optical parts of the Galaxy. The true orientation of the stellar ellisoid could be measured directly from suitable precise astrometric data. Data for a large number of stars at distances of up to 2kpc, with a space motion precision of about 1km/s are required. Sub-milliarcsec astrometry can realize this precision. We would then have a precise, reliable, and assumption-free direct determination of that part of the total Galactic mass density associated with the Galactic disk, and that part distributed in a (dark) halo.

3. The Shape of the Dark Halo

Arguments similar to those above, but applied to the radial Jeans equation (5) have been outlined by van der Marel (1991). He applied the methodology to determination of the shape of the dark matter distribution. The observational constraints are radial velocity data at several locations and star count determinations of the shape of the stellar distribution. The most recent and extensive determination of the shape of the stellar halo, which is in good agreement with most earlier determinations, provides $c/a = 0.55$ (Larsen & Humphries 1994).

Van der Marel shows that this observed shape of the stellar halo implies a corresponding axis ratio $c/a \approx 0.25$ for the case of an adopted spherical alignment of the velocity ellipsoid, and an axis ratio $c/a \approx 0.55$ for the case of an adopted cylindrical alignment of the stellar velocity ellipsoid. That is, there is more than a factor of two uncertainty generated by the absence of 3-D kinematics. This fractional error has the same effect as an error of a factor of two in the observed shape of the stellar halo, an error bound substantially outside current observational limits. That is, our best determinations of the shape of the dark matter halo in the Galaxy from a mix of stellar distribution and (radial velocity) kinematic data are entirely dominated by missing suitably precise astrometric data. These limits could be removed by provision of data with a precision of a few km/s for stars at distances of 5-10kpc.

4. Tests of Galactic Formation and Evolution

Modern models of Galaxy formation make fairly specific predictions which are amenable to detailed test with Galactic kinematic and chemical abundance data. For example, popular Cold Dark Matter models 'predict' growth of the Galaxy about a central core, which should contain the oldest stars. Later accretion of material forms the outer halo and the disks, while continuing accretion will continue to affect the kinematic structure of both the outer halo and the thin disk.

This blend-and-stir process will have been common at high redshifts, when a rain of dwarf 'proto-galaxies' was normal weather for a budding giant. It continues today, at a rate which may still be significant for some galaxies. The term 'significant' here is worth some thought: in the central regions of galaxies masses are large, timescales are short and dynamical friction effective. Thus significant changes to a galaxy require mergers of components of comparable mass. It has been suggested that such mergers would destroy the thin disk of a galaxy like the Milky Way, as argued recently by Toth & Ostriker(1992). If this argument were correct then normal late-type spirals must have completed the bulk of their merger events at very early times.

In the outer parts of galaxies mass densities are low, the fraction of the total luminous galaxy which is seen is very small, and timescales are comparable to a Hubble time. Thus one expects relatively little fossil kinematic structure to be observble in the central regions of normal galaxies, but it is probable that a large fraction of the outer parts of a large galaxy is a recent (on kinematic timescales) acquisition from afar. Fundamentally, the central regions of a galaxy need not be related in any obvious way to the outer parts of that same galaxy.

This picture, which contains aspects of both the monolithic ('ELS') and the multi-fragment ('Searle-Zinn') pictures often discussed in chemical evolution models, makes some specific predictions which are amenable to test. One specific example of current interest is the 'prediction' that mergers of small satellites are an essential feature of galactic evolution. This leads one to look for kinematic and spatial structures, and 'moving groups', as a primary test of such models. That is, the galaxy formation concepts outlined above suggest that a considerable amount of structure in the phase space distribution function for those stars (and DM particles) which inhabit the outer reaches of the galaxy is to be expected. This structure will be the remnant dispersion orbits occupied by the debris of former galactic satellites and near-neighbours which have now lost their former isolated identity. The existence of this structure provides a challenge in two ways: to devise dynamical analysis methods and/or sample selection methods which will still provide a 'fair sample' of the outer galaxy for dynamical studies; and to identify the fractional amount of phase space substructure, if any, and so test the (CDM) galaxy merger models.

The existence of such phase space structure in stars of the thick disk and halo, in addition to the younger stellar populations, has been persuasively argued by Eggen for many years (cf Eggen 1987 for a review). Such moving groups are of course a specific high contrast example of the structure being considered here.

Direct evidence for an ongoing merger event has been discovered recently in a study by Ibata, Gilmore & Irwin (1994). While investigating the kinematic structure of the Galactic bulge, they discovered a large phase-space structure consisting of > 100 K giant, M giant and carbon stars in three low Galactic latitude fields. The group has a velocity dispersion of < 10 kms^{-1}, and a mean heliocentric radial velocity of 140 kms^{-1}, that varies by less than 5 kms^{-1} over the 8° wide region of sky that the three kinematic fields cover. In a subsequent determination of the colour-magnitude relationship for that line of sight, a giant branch, red horizontal clump and horizontal branch are clearly visible, superimposed on the general distribution made up of stars in the bulge, and in the Galactic foreground. Stars belonging to the low velocity dispersion group lie on the upper giant branch of the colour-magnitude relation. From the magnitude of the horizontal branch, Ibata, Gilmore & Irwin find that the object is situated 15 ± 2 kpc from the Galactic centre.

An isodensity map shows an object which is elongated (with axial ratio ≈ 3), spanning $> 10°$ on the sky in a direction perpendicular to the Galactic plane. The interpretation is clear: this is a discovery, in kinematic phase space, of a dwarf (former) satellite galaxy currently well inside the galactic optical boundary. The tidal radius of the dwarf galaxy is then approximately an order of magnitude smaller than its apparent size on the sky, so most of its members will disperse into the Galactic halo over the next $\approx 10^8$ years. This finding clearly supports galaxy formation scenarios of with significant merging events happening right up to the present epoch.

The implications for the present are substantial. Kinematics can and has discovered phase space structure; presumably much more remains to be found. Kinematics can determine the present and future orbits of the (former) member stars of this dwarf galaxy. Kinematics can map out the merger history of the Milky Way. When astrometric data can provide distances to a few percent, and kinematics to a few km/s, at distances up to 20kpc from the Sun, then we will be able to determine in detail the evolutionary history, and the three-dimensional distribution of mass, in the Galaxy.

References

Cuddeford, P. & Amendt P. (1992) *MNRAS* **256** 166
Eggen, O.J. (1987) in *The Galaxy*, eds G. Gilmore & R. Carswell, p211 (Reidel)
Fux, R. & Martinet, L. (1994), *A&A*, **287** L21
Ibata, R., Gilmore, G., & Irwin, M. (1994) *Nature* **370** 194
Gilmore, G., Wyse, R.F.G., & Kuijken, K. (1989) *ARAA* **27** 555
Kuijken, K. & Gilmore, G. (1989a) *MNRAS* **239** 571
Kuijken, K. & Gilmore, G. (1989b) *MNRAS* **239** 605
Kuijken, K. & Gilmore, G. (1989c) *MNRAS* **239** 651
Kuijken, K. & Gilmore, G. (1991) *Ap.J.* **367** L9
Larsen, J.A. & Humphries, R. M. (1994) *preprint*
van der Marel, R.P. (1991), *MNRAS*, **248** 515
Toth, G., & Ostriker, J. (1992) *Ap.J.* **389** 5.

MOTIONS OF GLOBULAR CLUSTERS

P. BROSCHE, M. ODENKIRCHEN, H.-J. TUCHOLKE,
M. GEFFERT
Universitäts-Sternwarte, Auf dem Hügel 71
D–53121 Bonn
Germany

ABSTRACT. Absolute proper motions of seven globular clusters have been determined with respect to extragalactic references and with accuracies of ~ 1 mas/yr. Derived quantities and qualitative implications are described.

1. Introduction

A conservative manner for elucidating the objectives of sub-milliarcsecond optical astrometry consists in the consideration of already existing results with an accuracy not too far from the envisaged one. Since the *radial velocities* of almost all galactic globular clusters are known with a typical accuracy of ± 10 km s^{-1} (and since we shall not discuss motions *within* globular clusters), the bottleneck of our topic is the determination of *absolute proper motions* of the clusters as a whole. The informational value of such a determination will be seen if one compares the possible volumes in velocity space occupied by a cluster with and without the knowledge of a proper motion: while the third dimension is given by the radial velocity, in the tangential components a circle with a radius of the order of 300 km s^{-1} is a priori possible (letting aside transient intergalactic visitors). Even a crude determination of a proper motion which pins down the cluster just within 100 km s^{-1} bins (but in two dimensions!) leaves a posteriori only $\sim 1/30$ of the a priori volume as possible locus of the cluster in velocity space.

In what follows we provide examples of the results obtained by the Bonn group, partly in cooperation with other colleagues, for the globular clusters NGC 4147, NGC 5466, M 12 (Brosche et al. 1991), M 15 (Geffert et al. 1993), M 3 (Scholz et al. 1993, Tucholke et al. 1994), M 92 (Scholz et al. 1994), and M 5 (Scholz et al. 1995).

2. Photographic Astrometry, Direct Results and Invariants

Earlier attempts to find the proper motions of globular clusters have been corrupted by the errors in the motions of reference stars which entered directly into the results. In order to make use of the potentially higher accuracy of photographic data it is absolutely necessary to eliminate the influence of classical catalogue errors from the final results.

First, one would think that instead of using the few *individual* brighter stars lying around

E. Høg and P. K. Seidelmann (eds.),
Astronomical and Astrophysical Objectives of Sub-Milliarcsecond Optical Astrometry, 259–263.

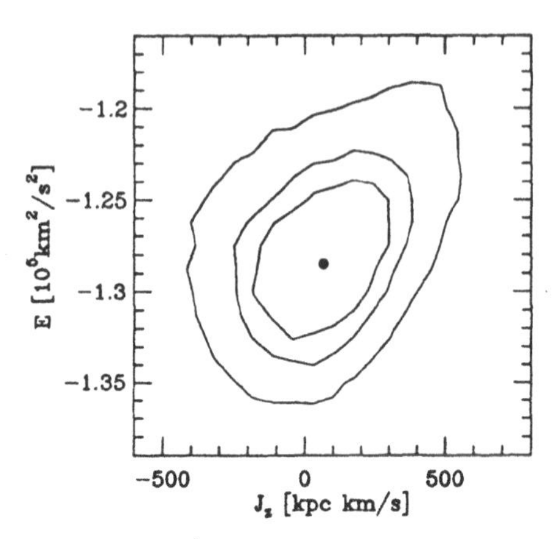

Fig. 1: Uncertainty in the determination of the constants of motion E and J_z of the globular cluster M 92. The dot corresponds to the values at the mean orbit. The contours are 50, 66 and 90% confidence limits.

a cluster, one should use merely the *essence* of the catalogues, i.e., a complete model of galactic motions. Such a model would describe the collective motions of many faint stars around a cluster, thus arriving at a statistically sound relative motion between cluster and stars. And because the motions of the stars are known from the model, so we had the motion of the cluster too, if the model were correct. However, as Brosche & Schwan (e.g. 1986) have demonstrated, the standard model is not sufficient to describe all systematic motions, neither for the FK 4-, nor N30-, nor FK 5-system.

Therefore we preferred to determine absolute proper motions with extragalactic reference, either directly relative to quasars or galaxies or indirectly through stellar proper motions measured with respect to galaxies provided by our colleagues A.R. Klemola (Lick Observatory) and R.-D. Scholz (Potsdam).

The most direct result consists in the two components of the absolute proper motion of a globular cluster, say in galactic coordinates. Adding its radial velocity and using the space velocity of the Sun with respect to the galactic center, we can derive the absolute space motion of the cluster. While such an 'instantaneous' information is interesting in itself, one would like also to obtain parameters of a less transient meaning, especially invariants of the motion. For an axisymmetric structure (which the Galaxy should possess approximately) there exist J_z, the specific angular momentum around the axis of symmetry, and E, the specific orbital energy (taken as vanishing at infinity). While the first can be obtained without further input directly from the space motion and the spatial position, the latter involves the knowledge of the galactic potential distribution. So far we have applied the one of Allen & Martos (1986). The two invariants depend on the input data in a complex manner. Although their errors are dominated by the errors of the proper motion components (and sometimes by distance errors) of the globular cluster, the magnitude of the errors often does not allow a linear treatment. Instead, the confidence area in the proper motions has to

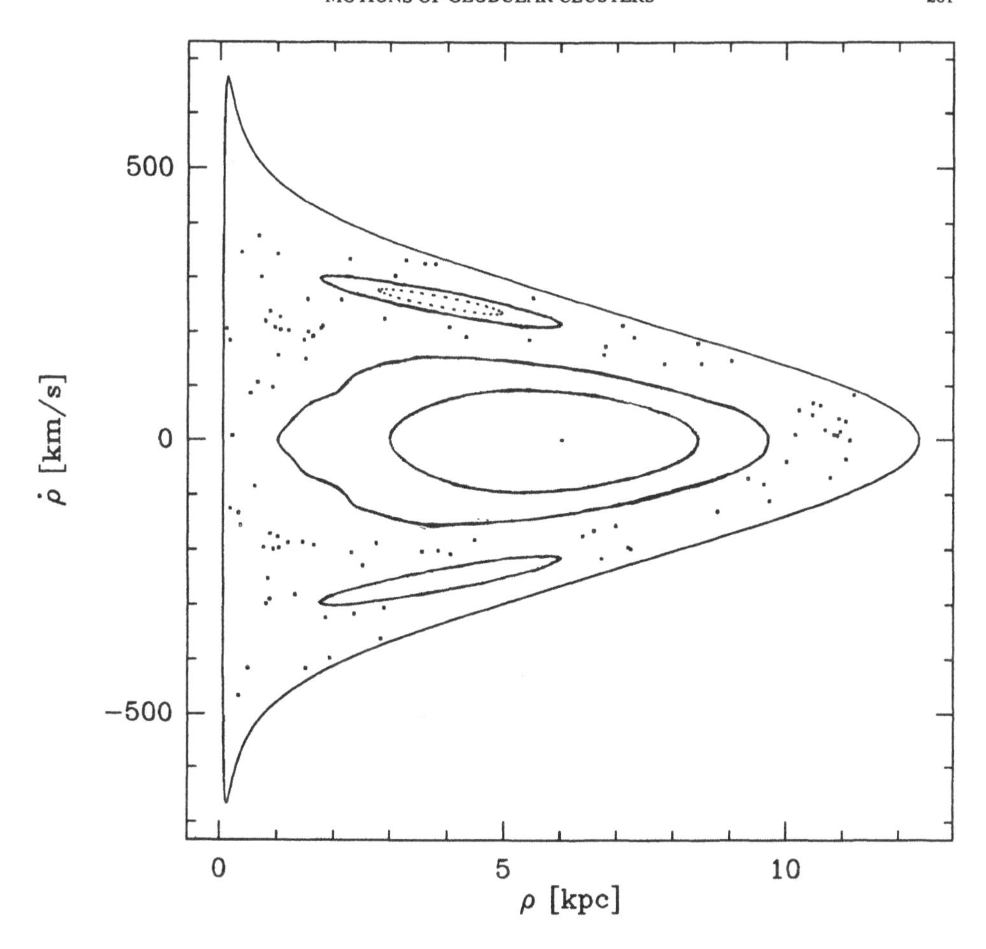

Fig. 2: Poincaré section for the mean values E and J_z of M 92. ρ denotes the radial distance to the galactic centre in a cylindrical coordinate system, while $\dot{\rho}$ is its time derivative. For the Poincaré section their values at the passages through the galactic plane are used.
Outer envelope: zero-velocity surface
Central ovals: family of box orbits
Upper oval: family of tube orbits containing the mean orbit of M 92 (dotted)
Isolated dots: stochastic orbit with initial conditions not far from those of M 92
Lower oval: another island of tube orbits.

be transferred into the surface of the two invariants. Fig. 1 shows a typical situation. Note the deviations from ellipses and the existence of correlations. We consider such figures to represent the overall concentrate of our results.

3. Further Interpretation and Conclusions

If the path of every globular cluster would totally fill the space allowed by its two invariants, the latter would be sufficient for the description. However, this is not generally true and the integration of the orbits in the potential quoted above reveals a rich variety of possibilities and of questions to be answered. Starting from short time scales, one can find the times, loci and velocities of the last transits through the galactic plane. If these values become more precise in the future, it may be useful to look for traces in the interstellar medium. Furthermore, at least the projections of the orbits allow the definition of some parameters prevailing for some time or oscillating in a characteristic manner (e.g. by precession): extremal distances and thereby tidal radii, or some kind of eccentricities and angles describing the position of an instantaneous orbital 'plane'. One can pose — and answer! — the question whether the present values of the parameters lie in a probable interval of their distribution for the whole history of the orbit.

It is possible to classify the orbits according to their behaviour in the Poincaré section (Fig. 2): stochastic or chaotic orbits leave a scattered distribution of crossing points while regular orbits produce pointwise closed contour lines of "islands of regularity". Inside the central island we find the usual type of (meridional) box orbit, which is symmetric with respect to the galactic plane (see Fig. 3 for an example). In outer islands we also obtain asymmetric orbits, usually confined in a rather narrow "tube" in the meridional section (Fig. 4). This type is not a rare exception, e.g. M 92 spends more than 80% of the time north of the galactic plane (Odenkirchen et al. 1994). Due to the proper motion errors, a probability can only be ascribed to each cluster for belonging to one or the other type of orbits. In some favourable cases like M 12, however, there is a high probability for a regular orbit. Taken together with the long periods of revolution, this enables us to integrate backwards a cluster orbit for the whole stationary life of the Galaxy.

With increasing numbers, one will be able to tackle the quest for correlations between orbital and astrophysical characteristics, e.g. maximal distance from the galactic plane and metallicity (Geffert et al. 1995).

An important problem not treated here is the sensitivity of the results to the underlying galactic model. Amongst the various competing ones, those based on the family designed by Miyamoto & Nagai (1975) seem especially attractive because of the explicit knowledge of both the potential and the mass distribution.

A most recent attempt to improve the outer part of the galactic potential from globular cluster space velocities has been performed by Dauphole & Colin (1995); clearly, the future use of velocities will strengthen the basis for such conclusions.

Due to the sparsity of our data basis, the main aim of this contribution consisted in the illustration of the potentialities of sub-mas-astrometry in the field of globular clusters.

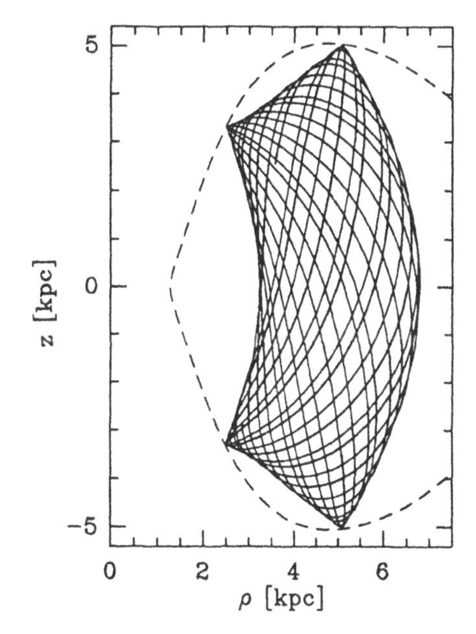

Fig. 3: Example for a box orbit: motion of the globular cluster M 12 in the meridional plane for the time interval [-2, 0] Gyr. ρ is the distance to the galactic centre, projected onto the galactic plane, and z the distance to the galactic plane. The dashed line shows the zero-velocity curve.

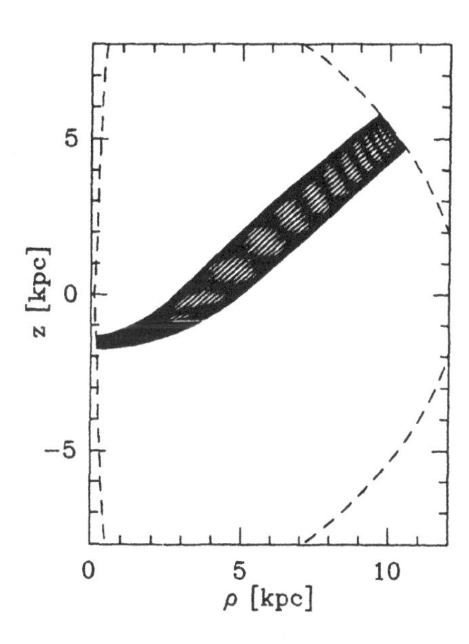

Fig. 4: Example for an asymmetric orbit: motion of the globular cluster M 92 in the meridional plane for the time interval [-20, 0] Gyr.

References

Allen C., Martos M.A., 1986, RMxAA 13, 137

Brosche P., Schwan H., 1986, IAU Symp. 109, 53

Brosche P., Tucholke H.–J., Klemola A.R., Ninković S., Geffert M., Doerenkamp P., 1991, AJ 102, 2022

Dauphole B., Colin J., 1995, IAU Symp. 169, in press

Geffert M., Colin J., LeCampion J.F., Odenkirchen M., 1993, AJ 106, 168

Geffert M., Odenkirchen M., Tucholke H.-J., Dauphole B., Colin J., 1995, IAU Symp. 164, in press

Miyamoto M., Nagai R., 1975, PASJ 27, 533

Odenkirchen M., Scholz R.-D., Irwin M.J., 1994, IAU Symp. 161, 453

Scholz R.–D., Odenkirchen M., Irwin M.J., 1993, MNRAS 264, 579

Scholz R.–D., Odenkirchen M., Irwin M.J., 1994, MNRAS 266, 925

Scholz R.–D., Hirte S., Irwin M.J., Odenkirchen M., 1995, IAU Symp. 164, in press

Tucholke H.–J., Scholz R.–D., Brosche P., 1994, A&AS 104, 161

2.3 EXTRAGALACTIC ASTROMETRY

BETTER PARALLAXES AND THE COSMIC DISTANCE SCALE

SIDNEY VAN DEN BERGH
Dominion Astrophysical Observatory
5071 West Saanich Road
Victoria, B.C., V8X 4M6
Canada

ABSTRACT. It is generally agreed that the Hubble parameter lies in the range $50 < H_0$ (km s^{-1} Mpc^{-1}) < 100. Recent observations, which are discussed in the present paper, favor $H_0 \geq 75$ km s^{-1} Mpc^{-1}. Hubble Space Telescope observations of Cepheids in Virgo cluster galaxies will probably reduce the uncertainty in H_0 to ~ 20%. It should be possible to lower this remaining uncertainty in the Hubble parameter by strengthening the calibration of the Cepheid period-luminosity relation, the M_V(RR) versus [Fe/H] relation of RR Lyrae stars, and the luminosity calibration of the subdwarf main sequence, by using parallaxes obtained with Hipparcos.

1. Introduction

A good mystery novel keeps one in suspense until the very last page, when we finally find out who did the evil deed. Papers on the extragalactic distance scale are generally less exciting because the names of the authors usually allow us to guess the value of the Hubble parameter that they will end up with.

Fig. 1, which is adapted from Okamura & Fukugita (1991), shows a plot of the distribution of recent determinations of H_0. In order to reveal my own prejudices the Hubble parameters that I obtained in review papers are shown as open circles. My most detailed discussion of this problem is in van den Bergh (1992), in which a value H_0(local) = 83 ± 6 km s^{-1} Mpc^{-1} was derived for the region of the Universe with $V < 10000$ km s^{-1}. This value represented an uneasy compromise between supernovae of Type Ia (SNe Ia) and of Type II (SNe II), which both appeared to yield low values of H_0, and planetary nebulae, the Tully-Fisher relation, and surface brightness fluctuations in early-type galaxies, which all yielded higher values of H_0. Finally observations of the luminosities and radial velocities of first-ranked galaxies in rich clusters (Hoessel, Gunn & Thuan 1980) indicated that H_0(global) = (0.92 ± 0.08) H_0(local), so that H_0(global) = 76 ± 9 km s^{-1} Mpc^{-1}.

E. Høg and P. K. Seidelmann (eds.),
Astronomical and Astrophysical Objectives of Sub-Milliarcsecond Optical Astrometry, 267–272.

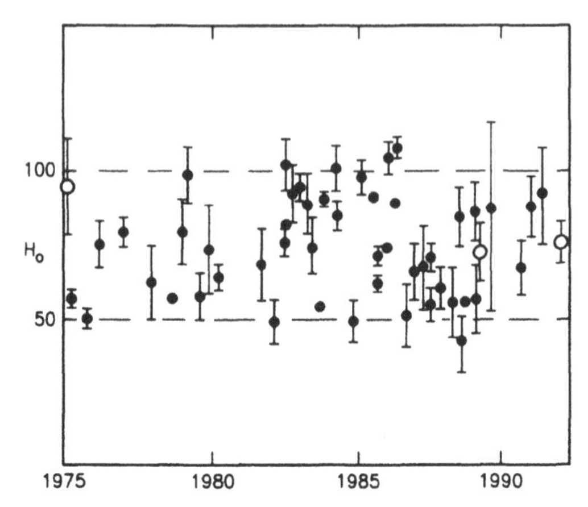

Fig. 1. Recent determinations of the Hubble parameter adapted from Okamura & Fukugita (1991). Results by the present author are shown as open circles.

2. Recent determinations of H_0.

The usefulness of supernovae of Type Ia as standard candles was thrown into doubt by the discovery of the subluminous object SN 1991bg, and by the probably superluminous supernova 1991T. Furthermore high quality observations obtained at Lick and at Cerro Tololo showed that SNe Ia exhibit a remarkable range in lightcurve morphologies and spectral characteristics that had not previously been appreciated. From a reevaluation of old observations of SN 1937C (Jacoby & Pierce 1994), and the maximum magnitude versus rate of decline relation of Phillips (1993), one obtains $H_0 = 75 \pm 12$ km s^{-1} Mpc^{-1}. However, this value is uncertain because observations of SN 1992bc and S Andromedae show that individual SNe Ia may deviate substantially from the mean relation adopted by Phillips.

Application of detailed model atmosphere calculations to the expanding photosphere (Baade-Wesselink) models of SNe II (Schmidt 1994) yields $H_0 = 73$ km s^{-1} Mpc^{-1}, which is significantly larger than values previously obtained by this method. For the two most distant objects so far observed (SN 1990ae, with V = 7800 km s^{-1}, and SN 1992am, with V = 14500 km s^{-1}) Schmidt et al. (1994) find $H_0 = 68$ km s^{-1} Mpc and $H_0 = 81$ km s^{-1} Mpc^{-1}, respectively.

From studies of two volume elements, in the direction of (but beyond) the Virgo cluster, and another in the opposite direction Lu et al. (1994) find $H_0 = 84 \pm 8$ km s^{-1} Mpc^{-1}. This value includes a small correction for Malmquist bias, and is independent of the adopted infall (retardation) velocity of the Local Group into the Virgo cluster.

Tonry & Schneider (1988) have shown that surface brightness fluctuations of galaxies containing old stellar populations can be used to derive their distances. Using this technique Jacoby et al. (1992) derive a distance of 15.0 ± 1.4 Mpc to the elliptical galaxies in the core of the Virgo cluster. Combining this distance with the cosmic velocity V = 1311 ± 132 km s^{-1}, that van den Bergh (1992) derives from the Coma cluster velocity, and the Coma/Virgo distance ratio yields $H_0 = 87 \pm 12$ km s^{-1} Mpc^{-1}.

From a careful study of the luminosity functions of planetary nebulae in early-type Virgo galaxies Méndez et al. (1993) derive a Virgo distance modulus of 30.9 ± 0.4 mag. In conjunction with the cosmological velocity of the Virgo cluster this yields $H_0 = 86 \pm$ km s^{-1} Mpc^{-1}.

From observations of planetary nebulae in three galaxies in the Fornax cluster (McMillan, Ciardullo & Jacoby 1993) $H_0 = 75 \pm 8$ km s^{-1} Mpc^{-1}. From surface brightness profiles of early-type dwarfs in Fornax, Young & Currie (1994) find $H_0 = 99 \pm 16$ km s^{-1} Mpc^{-1}.

Distance determinations based on intercomparisons of the peaks of the luminosity functions of distant globular clusters, with those of globulars in M31 and the Galaxy, are based on the (as yet unproven) hypothesis that globulars in nearby spiral galaxies are similar to those in distant spirals. This technique has been discussed by Sandage & Tammann (1994) and more recently by van den Bergh (1994), who obtains $H_0 \geq 73$ km s^{-1} Mpc^{-1}.

Additional estimates of H_0 based on observations of the gravitational lens system 0957 + 561 (Bernstein, Tyson & Kochanek 1993), the Sunyaev-Zel'dovich effect in the cluster Abell 2218 by Jones et al. (19930, and from VLBI observations of compact radio sources (Roland 1994) are listed in Table 1. Taken in their entirety the data in this table suggest that $H_0 \gtrsim 75$ km s^{-1} Mpc^{-1}.

Finally Lauer & Postman (1992), who studied brightest cluster galaxies, found no evidence for a systematic change in the value of H_0 over the range $0.01 \leq z \leq 0.05$.

Table 1. Compilation of recent H_0 determinations.

Method	H_0 (km s^{-1} Mpc^{-1})
SNe Ia	75 (± 12?)
SNe II	68 - 81
Tully-Fisher	84 ± 8
Surface brightness fluctuations	87 ± 12
Planetary nebulae (Virgo)	86 ± 18
Planetary nebulae (Fornax)	75 ± 8
Galaxy diameters	76 - 124
Globular clusters (Virgo)	≳ 73
Surface brightness profiles (Fornax)	99 ± 16
Gravitational lens	< 87
Sunyaev-Zel'dovich effect	~ 50
Compact radio sources	~ 100

3. Hipparcos and the distance scale

Hopefully the Hubble Space Telescope will start observing Cepheid variables in Virgo spiral galaxies later this year. To derive a value of the Hubble parameter H_0 from these observations it will be necessary to obtain reddening values of individual Cepheids from multi-color observations. Secondly, we shall have to improve our understanding of the rather complex structure of the Virgo region, so as to prevent contamination of the sample by background galaxies. Spirals with high negative velocities, relative to the mean cluster velocity, are most likely to be situated close to the cluster core. Finally the calibration of the zero point of the Cepheid period-luminosity relation (Turner 1992), and of the Cepheid period-color relation (Turner 1994 et al.) will have to be strengthened. This is where Hipparcos can make a major contribution.

Traditionally extragalactic distances have been determined by using the slope of the P-L relations of Cepheids in the Magellanic Clouds, in conjunction with zero point calibrations derived by fitting the main sequences of clusters and associations containing Cepheids, to that of the Pleiades. A difficulty with this approach is that (1) individual Galactic Cepheids may have metallicities that differ significantly from that of the Pleiades (Fry & Carney 1994). This difficulty can be circumvented by measuring the distances to a significant number of individual nearby Cepheids directly with Hipparcos. A problem that still remains to be resolved is whether the Cepheid period-luminosity relation is itself metallicity dependent. By *assuming* that the wavelength dependence of extinction is independent of distance from the center of M31 Freedman & Madore (1990) found that Cepheids at different distances from the nucleus of the Andromeda nebula (which presumably have differing metallicities) yield the same distance modulus $(m\text{-}M)_0$. However, using the same observational data Gould (1994) arrives at a different conclusion. He finds that $\Delta(m\text{-}M)_0 = (0.56 \pm 0.20)\ \Delta[Fe/H]$ *i.e.* distance modulus and metallicity are related at the 2.8 σ confidence level. A different parameterization of the relationship between distance modulus and metallicity is given by Caldwell & Coulson (1986).

Observations of Population II distance indicators, such as RR Lyrae stars, W Virginis variables, and globular clusters add significant weight to the distance determinations of nearby galaxies, that are used to calibrate the extragalactic distance scale. Hipparcos can contribute by providing accurate parallaxes of subdwarfs of various metallicities, to which globular cluster main sequences can be fit. Furthermore parallax observations of nearby RR Lyrae stars should allow one to chose between the (significantly different) M_V(RR) versus [Fe/H] relations (see Fig. 2) that have recently been proposed by Carney, Storm & Jones (1992), Sandage (1993) and Walker (1992).

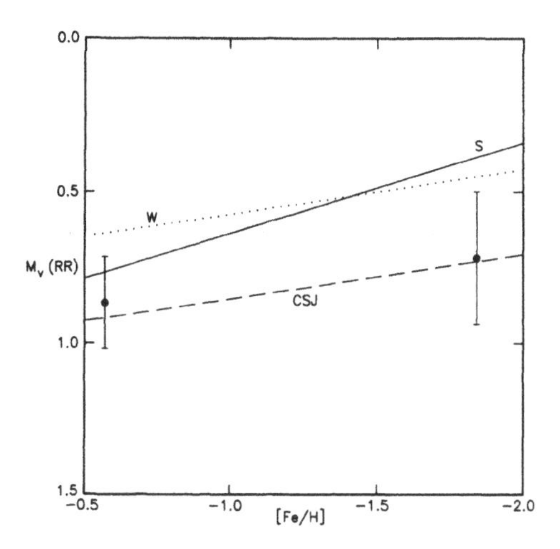

Fig. 2. M_V(RR) versus [Fe/H] relations proposed for RR Lyrae stars by CSJ (Carney, Storm & Jones 1992), S (Sandage 1993), and W (Walker 1992). The two observational points are the horizontal branch magnitudes of the M31 clusters K58 (left) and K219 (right) that have recently been measured with the Hubble Space Telescope (Ajhar et al. 1994). These data appear to favor the faint calibration of CSJ.

It is concluded that recalibration of the luminosities of Cepheids of Population I and of RR Lyrae stars of Population II with Hipparcos, in conjunction with further elucidation of the spatial structure of the Virgo cluster, can probably improve the accuracy with which H_0 is determined to ~ 10%.

References

Ajhar, E.A., Lauer, T.R., Lynds, C.R., O'Neil, E.J., Faber, S.M., Grillmair, C.J., Baum, W.A., Ewald, S.P., Light, R.M., & Holtzman, J.A. 1994, poster paper presented at 184th AAS meeting.
Bernstein, G.M., Guhathakurta, P., Raychaudhury, S., Giovanelli, R., Haynes, M.P., Herter, T., & Vogt, N.P. 1994, AJ, 107, 1962
Caldwell, J.A.R., & Coulson, I.M. 1986, MNRAS, 218, 223
Carney, B.W., Storm, J., & Jones, R.V. 1992, ApJ, 386, 663
Freedman, W.L., & Madore, B.F. 1990, ApJ, 365, 186
Fry, A.M., & Carney, B.W. 1994, BAAS, 26, 869
Gould, A. 1994, ApJ, 426, 542
Hoessel, J.G., Gunn, J.E., & Thuan, T.X. 1980, ApJ, 241, 486
Jacoby, G.H., Branch, D., Ciardullo, R., Davies, R., Harris, W.E., Pierce, M.J., Pritchet, C.J., Tonry, J.L., & Welch, D.L. 1992, PASP, 104, 599
Jacoby, G.H., & Pierce, M.J. 1994, NOAO Newsletter No. 38, p.1
Jones, M. et al. 1993, Nature, 365, 320
Lauer, T.R., & Postman, M. 1992, ApJ, 400, L47
Lu, N.Y., Salpeter, E.E., & Hoffman, G.L. 1994, ApJ, 426, 473
McMillan, R., Ciardullo, R., & Jacoby, G.H. 1993, ApJ, 416, 62
Méndez, R.H., Kudritzki, R.P., Ciardullo, R., & Jacoby, G.H. 1993, A&A, 275, 534
Okamura, S., & Fukugita, M. 1991 in Primordial Nucleosynthesis and Evolution of the Early Universe, eds. K. Sato and J. Audouze (Kluwer, Dordrecht), 45
Phillips, M.M. 1993, ApJ, 413, L105
Roland, J. 1994 in Proceedings of the 9th IAP Astrophysics Meeting, eds. F. Bouchet and M. Lachièze-Rey, Editions frontières, Gif-sur-Yvette, in press
Sandage, A. 1993, AJ, 106, 703
Sandage, A., & Tammann, G.A. 1994 preprint
Schmidt, B.P. 1994, private communication
Schmidt, B.P. et al. 1994, AJ, 107, 1444
Tonry, J., & Schneider, D.P. 1988, AJ, 96, 807
Turner, D.G. 1992, AJ 104, 1865
Turner, D.G., Mandushev, G.I., & Forbes, D. 1994, AJ, 107, 1796
van den Bergh, S. 1992, PASP, 104, 861
van den Bergh, S. 1994, in preparation
Walker, A.R. 1992, ApJ, 390, L81
Young, C.K., & Currie, M.J. 1994, MNRAS, 268, L11

Question:

E. Høg: Could you be more specific about various contributions to the quoted error of H_0 ?

S. van den Bergh: The three main sources of error in the determination of the Hubble parameter from observations of Cepheids in the Virgo cluster are: (1) Uncertainty in the zero-point of the Cepheid Period-Luminosity relation,which probably amounts to 0.1-0.2 mag. (2) Uncertainty in the location of the Cepheid parent galaxy relative to the core of the Virgo Cluster. This uncertainty probably amounts to a few tenths of a magnitude in the distance modulus difference for M100, but is probably much smaller for NGC4571 because its low hydrogen abundance indicates that it has recently been swept by intra-cluster gas in the Virgo cluster core. (3) Large-scale motions in nearby region of the Universe introduce uncertainty in the cosmic velocity of the Virgo cluster that are of the order of 10 percent. The errors resulting from (1) can be greatly reduced by HIPPARCOS observations of Cepheids. The uncertainty introduced by (3) can be minimized by deriving H_0 from the Coma cluster velocity and the Coma/Virgo distance ratio, which is known to better than 5 percent.

THE KINEMATICS OF THE MAGELLANIC CLOUDS

B.E.WESTERLUND
Astronomical Observatory, Uppsala University
Box 515, S-75120 Uppsala, Sweden

Abstract. It is essential for our understanding of the evolution of the Magellanic System, comprising the Large and the Small Magellanic Cloud, the Intercloud or Bridge region and the Magellanic Stream, to know its motions in the past. The Clouds have a common envelope of neutral hydrogen; this indicates that they have been bound to each others for a long time. The Magellanic System moves in the gravitational potential of our Galaxy; it is exposed to ram pressure through its movement in the galactic halo. Both effects ought to be noticeable in their present structure and kinematics. It is generally assumed, but not definitely proven, that the Clouds have been bound to our Galaxy for at least the last 7 Gyr. Most models assume that the Clouds lead the Magellanic Stream. The interaction between the Clouds has influenced their structure and kinematics severely. The effects should be possible to trace in the motions of their stellar and gaseous components as pronounced disturbances. Recent astrometric contributions in this field show a great promise for the future if still higher accuracy can be achieved.

1. Introduction

The past history of the Magellanic Clouds is still veiled in obscurity. As all members of the Local Group they may have emerged from its barycentre. Whereas many of the dwarf galaxies in the Local Group may have had close encounters with either the Galaxy or M 31 but escaped (Mishra 1985), several, including the Large (LMC) and the Small (SMC) Cloud, were caught and are now satellites to either of the two giant galaxies in the Group. Shuter (1992) has suggested that the Clouds set out in a Hubble flow together with M 31 and the Galaxy, were subsequently attracted to M 31, and suffered a close collision with it $\sim$ 6 Gyr ago. The tidal acceleration in this collision projected them back toward the Galaxy and produced the velocity differentials required to establish the Magellanic Stream. Byrd *et*

E. Høg and P. K. Seidelmann (eds.),
Astronomical and Astrophysical Objectives of Sub-Milliarcsecond Optical Astrometry, 273–282.

al. (1994) consider that the Magellanic Clouds left the M 31 neighbourhood $\sim$ 10 Gyr ago and were captured by our Galaxy $\sim$ 6 Gyr ago. According to Murai & Fujimoto (See Gardiner *et al.* 1994 for references) the Clouds may have been satellites to our Galaxy for up to 10 Gyr and always have been separate objects.

The Bridge between the SMC and the LMC contains HI gas but also young stars and associations with a common age of about 0.1 Gyr (Irwin *et al.* 1989). The Magellanic Stream extends from the Magellanic Clouds as an essentially continuous filament of HI about 100° along a great circle as seen from the Galactic centre. There are six main concentrations, MS I to MS VI (Cf.Mathewson *et al.* 1987). No evidence for stars in the Stream have been found. CaII has been detected as well as interstellar lines(HI, CII, CIV, OI, MgII, SiII, Fe II) (cf. Penston 1982).

Several theories exist for the origin of the Magellanic Stream (See Mathewson *et al.* 1987, Liu 1992, Shuter 1992, Sofou 1994, Gardiner *et al.* 1994). In the ram pressure model the Intercloud Gas and the Magellanic Stream were produced in a collision between the LMC and the SMC about 0.4 Gyr ago. In the tidal model the Magellanic Stream was torn off from the SMC in an encounter with the LMC $\sim$ 1.5 Gyr ago which also coincided with a perigalactic passage.

2. The Distances of the Components of the Magellanic System

The Magellanic System covers a large area of the sky and a considerable distance in depth. Individual distances must be measured for its components. The uncertainties in the distance determinations arise mainly through the dependence of the criteria on the metallicities and ages of the involved objects and on the interstellar absorption in the line of sight.

2.1. THE INTERSTELLAR REDDENING, FOREGROUND AND INTRINSIC

Schwering & Israel (1991) have investigated the Galactic foreground colour excess towards the Magellanic Clouds on scales of 48 arcmin. Over the LMC the foreground reddening falls in the range0.07 $\leq E_{B-V} \leq$ 0.17 mag. For the SMC the range is 0.07 to 0.09 mag. Bessell (1991), using various reddening determinations, derived a foreground excess in the SMC of 0.04 $\leq E_{B-V} \leq$ 0.06 mag and in the LMC of 0.04 to 0.09 mag. The average reddening within the SMC is probably $\sim$ 0.06 mag; there are regions with reddenings up to 0.3 mag. The average reddening within the LMC is similar but the variations are larger.

2.2. THE GEOMETRY OF THE MAGELLANIC CLOUDS

The LMC has long been regarded as a thin flat disk seen nearly face-on. Its east side is closer to us than the west. For the determination of the inclination (i) of its main body low-frequency observations, referring to non-thermal emission, are the most reliable. The thermal radio continuum emission receives strong contributions from discrete sources which are not necessarily in the main plane. The use of HI for determinations of i is questionable as HI has a complicated structure in and away from the main plane. From data summarized by Westerlund (1990) and from Schmidt-Kaler & Gochermann (1992) $35° \leq i \leq 45°$ appears as the most likely range for the tilt.

The SMC is extended in depth with the Wing and the NE section closer to us than the southern part so that its possible tilt and line of nodes have little relevance.

2.3. THE DISTANCES OF THE MAGELLANIC CLOUDS

TABLE 1. Determinations of the distance moduli of the Magellanic Clouds

LMC	SMC	Ref.
18.5	18.9	Westerlund (1990), review
18.19 ± 0.22	19.33 ± 0.31	Arellano Ferro *et al.* (1991): F supergiants
	18.9 ± 0.2	Barnes III *et al.* (1993): Cepheid HV 829
18.66 ± 0.05		Hughes & Wood (1990): old LPVs, P 200 d.
18.70 ± 0.23		Capaccioli *et al.* (1990): novae
18.4 ± 0.18		Hanuschik & Schmidt-Kaler (1991): SN1987A
18.50 ± 0.13		Panagia *et al.* (1991): SN1987A shell
18.38 ± 0.03		de Vaucouleurs (1993): Mean of 55 estimates
18.3 – 18.5		Reid & Freedman (1994): RR Lyr, NGC 2210

Table 1 summarizes recent determinations of the distance moduli of the LMC and the SMC. For estimates of the uncertainties in the moduli I refer to the table with individual data given by de Vaucouleurs (1993) and to Reid and Freedman (1994).

3. The Kinematics of the Magellanic System

Information about the conditions in the Magellanic System is now available from radio, infrared, optical, UV and X-ray observations. For the kinematics of the System the 21-cm HI observations have been particularly important.

3.1. TRANSVERSE MOTIONS OF THE LMC AND THE SMC

3.1.1. *Proper Motions of the LMC and the SMC*

All models proposed so far assume that the LMC and the SMC lead the Magellanic Stream and predict a proper motion for the LMC of between 1.5 and 2.0 mas yr^{-1}.

TABLE 2. Model motions of the LMC

Model	μ mas yr^{-1}	$v_{t,g}$ km s^{-1}	$V_{c,\odot}$ km s^{-1}	$R_{\odot}$ kpc
Murai & Fujimoto (1980)	1.7	288	250	10
Lin & Lynden Bell (1982)	2.0	373	244	9
Shuter(1992)	1.9	355	220	8.5
Liu (1992)	1.7	310	220	8.5
Gardiner*et al.* (1994)	1.8	287	220	8.5

TABLE 3. Proper motions of the Magellanic Clouds

μ_{α} $cos\delta$ mas yr^{-1}	μ_{δ} mas yr^{-1}	Reference
0.91 ± 2.34	-0.23 ± 2.77	Tucholke, Hiesgen (1991), LMC
1.3 ± 0.6	$+1.1 \pm 0.7$	Kroupa *et al.* (1994), LMC, SMC
1.37 ± 0.25	-0.18 ± 0.25	Jones *et al.* (1994), LMC

Tucholke & Hiesgen (1991) measured absolute proper motions (498 reference galaxies, epoch span 15 yr). Kroupa *et al.* (1994) identified 35 PPM stars as members of the LMC and 8 of the SMC and derived proper motions from data spanning almost a century. The present values are consistent with bound as well as unbound orbits of the Magellanic Clouds. Jones *et al.* (1994) determined proper motions for 251 LMC members (92 reference galaxies, epoch span 14 yr) near NGC 2257, 8.5° from the LMC center, PA = 61°. The mean absolute proper motion of the LMC stars in this region is $\mu_{\alpha} = 0.120 \pm 0.028$ arcsec $century^{-1}$, $\mu_{\delta} = 0.026 \pm 0.027$ arcsec $century^{-1}$.

3.1.2. *Transverse Motions from Radial Velocities of the LMC and the SMC*

Feitzinger *et al.* (1977) used the difference in position angle between the kinematical (188.0° ± 2.6°) and photometric lines (168° ± 4°) of nodes to determine the transverse velocity. The same method was used by Meatheringham *et al.* (1988) and by Prévot *et al.* (1989). Hughes *et al.* (1991) found that the dynamics of the LMC is dominated by a single rotating disk and that all major populations of the Bar have solid- body rotation.

TABLE 4. Transverse motion of the LMC

Heliocentric velocity km s^{-1}	Galactocentric velocity km s^{-1}	Reference
	100	Feast *et al.* 1961; RV of stars and nebulae
275	143	Feitzinger *et al.* 1977
275 ± 65		Meatheringham *et al.*1988; gradients in RV
	150	Prévot *et al.* 1989; late- type supergiants
	200	Hughes *et al.*1991; HI, CO, PN, CHs, clusters
	236	Lin 1993; proper motions
	215 ± 48	Jones*et al.*1994; proper motions

The individual radial-velocity measurements are accurate to $\sim$ 10 –15 km s^{-1}; the individual proper motions are accurate to $\sim$ 3 mas yr^{-1}. At the distance of the LMC the latter corresponds to $\sim$ 700 km s^{-1}. If a 20 microarcsec accuracy can be achieved (see the GAIA concept at this conference) annual proper motions could be determined for LMC stars to better than $\sim$ 5 km s^{-1}.

3.2. THE INTERNAL MOTIONS IN THE LMC

Both the LMC and the SMC have been severely disturbed by interactions. Observed radial velocities may therefore not necessarily be interpreted as effects of rotation only.

3.2.1. *Local Disturbances in the Observed Radial Velocities in the LMC*

In their 21-cm line map of the LMC Rohlfs *et al.* (1984) note that the central region, $r < 1.5°$, is strongly perturbed by non-circular motions. There is a kink in the rotation curve; due to the HI void in the SGS LMC-4 area. The large cloud around the 30 Dor complex and extending south for about 2° is visible as a strong perturbation of the velocity field with a second component of lower radial velocity. A similar component is seen at the NW end of the Bar. These components may be connected with a warp of the LMC disk.In an extension of the Rohlfs *et al.* survey Luks & Rohlfs (1992) interpret the HI gas distribution as two rotating disks and no warps.

In the supergiant shell SGS LMC-4 the HI gas has three components (Dopita *et al.* 1985b): one in the plane, one above and one below it; the ejected gas has a velocity of 36 km s^{-1}. Supergiants have been found outside the LMC HI plane; they group around a velocity 35 km s^{-1} larger than the HI velocity in their neighbourhood or around a 34 km s^{-1} smaller velocity (McGee 1964), i.e. with velocities similar to the expansion velocity of SGS LMC-4. This may be typical for the motion of young objects forced out of the plane.

3.2.2. *The Kinematics of the Youngest Population in the LMC*

Images of the LMC in any wavelength region are dominated by radiation from its Extreme Population I constituent (stellar associations, supergiants, etc.) or the connected gas (HII regions, HI complexes, molecular clouds) and dust. They display the regions of recent star formation as an asymmetric pattern, not completely at random but with some structure. During the past 30 years many attempts have been made to interpret the observations of the youngest population as showing a spiral-arm structure. Two lines of thought regarding the structure of the LMC appear to develop independent of each others: (1).– The radio centre is given the role played by the nucleus in our Galaxy, i.e. as the centre of rotation and mass for all classes of objects.– In a recent investigation Luks & Rohlfs (1992) found two rotating disks in their HI maps: one extending over all of the LMC with 72 % of the HI gas and a lower-velocity component with 19 %. The rotation curve of the major component is symmetric. It has its centre at $5^h12^m.8$, - 69°.1 (1950), i.e. 1.2° from the centre of the Bar.The low-velocity component contains the complex south of 30 Dor and a lobe N and W of it with the 30 Dor nebula as a link between the two lobes. Its distance above the main disk is at least 250 – 400 pc. (2).– The 30 Doradus nebula is considered as the origin of the spiral arms.– A number of well separated sources in the 1.4 GHz radio continuum map form long ridges suggested to originate in the 30 Doradus nebula (Feitzinger *et al.* 1987). The ridges correlate well with a series of blobs in the 100μm emission in the IRAS maps and with the UV brightness distributions. Most outstanding is the chain of OB associations and HII regions, " a bright and sharply defined spiral", through the Bar towards the NW.

3.2.3. *The Kinematics of the Old and Intermediate-Age Clusters in the LMC*

Doubts on the LMC as a single uniformly rotating disk arose from an investigation of the RVs of 59 clusters by Freeman *et al.* (1983). Clusters younger than 1 Gyr had motions similar to the gas in their vicinity and shared the rotation solution found from HI and HII region velocities (line-of-sight velocity dispersion 15 km s^{-1}, rotation amplitude 37 $\pm$5 km s^{-1}, galactocentric systemic velocity 40 $\pm$ 3 km s^{-1}, and a line of nodes in PA 1° $\pm$ 5°. Also the intermediate-age clusters formed a flattened system. The oldest clusters, age $\sim$ 10 Gyr, had a line-of-sight dispersion of 16 km s^{-1}, a rotation amplitude of 54$\pm$7 km s^{-1}, a systemic velocity of 38 $\pm$ 4 km s^{-1}, and a line of nodes in PA 44° $\pm$ 6°. The oldest LMC clusters appear to rotate in a disk separate from that of the other populations but do not form a spherical halo population.

Schommer *et al.* (1992) analyzed RVs for 83 star clusters in the LMC. About one half of the clusters are more than 5° from the center. The outer cluster sample and the inner intermediate age clusters form a disk aligned with the inner HI kinematics and the outer LMC isophote major-axis position angle.The oldest clusters rotate with an amplitude comparable to that of the younger disk with a small velocity dispersion. A single rotating disk solution fits the old and intermediate-age clusters and other tracers (no need for an additional "tilted disk" system). In the inner 2° the old clusters exhibit peculiar velocities, as do the CH stars and the old LPVs, possibly due to perturbations from the Bar. The rotation curve does not show signs of a Keplerian falloff out to at least 5 - 6 disk scale lengths, implying the existence of dark matter associated with the LMC.

3.2.4. *The Kinematics of the Planetary Nebulae and the Oldest Stellar Field Populations in the LMC*

Meatheringham *et al.* (1988) have derived a rotation solution for the PN in the LMC essentially identical with that of the HI but the vertical velocity dispersion of 19.1 km s^{-1} is much greater than the 5.4 km s^{-1} found for HI. The larger dispersion of the PN is consistent with the action of orbital diffusion over the lifetime of the PN. The bulk of the PN cannot represent a halo population.

Hughes *et al.* (1991) have obtained RVs for a significant sample of LPVs and applied a kinematic analysis to a wide range of LMC populations (HI, CO, PN, CH stars, clusters, LPVs). The oldest LPVs ($\sim$ 10 Gyr) were found to have a high velocity dispersion and a low rotational velocity, proving that they belong to a flattened spheroid population (maximum height $\leq$ 2.8 kpc); may be part of a disk population $\sim$ 4 Gyr old. The dynamics of the LMC is dominated by a single rotating disk.

The velocity distribution derived by Hartwick & Cowley (1991) from radial velocities for 81 CH stars appears assymetrical. Two groups stars may exist, one of which may have about the same age as the oldest LPVs and belong to the spheroid population. The other may be associated with a flattened-disk system; it contains stars with luminosities up to M_{bol} = -6 or brighter. They may belong to an AGB population as young as $\sim$ 0.1

Gyr(Suntzeff *et al.* 1993) or be products of binary mergers, age a few Gyr (Feast 1992).

3.3. THE INTERNAL MOTIONS IN THE SMC

The early attempts to interpret the SMC as a rotating galaxy with a spiral structure have been replaced by questions about its extension in depth and/or about the kind and extent of its fragmentation.

3.3.1. *The Extension of the SMC in Depth*

The SMC was first suspected to have an appreciable extent in depth by Johnson (1961). Azzopardi (1982) showed that the A, B, O supergiants in the SMC have an appreciable spread in distance, up to 7 kpc. At least two stellar groupings may thus exist in the SMC. Feast *et al.* (1961) found no convincing rotational effects and concluded that their results could be represented by a random scatter around a mean velocity of 166 $\pm$ 3 km s^{-1}.

Analyses of the structure of the SMC based on new extensive radio, optical and UV observations show that several velocity groups exist. In a series of investigations Mathewson *et al.* (See Mathewson *et al.* 1988 for references) have shown that two separate entities exist in the HI distribution in the SMC, each with its own stellar and nebular populations. They may have formed when the SMC was torn apart in the encounter with the LMC $\sim$ 0.2 Gyr ago into a low-velocity fragment, the SMC Remnant (SMCR), in front of a higher-velocity fragment, the Mini-Magellanic Cloud (MMC). Cepheids were found in the range 43 and 75 kpc with two 6 kpc deep components centered 12 kpc apart. A further detailed study of Cepheids in the SMC Bar combined with a high resolution HI survey of this region confirmed that the SMC has a depth of $\sim$ 20 kpc and that the NE section of the Bar is 10 –15 kpc closer than the southern. An extension of the SMC in depth has also been proposed from observations of red stars aged more than 1 Gyr (see Hatzidimitriou & Cannon 1993).

Martin *et al.* (1989) carried out an extensive discussion of the structure and motions of the SMC based on the HI velocity distribution, on accurate RVs of 307 young stars and 35 HII regions, and on very high spectral resolution profiles of interstellar absorption lines. The HI in the Bar was found to consist of 4 components: a very low (VL) component in the SW, a low (L) major component covering the southern half of the SMC, a high velocity (H) component everywhere except in the SW, and a weak very high (VH) velocity component mainly in the NE. The L and H components correspond more or less to the SMCR and MMC. The main H complex is behind the L complex. Most of the young SMC stars lie within a depth of

$\leq$ 10 kpc. Interstellar CaII is seen at higher and lower velocities than the HI lines. (cf. Fitzpatrick & Savage 1985).

3.3.2. *The Kinematics of the Planetary Nebulae, C Stars and Halo Giants in the SMC*

The PN in the SMC form an unstructured spheroidal population apparently associated with the Bar and with the centroid at $0^h49^m.7$, $-73°.5$ (1950) and a mean $RV_{GSR} = -17$ kms^{-1} (Dopita *et al.* 1985a). No evidence was found of any organized rotation, nor any bimodal velocity distribution of the kind observed by Torres & Caranza (1987) and others. Infrared spectroscopy by Hardy *et al.* (1989) with an individual precision of $\sim$ 1.8 km s^{-1} show that the C stars behave kinematically like the PN with which they share the mean velocity as well as a velocity dispersion of $\sim$ 27 km s^{-1}. No evidence was found of a velocity splitting, nor of rotation of the main body. C stars and PN may belong to a spheroidal-like system.

In their study of the metallicity and RV of SMC halo giants Suntzeff *et al.* (1986) compared the kinematics of the NGC 121 field with the SMC in general They found a velocity of -29 km s^{-1} for these stars. This is significantly smaller than the velocities of $+2 \pm 3$ km s^{-1} for the main body of the SMC and $+$ 21 km s^{-1} for the K1 region east of the SMC Bar found by Ardeberg & Maurice (1979). The three values give together the impression of a small gradient in stellar RV about orthogonal to the HI major axis. However, they show more likely the double-peaked velocity structure in the SMC.

The present data for PN, C stars, and metal-poor giants appear to show that the near collision with the LMC $\sim$ 0.2 Gyr ago did leave the older stellar component roughly spheroidal whereas the gaseous component was drawn out along our line of sight.

3.4. THE MOTIONS IN THE BRIDGE REGION AND IN THE MAGELLANIC STREAM

The RV of the Stream with respect to the Galactic Centre becomes increasingly more negative from 0 km s^{-1} at MS I to -200 km s^{-1} at MS VI. There is a sharp discontinuity in velocity of $\sim$ 100 km s^{-1} between the top of the Intercloud region and MS I (Mathewson *et al.* 1987). Other components are also seen in HI as overlapping the main Stream. Several of them appear to be typical high-velocity clouds (Morras 1985). The Stream is following behind the Magellanic Clouds.

References

Ardeberg A., Maurice E., 1979, *A&A* **77**, 277

Arellano Ferro A., Mantegazza L., Antonello E., 1991, *A&A* **246**, 341
Azzopardi M., 1982, *C.R. Journées Strasbourg* **4**, p.20
Barnes III Th.G., Moffett Th.J., Gieren W.P., 1993, *ApJ* **405**, L51
Bessell M.S., 1991, *A&A* **242**, L17
Byrd G., Valtonen M., McCall M., Innanen K., 1994, *AJ* **107**, 2055
Capaccioli M., Della Valle M., D'Onofrio M., Rosino L., 1990, *ApJ* **360**, 63
Cowley A.P., Hartwick F.D.A., 1991,*ApJ* **373**, 80
De Vaucouleurs G., 1993, *ApJ* **415**, 10
Dopita M.A., Ford H.C., Lawrence C.J., Webster B.L., 1985a, *ApJ* **296**, 390
Dopita M.A., Mathewson D.S., Ford V.L., 1985b, *ApJ* **297**, 599
Feast M.W., 1992, *New Aspects of Magellanic Cloud Research* , p. 239
Feast M.W., Thackeray A.D., Wesselink A.J., 1961,*MNRAS* **122**, 433
Feitzinger J.V., Isserstedt J., Schmidt-Kaler Th., 1977, *A&A* **57**, 265
Feitzinger J.V., Haynes R.F., Klein U., Wielebinski R., Perschke M.,1987, *Vistas Astr.* **30**, 243
Fitzpatrick E.L., Savage B.D., 1985, *ApJ* **292**, 122
Freeman K.C., Illingworth G., Oemler Jr.A., 1983, *ApJ* **272**, 488
Gardiner L.T., Sawa T., Fujimoto M., 1994, *MNRAS* f 266, 567
Hanuschik R.W., Schmidt-Kaler Th., 1991, *A&A* **249**, 36
Hardy E., Suntzeff N.B., Azzopardi M., 1989, *ApJ* **344**, 210
Hatzidimitriou D., Cannon R.D., 1993, *New Aspects of Magellanic Cloud Research* , p. 17
Hughes S.M.G., Wood P.R., 1990, *AJ* **99**, 784
Hughes S.M.G., Wood P.R., Reid N., 1991, *AJ* **101**, 1304
Irwin M.J., Demers S., Kunkel W.E., 1990, *AJ* **99**, 191
Jones B.F., Klemola A.R., Lin D.N.C., 1994, *AJ* **107**, 1333
Kroupa P., Roeser S., Bastian U., 1994, *MNRAS* **266**, 412
Lin D.N.C., 1993, *BAAS* **25**, 783
Lin D.N.C., Lynden-Bell, D., 1982, *MNRAS* **198**, 707
Liu Y.-z., 1992, *A&A* **257**, 505
Luks Th., Rohlfs K., 1992, *A&A* **263**, 41
McGee R.X., 1964, *Aust. J.Phys.* **17**, 515
Mathewson D.S., Wayte S.R., Ford V.L, Ruan K., 1987, *Proc.Astron. Soc. Aust.* **7**, 19
Mathewson D.S., Ford V.L., Visvanathan N., 1988, *ApJ* **333**, 617
Meatheringham S.J., Dopita M.A., Ford H.C., Webster B.L., 1988, *ApJ* **327**, 651
Mishra R., 1985, *MNRAS* **212**, 163
Morras R., 1985, *AJ* **90**, 1801
Panagia N., Gilmozzi R., Macchetto F., Adorf H.-M., Kirshner R.P., 1991, *ApJ* **380**, L23
Penston M.V., 1982, *Observatory* **102**, 174
Prévot L., Rousseau J., Martin N., 1989, *A&A* **225**, 303
Reid N., Freedman W., 1994, *MNRAS* **267**, 821
Roeser S., Bastian U., 1993, *Bull. Inf. CDS* **42**, 11
Schmidt-Kaler Th., Gochermann J., 1992, *ASP Conf. Ser.* **30**, 203
Schommer R.A., Olszewski E.W., Suntzeff N.B., Harris H.C., 1992, *AJ* **103**, 447
Schwering P.B.W., Israel F.P., 1991, *A&A* **246**, 231
Shuter W.L., 1992, *ApJ* **386**, 101
Sofue Y., 1994, *Univ. Tokyo Preprint No. 94-07*
Suntzeff N.B., Friel E., Klemola A., Kraft R.P., Graham J.A., 1986, *AJ* **91**, 275
Suntzeff N.B., Phillips M.M., Elias J.H., Cowley A.P., Hartwick F.D.A., Bouchet P., 1993, *PASP* **105**, 350
Torres G., Carranza G.J., 1987, *MNRAS* **226**, 513
Tucholke H.-J., Hiesgen M., 1991, *Proc IAU Symp. 148* (eds. R.Haynes, D.Milne), p.491
Westerlund B.E., 1990, *A&AR* **2**, 29

SECULAR MOTIONS OF EXTRAGALACTIC RADIO-SOURCES AND THE STABILITY OF THE RADIO REFERENCE FRAME

T.M. EUBANKS, D.N. MATSAKIS, F.J. JOSTIES,
B.A. ARCHINAL, K.A. KINGHAM, J.O. MARTIN,
D.D. McCARTHY, S.A. KLIONER, T.A. HERRING

Abstract. The best current approximation to an inertial reference frame is provided by Very Long Baseline Interferometry (VLBI) observations of extragalactic radio sources with red shifts (z) up to 3.8. The stability of the resulting reference frame directly depends on the amount of any secular changes in the observed source positions.

Two types of potentially observable secular motions should be present in extragalactic source positions. Gravitational accelerations of the solar system will cause secular motions through aberration, amounting to, e.g., about 4 microarcsec (μas) year^{-1} due to the mass of the galaxy. Extragalactic mass concentrations will cause gravitational deflections in the apparent positions of more distant radio sources, and these will change with time as the mass concentrations evolve. This effect could easily cause secular motions of order 1 μas year^{-1} in some, or even most, radio sources with $z \geq 1$.

The present astrometric VLBI data set contains about one million observations over a 15 year period, with current source proper motion formal errors being as small as 2.5 μas year^{-1}. Proper motion estimates from these data reveal many sources with statistically significant proper motion estimates of order 30 μas year^{-1}, about an order of magnitude larger than expected. Work continues to determine if the observed motions are due to systematic errors or reflect true secular changes in source positions. The results from a continued proper motion analysis of the complete astrometric VLBI data set will be presented.

E. Høg and P. K. Seidelmann (eds.),
Astronomical and Astrophysical Objectives of Sub-Milliarcsecond Optical Astrometry, 283.

IMPROVEMENT AND MAINTENANCE OF THE EXTRAGALACTIC REFERENCE FRAME

E.F. ARIAS and A.M. GONTIER

Abstract. Very Long Baseline Interferometry (VLBI) is at present the most powerful technique to construct the best approximation to an inertial reference frame. After more than a decade of VLBI observations several hundreds of extragalactic objects have positions known within ±0.0003". Since 1988 the International Earth Rotation Service (IERS) elaborates a global extragalactic celestial reference frame (IERS Celestial Reference Frame, ICRF) that is tied to the international terrestrial reference frame through the high precision monitoring of the Earth's rotation. The direction of the ICRF axes relative to the IAU definitions are known within better than ±0.001" for the polar axis and ±0.003" for the origin of right ascensions. The FK5 axes are consistent with the ICRF ones within their uncertainties (0.050"-0.100"). The maintenance of this high accuracy extragalactic frame will be necessary for the long term programs, such as the future monitoring of the tie of the Hipparcos galactic frame as well as of the dynamical planetary frame (millisecond pulsars, lunar laser ranging).

E. Høg and P. K. Seidelmann (eds.),
Astronomical and Astrophysical Objectives of Sub-Milliarcsecond Optical Astrometry, 284.

2.4 REFERENCE FRAMES AND SOLAR SYSTEM

EARTH ORIENTATION - THE CURRENT AND FUTURE SITUATION

DENNIS D. MCCARTHY
U. S. Naval Observatory
Washington, DC 20392-5420

Abstract.
Sub-milliarcsecond astrometry often requires an accurate characterization of the orientation of the Earth in a quasi-inertial reference frame. The International Earth Rotation Service (IERS) standards provide the current state of the art in the transformation between celestial and terrestrial reference systems. Improvements in the determination of Earth orientation parameters which describe this transformation continue to be made. Current and future capabilities are given.

1. Introduction

Earth orientation refers to the direction in space of axes which have been defined on the Earth. It is measured using five quantities: two angles which identify the direction of the Earth's rotation axis within the Earth (polar coordinates x and y), an angle describing the rotational motion of the Earth (UT1-UTC), and two angles which characterize the direction of the Earth's rotation axis in space (longitude of the ascending node of the Earth's orbit and obliquity of the ecliptic). With these data the orientation of the Earth in space is fully described. Anyone who must relate changing aspects of reference systems on Earth to a system in space requires information on the Earth's orientation.

2. Current Situation

Astronomical observations are made routinely by a number of observatories located around the world to monitor variations in the Earth's orientation. The International Earth Rotation Service (IERS) is the international or-

E. Høg and P. K. Seidelmann (eds.),
Astronomical and Astrophysical Objectives of Sub-Milliarcsecond Optical Astrometry, 287–291.

ganization responsible for the coordination of observations of polar motion and nutation as well as astronomical time. The structure and functions of the IERS are outlined in the annual reports of the IERS, and much of the following information is taken from the *1993 IERS Annual Report* (IERS, 1994). The IERS routinely provides the information necessary to define the Conventional Terrestrial Reference System and a Conventional Celestial Reference System and relate them and their frames to each other and to other reference systems used in the determination of the Earth orientation parameters. Responsibilities of the IERS include defining and maintaining a conventional terrestrial reference system based on observing stations using high-precision techniques in space geodesy, defining and maintaining a conventional celestial reference system based on extragalactic radio sources, and relating it to other celestial reference systems, determining the Earth orientation parameters connecting these systems, collecting and archiving appropriate data and results, and disseminating the results to meet the needs of users.

The IERS consists of a Central Bureau and Coordinating Center for each of the principal observing techniques, Very Long Baseline Interferometry (VLBI), Lunar and Satellite Laser Ranging (LLR, SLR), Global Positioning System (GPS), and DORIS and is supported by organizations that contribute observations and their analyses. Coordinating Centers are responsible for developing and organizing the activities in each technique to meet the objectives of the service. The Central Bureau combines the various types of data collected by the service, and disseminates to the user community the appropriate information on Earth orientation and the terrestrial and celestial reference systems. It includes sub-bureaus for the accomplishment of specific tasks.

2.1. THE IERS REFERENCE SYSTEM

The IERS Reference System is composed of the IERS Standards and the IERS reference frames. The IERS Standards (McCarthy, 1992) are a set of constants and models used by IERS Analysis Centers for and by the Central Bureau in the combination of results. The values of the constants are adopted from recent analyses, and, in some cases, differ from the current IAU conventions. The models represent, in general, the state of the art in the field concerned.

The IERS reference frames consist of the IERS Terrestrial Reference Frame (ITRF) and IERS Celestial Reference Frame (ICRF). Both frames are realized through lists of coordinates of fiducial points, terrestrial sites or compact extragalactic radio sources. The origin, the reference directions and the scale of ITRF are implicitly defined by the coordinates adopted for the terrestrial sites. Its origin is located at the center of mass of the Earth with an uncertainty of ±10cm. The unit of length is the meter (SI). The IERS Reference Pole (IRP) and Reference Meridian (IRM) are consistent with the corresponding directions in the BIH Terrestrial System (BTS) within $\pm 0\rlap{.}''003$. The BIH reference pole was adjusted to the Conventional International Origin (CIO) in 1967, and it was then kept stable independently until 1987. The uncertainty of the tie of the BIH reference pole with the CIO was $\pm 0\rlap{.}''003$.

The origin of the ICRF is at the barycenter of the solar system. The direction of the polar axis is that given for epoch J2000.0 by the IAU 1976 Precession and the IAU 1980 Theory of Nutation. The origin of right ascensions is in agreement with that of the FK5 within $\pm 0\rlap{.}''001$.

2.2. THE EARTH ORIENTATION PARAMETERS

The IERS Earth Orientation Parameters (EOP) are the parameters which describe the rotation of the ITRF with respect to the ICRF, in conjunction with the conventional precession/nutation model. They model the unpredictable part of the Earth's rotation. x and y are the coordinates of the Celestial Ephemeris Pole (CEP) relative to the IRP, the IERS Reference Pole. The CEP differs from the instantaneous rotation axis by quasi-diurnal terms with amplitudes under $0\rlap{.}''001$. The x-axis is in the direction of IRM; the y-axis is in the direction 90 degrees West longitude. $d\psi$, $d\epsilon$ are the offsets in longitude and in obliquity of the celestial pole with respect to its position defined by the conventional IAU precession/nutation models. UT1 is related to the Greenwich mean sidereal time (GMST) by a conventional relationship (Aoki *et al.*, 1982); it gives access to the direction of the IRM in the ICRF, reckoned around the CEP axis. It is expressed as the difference UT1-TAI or UT1-UTC.

The precision of the published results depends on the delay of their availability. For the operational solutions of Earth rotation (weekly and monthly bulletins) it is of the order of one millisecond of arc. The prediction accuracy is in the range of $\pm 0\rlap{.}''005$ to $\pm 0\rlap{.}''020$ for x and y, $0\rlap{.}^{s}002$ to $0\rlap{.}^{s}15$ for UT and $0\rlap{.}''002$ for $d\psi$, $d\epsilon$ (prediction lags of 10 and 90 days). For the scientific solution of reference frames and Earth orientation, the inaccuracy is lower than $0\rlap{.}''001$ (3 cm).

Currently, accuracy appears to be limited by systematic errors, mainly in the definition of reference frames and in the observation and reduction procedures. Contributors to the IERS make use of different observing sites and perhaps different philosophies in arranging their observing programs. These differences can lead to systematic variations among the contributors' results. Also, differing models employed in the analyses of the observations by various analysts can lead to differences among the data provided to the IERS. The IERS Standards (McCarthy, 1992) are intended to provide a set of constants and models in order to minimize this problem. However, subtle differences may still lead to systematic effects. Another non-scientific concern is that, in times of dwindling budgets, resources required to operate expensive observational programs may be reduced. This situation may, in turn, cause problems which affect accuracy.

3. Future Situation

In the future we should expect that the current systematic error problems will lead to improved definitions of the reference frames. This will be accomplished in the case of the terrestrial frame by increased emphasis on collocation of observations with multiple techniques at the same locations. The celestial frame will be refined with more numerous southern hemisphere observations and efforts, already planned, for an improved radio reference frame which will be continuously updated. Improved models will also be forthcoming from efforts to understand differences between analyses. For example, a new precession/nutation model will certainly be discussed and adopted in the near future. Other refinements may be expected in modelling of site displacements due to crustal motion, glacial rebound and atmospheric loading.

Possible new concepts may also be called for as a result of advancements in accuracy. The use of the "non-rotating origin" formalism (Guinot, 1979; Capitaine, 1990) may find increased popularity as the demand for more precise definition of reference frames grows. Along with this development, a new definition of the relationship between sidereal time and solar time is likely to be accepted.

The necessity for the increasing frequency of leap seconds may lead to demands for a new definition of the second or at least a change in the relationship between UT1 and uniform time. Currently leap seconds must be inserted in the UTC time scale at the rate of approximately once per year. In the 21st century we expect this frequency to increase to almost twice per year. Users may find the current system too cumbersome for practical use and demands may grow for an alternate procedure for handling the growing difference between rotational and uniform time scales.

With improving accuracy and increased time resolution of Earth orientation parameters, refinements in geophysics will progress. Advancements in models of the Earth's interior structure will occur, and a better understanding of the relationship between the rotation of the Earth and the oceans and atmosphere will be developed. These improvements may then lead to enhancements in our ability to forecast variations in the Earth's orientation.

New observing techniques may also be developed within the next ten years providing improved determinations of the Earth's orientation. Only recently has the use of analyses of the orbits of the satellites of the Global Positioning System (GPS) provided an important improvement in the accuracy and time resolution of our determination of polar motion.

With improvements such as those listed above, it can be expected that accuracy approaching 0.01 millisecond of arc will be achievable within the next ten years.

References

Aoki, S., Guinot, B., Kaplan, G., Kinoshita, H., McCarthy, D., Seidelmann, P. K., 1982, "The New Definition of Universal Time," *Astron. Astrophys.*, **105**, 359-362.

Capitaine, N., 1990, "The Celestial Pole Coordinates," *Celest. Mech. Dyn. Astr.*, **48**, 127-143.

Guinot, B., 1979, "Basic Problems in the Kinematics of the Rotation of the Earth," in *Time and the Earth's Rotation*, D. D. McCarthy and J. D. Pilkington, eds., D. Reidel Publishing Co., 7-18.

IERS, 1994, *IERS Annual Report for 1993*, Observatoire de Paris, Paris.

McCarthy, D. D. (ed.), 1992, IERS Standards (1992), *IERS Technical Note 13*, Observatoire de Paris, Paris.

EARTH ROTATION VELOCITY IN RELATION WITH DIFFERENT REFERENCE FRAMES

V. A. BRUMBERG

Institute of Applied Astronomy,
197042 St.-Petersburg, Russia

Abstract. The high precision of present observations makes it reasonable to clear up a question about GRT (general relativity theory) corrections in the problem of Earth's rotation. The answer is that one may almost forget about GRT corrections when dealing in an adequate reference system (RS). The problem of Earth's rotation may be related to the relativistic hierarchy of RS started in (Brumberg and Kopejkin, 1989) and completed in (Klioner, 1993). Let letters B, G and T be related to barycentric, geocentric and topocentric RS, respectively. Let DRS and KRS be dynamically nonrotating or kinematically nonrotating RS, respectively. From the dynamical equations of rotation it follows that the most adequate system for studying the Earth's rotation is DGRS. Apart from the geophysical factors the rotation of the Earth in this system is fairly well approximated by the rigid-body rotation with some angular velocity $\hat{\omega}^i$. The same rotation of the Earth as considered in BRS and DTRS may be also approximated by the rigid-body rotation but with some additive relativistic corrections and with other angular velocities ω^i and $\tilde{\omega}^i$, respectively. Substituting these three rotation relations into four-dimensional BRS–DGRS and DGRS–DTRS transformations one may express ω^i and $\tilde{\omega}^i$ in terms of $\hat{\omega}^i$ and determine the additive relativistic corrections in BRS and BTRS. These corrections are of importance for treating kinematics problems in various coordinate systems and for obtaining physically meaningful solutions of the dynamical equations of rotation in the barycentric reference system.

The complete text will be published in Journal of Geodynamics.

References

Brumberg V.A. and Kopejkin S.M., 1989, in: Kovalevsky J., Mueller I.I., Kolaczek B. (eds.) Reference Frames. Kluwer, Dordrecht, p. 115
Klioner S.A., 1993, A&A 279, 273

E. Høg and P. K. Seidelmann (eds.),
Astronomical and Astrophysical Objectives of Sub-Milliarcsecond Optical Astrometry, 293.

THE GENERAL RELATIVISTIC POTENTIAL OF ASTROMETRIC STUDIES AT MICROARCSECOND LEVEL

V. I. ZHDANOV
Astronomical Observatory of Kiev University
Observatorna St., 3 Kiev 254053 Ukraine

ABSTRACT. The relativistic effects in positioning of distant objects at microarcsecond level are studied. The main points are: statistics of random variations of the image position of a distant radiation source due to the gravitational field of moving stars, motion of the image due to individual invisible gravitators, and possibilities to obtain information on their masses and velocities. The gravitators shifting the object image may be stars of our Galaxy or of a lensing galaxy.

1. Preliminary remarks

1.1. INTRODUCTION

The gravitational lensing theory contains a number of results concerning the impact of relativistic gravity upon electromagnetic wave propagation, which may be useful to obtain independent astrophysical and cosmological information (Schneider et al. 1992; Refsdal & Surdej 1993). The advances in astrometric techniques for precise positioning of a radiation source image open new possibilities in this direction (Hosokawa et al. 1993; Høg et al. 1994). Following these authors we discuss in this paper the variations of the image position (IP) due to the gravitational field of stars. For typical positions of the source we consider the statistical action of a large number of stars on IP; then we proceed to the cases of the IP location near individual gravitators, that are less probable, but that may produce a more sensible relativistic effect. The gravitators are supposed to be unobservable. This may be of interest when the source and the deflecting star can be observed in different wave bands (radio and optical), and when the deflector is a dark object or a star of another galaxy.

The order of magnitude of the effects discussed would be the same (for comparable values of the impact parameter of the ray from the source with respect to gravitators) in either the case **(a)** of a distant source image microlensed by stars of a foreground lensing galaxy, or **(b)** the case of our Galaxy stars.

We deal with the image displacements at the microarcsecond level; these effects are at present almost unattainable, however their consideration demonstrates the needs for the precision astrometry.

1.2. BASIC RELATIONS AND NOTATIONS

In case (a) our starting point is the relation for the two-dimensional vector of the IP angular shift due to the gravitational field of a deflector (see, e.g., Schneider et al. 1992)

$$\vec{\psi} = -4 k_{ds} m \mathbf{r}_m / r_m^2, \qquad (1)$$

E. Høg and P. K. Seidelmann (eds.),
Astronomical and Astrophysical Objectives of Sub-Milliarcsecond Optical Astrometry, 295–300.

$k_{ds} = D_{ds} / D_s$. Here D_{ds}, D_d, D_s are the angular distances from the deflector to the source, from the observer to the deflector, and from the observer to the source, respectively; $m = GM / c^2$, M is the deflector mass with the position $\mathbf{r}_p$ with respect to the unperturbed ray trajectory, $\mathbf{r}_p$ being orthogonal to this trajectory and directed towards the deflector, $\mathbf{r}_m = r_m \mathbf{r}_p / r_p$, $r_p = |\mathbf{r}_p|$, r_m is related to r_p by the equation:

$$r_m^2 - r_p r_m - R_E^2 = 0 \quad , \tag{2}$$

$R_E = (4mk_{ds}D_d)^{1/2}$ being the Einstein radius of the deflector. Equation (2) has two roots that describe two source images. In case of linear microlensing (that is $r_p >> R_E$) one of these images is faint and, for the main image, one has $\mathbf{r}_m \approx \mathbf{r}_p$. Formula (1) refers to the case when $r_m << D_d$ and $r_m << D_{ds}$, and this is not the case (b) when $k_{ds} \approx 1$ and the contribution of stars far from the ray path may be essential. In the latter case the following relation must be used:

$$\bar{\psi} = -2(m\mathbf{r}_m / r_m^2)(1 + z \cdot (z^2 + r_p^2)^{-1/2}), \qquad z << D_s \tag{3}$$

for the mass at the point $(\mathbf{r}_p, z)$. We choose z-axis in the direction $\mathbf{N}$, where $\mathbf{N}$ is the unit vector pointing to the position of the source (unperturbed IP), the observer being at the origin. Formula (3) is valid also when r_p may be of the same order of magnitude as $z = D_d << D_s$. Motion of gravitators results in nonzero angular velocity and acceleration that can be recovered from (1) and (3).

2. Statistical dragging of distant source image by gravitational field of moving stars

2.1. PROBABILITY DISTRIBUTION FOR ANGULAR VELOCITIES OF THE IMAGE

Here we partially use the results of Zhdanov & Zhdanova (1994) dealing with the angular variations of a distant source IP induced by random gravitators under some restrictions on the gravitator distribution. The relations below generalize these results to rather a general case incorporating both situations (a) and (b). The gravitator distribution is described by the volume density of stars $f_v(\mathbf{r}, z, \mathbf{v}, w, M)$ with the transverse velocity $\mathbf{v}$, longitudinal velocity w and the mass M at the point $(\mathbf{r}, z)$. In the case (a) the dependence upon w and z is not essential.

Under the supposition of linear microlensing the total IP shift is obtained by summing up the contributions of all gravitators. Then, in view of the gravitators' motion, differentiating (1) and (3) over the time we obtain the relation for the average velocity of the image

$$\mathbf{u}_t = -2k\int dM\int dw\int d^2\mathbf{v}\int dz\int d^2\mathbf{r}\; m\; f_v(\mathbf{r},z,\mathbf{v},w,M)\cdot$$
$$\cdot\Big\{r^{-2}[\mathbf{v}-2\mathbf{r}(\mathbf{r}\cdot\mathbf{v})/r^2]\cdot[1+z(z^2+r^2)^{-1/2}+$$
$$+(zr^2/2)\cdot(z^2+r^2)^{-3/2}]+(w\mathbf{r}-z\mathbf{v}/2)\cdot(z^2+r^2)^{-3/2}\Big\},$$

which can be simplified in the case (a) ($k = k_{ds}$) by assuming $r << z$ and recalculating the result in terms of the surface density of stars in the gravitator plane $z = D_d$. In the case (b) we assume $k = 1$; here the dependencies upon longitudinal velocity components and upon z are significant. The singular points in the integral do not lead to any troubles.

We obtained the probability distribution of angular image velocities describing the action of a large number of gravitators. The Markov method has been used to obtain the characteristic function of this distribution. For a typical range of its independent variable, $\mathbf{y}$, it is approximated by the expression

$$\exp[-u_0 y + i(\mathbf{u}_t\mathbf{y})]\;,\quad \text{where}\quad y = |\mathbf{y}|,\quad u_0 = 4\pi k\int dM\int d^2\mathbf{v}\; m\cdot|\mathbf{v}|\cdot f_s(\mathbf{v},M)\;,$$

$$f_s(\mathbf{v},M) = \int_0^{D_s} dz\int_{-\infty}^{\infty} dw\; f_v(0,z,\mathbf{v},w,M).$$

Therefore, for a realistic f_v we obtain the velocity distribution valid for both cases (a) and (b):

$$\mathbf{P}(\mathbf{u}) = \frac{u_0}{2\pi}[u_0^2+(\mathbf{u}-\mathbf{u}_t)^2]^{-3/2} \tag{4}$$

As it can be seen from the relations, in the case of a nonuniform gravitator distribution and a nonzero bulk velocity, there is an average dragging of the image with the velocity $\mathbf{u}_t$, depending upon the input of stars far from the ray. The other parameter u_0 is determined by the masses in the close vicinity of the ray path; its order of magnitude for a gravitator volume density 0.1pc^{-3} along the ray path of 10kpc may be estimated as:

$u_0 \approx 1''\cdot 10^{-7}(10\text{yr})^{-1} kM_{ch}v_{ch}/(M_\odot\cdot 100\text{km}/\text{s})$,
that involves the characteristic values of the gravitator mass and velocity. It is rather small, but distribution (4) decreases slowly, and there is, e.g., 1% of events with $u \geq 100u_0$. The parameter $\mathbf{u}_t$ is roughly of the same order of magnitude, but it depends on the form of the star density distribution. Then, as it can be seen from (4), the presence of bulk gravitator motions amplifies the probability of dragging with the characteristic values of u, but it is not important for distribution at large u.

For (b), the Galaxy rotation produces an anisotropy in the velocity distribution of the remote radiation sources close to the Galaxy plane. The dragging due to the Galaxy stars seems to play a negative role. Indeed, the above characteristic values appear to be comparable to the estimates of proper angular velocities of quasars and, provided that the corresponding accuracy can be achieved, it would be difficult to separate them in observations. However, for (a), the effect might be interesting in precise positioning of different images of the same quasar, when the characteristic parameters may be increased by one or two orders of magnitude because of a larger bulk velocity of the lensing galaxy motion and a more dense region near the ray path. This is the case of some gravitational lenses, e.g., the "Einstein Cross" 2237+0305. On the other hand, the proper motion of the source may be ruled out in relative angular velocity measurements for different source images.

2.2. PROBABILITY DISTRIBUTION FOR ACCELERATIONS

The motion of foreground stars causes the angular acceleration $\mathbf{a}$ of the distant source image. To obtain the distribution $\mathbf{P}_A(\mathbf{a})$, one may neglect the accelerations of the stars. As distinct from (4), the acceleration distribution turns out to be determined only by the stars near the ray path and relation (1) may be used from the very beginning. The characteristic function of the two-dimensional variable $\mathbf{y}$ is $\exp[-(a_0|\mathbf{y}|)^{2/3}]$, where

$$a_0 = k\{6\sqrt{\pi}\,\Gamma(5/6)\int dM \int d^2\mathbf{v}\, m^{2/3}|\mathbf{v}|^{4/3} f_s(\mathbf{v},M)\}^{3/2}.$$

This yields the distribution written here in the form suitable for the estimates for large values of $a = |\mathbf{a}|$:

$$\mathbf{P}_a(\mathbf{a}) = \frac{1}{3\pi}\sum_{n=1}^{\infty}\sin(\pi n/3)\cdot[n\Gamma(n/3)]^2\frac{(-1)^{n+1}(2a_0)^{2n/3}}{n!\,a^{2n/3+2}} \quad . \tag{5}$$

The numerical estimate, for the same values of the volume density and the ray path as in sect. 2.1, is given by the formula

$$a_0 \approx 1''\cdot 10^{-8}(10yr)^{-2}\{kM_{ch}v_{ch}^2/(M_\odot\cdot(100km/s)^2)\}.$$

As before, this value could be more significant in the case (a). From (5) we infer that there is about 3% of events with $a \geq 100a_0$. Note that in case (b) of a differential measurement with two distant sources, these must be separated by an angular distance exceeding the characteristic one between the Galaxy stars to avoid correlations. If the IP variation is measured with respect to some reference Galaxy star close to IP, then the integrations over Z in the above expressions must be performed from the positions of this star.

3. Gravitational image motion due to an individual deflector, and its mass, distance and velocity

The observations of the relative motion of the source image with respect to the deflecting star due to the gravitational parallactic effect would allow one to measure the star mass and distance (Hosokawa et.al 1993). These authors pointed out that this effect must be studied jointly with the image motion owing to the motion of a gravitator considered by Chollet (1979) and Kovalevsky et al. (1986). Here we present a simple argument that observation of both effects could yield the values of the mass and the velocity of the invisible gravitator (linear microlensing, $\mathbf{r}_m \approx \mathbf{r}_p$). A more detailed consideration of the question involves evaluations by the nonlinear least square method. We restrict our consideration to show that the above parameters may be determined in principle. The way used below is not the best one for the real data treatment, but it makes evident the formal solvability of the problem.

Here the source is assumed to be at rest. The observer's shift by the vector $\mathbf{p}$ in a transverse plane orthogonal to $\mathbf{N}$ gives rise to the corresponding change in the relative position of the gravitator with respect to the unperturbed ray trajectory $\mathbf{r}_p \to \mathbf{r}_p + k^*_{ds}\mathbf{p}$, the coefficient k^*_{ds} depending upon background space-time (in flat space-time $k^*_{ds} = k_{ds}$). For the gravitator velocity, $\mathbf{v}_g$, in the plane orthogonal to $\mathbf{N}$, one can write $\mathbf{r}_p = \mathbf{r}_{\rm ort} + (\mathbf{v}_g + k^*_{ds}\mathbf{v}_S)(t - t_m) + k^*_{ds}\mathbf{p}_{\rm orb}$, where we have separated the linear motion due to the velocity of the Sun, $\mathbf{v}_S$, and the periodic orbital motion of the observer; the vector $\mathbf{r}_{\rm ort}$ is orthogonal to $\mathbf{v}_g + k^*_{ds}\mathbf{v}_S$ and corresponds to the minimum distance from the ray to the gravitator at an epoch t_m in absence of the orbital motion. Under the above condition of the linear microlensing, the straight line motions of the gravitator and of the observer induce the apparent motion of the image exactly along a circle. The actual source position is represented by the limiting point, $\mathbf{S}_0$, on the circle that corresponds to an infinite position of the gravitator. The gravitational shift of the IP from the unperturbed position $\mathbf{S}_0$ is given by the angle $\vec{\psi}$ from (1), and the epoch t_m is determined at the maximum value of this angle $\psi_{\rm max} = 4m / r_{\rm ort}$.

Using (1) we have

$$\vec{\psi} / |\vec{\psi}|^2 = [\mathbf{r}_{\rm ort} + (\mathbf{v}_g + k^*_{ds}\mathbf{v}_S)(t - t_m) + k^*_{ds}\mathbf{p}_{\rm orb}] / (4mk_{ds}).$$

Separating the linear terms from the periodic ones (with known $\mathbf{p}_{\rm orb}$) one obtains the combination $k^*_{ds} / (k_{ds}M)$ and, for $k^*_{ds} = k_{ds}$, the inverse mass. For an infinite source we have $k^*_{ds} = k_{ds} = 1$ and from the above equation we find $\mathbf{v}_g + k^*_{ds}\mathbf{v}_S$ and $\mathbf{r}_{\rm ort}$ which defines the IP circle diameter. For known $k^*_{ds}\mathbf{v}_S$ this allows us to find the transversal velocity of the gravitator. In the example by Hosokawa et al. the condition of linear microlensing is satisfied, $\psi_{\rm max} \approx 0.7$milliarcsec, and the gravitational IP shift due to the observer's annual motion is estimated to be about 10 microarcsec.

If r_p and R_E are of the same order of magnitude, the following points should be mentioned: (i) due to microlensing there is a high amplification of the source images, which yields an additional source of information; (ii) the Einstein radius R_E appears in photometric and astrometric effects. This leads to the results of Høg et al. (1994), who have shown that combination of photometric and astrometric measurements during microlensing of a star in the Large Magellanic Cloud (LMC) by a massive compact halo object is sufficient to determine the mass, proper motion and distance of this gravitator, provided that the position, distance and relative motion of the source with respect to the observer are known. The probability of this event is very small, but the examination of millions of stars in the LMC, proposed by Paczinsky (1986), makes these observations possible, and such events have already been detected (Alcock, 1993; Aubourg, 1993). The order of magnitude of ψ_{max} during the event is milliarcsecond.

The brightness of two source images during microlensing is comparable to each other and the interpretation of observations depends on whether it is possible to resolve them. In the case of a positive answer, there is one more opportunity to obtain information: (iii) one may hope to estimate the angular separation of the images $\Delta\psi = (r_p^2 + R_E^2)^{1/2} / D_d$, which (in this case) is roughly of the same order of magnitude as ψ_{max}.

There is a considerable probability of microlensing events in gravitational lenses (Schneider et al. 1992; Refsdal & Surdej, 1994). The angular separation of the images due to the gravitational field of an isolated star and the corresponding ψ_{max} are of the order of

$$R_E / D_d \approx 1'' \cdot 10^{-5} \cdot [k_{ds} \cdot M \cdot 10^8 \mathrm{pc} / (M_\odot D_d)]^{1/2} .$$

This value may be enhanced by gravitational field of the lensing galaxy.

References

Alcock, C., Akerloff, C.W., Allsman, R.A., et al., 1993, Nature 365, 621
Aubourg, E., Bareyre, P., Brehin, S., et al., 1993, Nature 365, 623
Chollet, F., 1979, C.R. Acad. Sci. Paris, Ser.B 288, 163
Høg, E., Novikov, I.D., Polnarev, A.G., 1994, MACHO photometry and astrometry. NORDITA prepr. 94/26 A
Hosokawa, M., Ohnishi, K., Fukushima, T., Takeuti, M., 1993, A&A 278, L27
Katz, N., Balbus, S., Paczinsky, B., 1986, ApJ 306, 2
Kovalevsky, J., Mignard, F., Froschle, M. 1979. Proc. IAU Symp.114, Eds.: J.Kovalevsky and V.A.Brumberg, p.369.
Paczinsky, B. 1986, ApJ 304, 1
Refsdal, S., Surdej J., 1994, Repts. Progr. Phys. 56, 117.
Schneider, P., Ehlers, J., Falco, E.E., 1992, Gravitational Lensing. Springer, New York
Zhdanov, V.I., Zhdanova, V.V., 1994, Analytical relations for time-dependent statistical microlensing. Submitted to A&A.

HIGH ACCURACY ASTROMETRY BENEFITS FOR CELESTIAL MECHANICS

P. K. SEIDELMANN
U. S. Naval Observatory
Washington, D.C.

Abstract.

Achieving angular accuracies at milli- and microarcsecond levels for objects within the solar system presents real challenges concerning the meaning of the observations due to the effects of phase and due to the sizes of the objects being observed. The use of retroreflectors or transmitters located on solar system objects may be the most promising technique.

However, if these accurate observations can be achieved, this provides the opportunity for significant improvements in the orbits of the planets, satellites, and minor planets, and will provide, or require, improvements in the models for the tidal interactions, secular accelerations, and masses. The observations will provide a test, or improvement, of the ties between radar, optical, and radio observations. This should provide an opportunity for improving the reductions of pulsar timing observations and improve our determination of the solar system reference system. This will also test the accuracy of the equivalence between Atomic Time and Dynamical Time. These accuracy levels should permit the discovery of other solar systems and the resulting increase in our knowledge, based on more than a single solar system.

Improvements in the accuracy level of observations provide increases in our knowledge and raises new challenges to explain the signatures detected by the observations.

1. Introduction

For solar system objects, there is a continuing interaction between observational accuracy and theoretical improvements. In principle, the method of computing the orbit, or the theoretical orbit computations, should be an

E. Høg and P. K. Seidelmann (eds.),
Astronomical and Astrophysical Objectives of Sub-Milliarcsecond Optical Astrometry, 301–304.

order of magnitude more accurate than the observational data. Once the observations become more accurate, it requires improved methods of orbit computation or it reveals the errors in the previous method. So, there is a continuing cycle of observational improvement, computational improvement, observational improvement, etc. Thus, as we contemplate observational improvement, we can anticipate the improvement in theory and the possibilities of new effects which must be included in our models of the solar system.

2. Observational Challenge

The benefits to celestial mechanics of milli- to microarcsecond angular astrometry will come primarily from observations of solar system bodies to that accuracy level. However, achieving optical angular accuracies will be a challenge. The solar system bodies present a very different image than a stellar point source. The solar system bodies are extended sources. They have phase effects. Thus the Navy Prototype Optical Interferometer has the problem that, if the objects are bright enough, then they are also large enough that the interferometer will resolve the body. Larger siderostats would permit unresolved observations of small minor planets.

The highly accurate observations of the solar system today are not made by angular measures. The high accurate observations are lunar laser ranging, radar ranging, radar illumination and VLBI measurement, millisecond pulsar recordings, spacecraft, and mutual events; specifically occultations, eclipses, or objects observed when they are very close to each other, being equivalent to a mutual event. Since one milliarcsecond at one AU is 0.7 km and one microarcsecond at one AU is 0.7 m, radar to inner planets is at mas level, some spacecraft are at 10 μas and lunar laser ranging is at 50 μas level for some parameters.

What are the future possibilities for accurate observations? The ones that come to mind are putting a transponder on a solar system body, or placing a small emitting object in orbit around the planet. Alternatively, can we come up with a new technique? Accurate observations of selected asteroids when passing near other asteroids would provide a means of determining the mass of the asteroid. This is essential to improving the model of the solar system for computations of improved ephemerides. This would, with the diameter, provide density and composition information about the asteroid.

3. Theoretical Challenges

As mentioned, improvements in observations require improvements in orbit computational methods. Some of the limitations on the theoretical methods are recognized and known. The current computation of the ephemerides does not include a complete model of the asteroid belt, or alternatively, all the individual asteroids. Likewise, the Kuiper belt and the Oort cloud are not included in the present model. Galactic effects are not included in the computed orbits as a direct effect. The tidal interactions between the extended bodies are not included, except in the Earth/Moon case. Improved accuracies will require higher order relativistic terms to be included in the equations of motion, or post-post-Newtonian approximations.

4. Observational Benefits

What are the benefits of more accurate observations? It should be stated here that individual, more accurate, observations will not be of any real benefit to the solar system. To achieve benefits it is necessary to have a continuing supply of more accurate observations in a systematic manner. With this type of observation, more accurate orbits can be determined, and in turn, more accurate astronomical constants. Assuming these observations are of different types, then more accurate ties between radio, radar, or optical frames can be determined. The secular accelerations can be determined, in addition to those for the Moon, Phobos, and Deimos. Improved values of the masses of satellites and asteroids can be determined. Improved orbits of the asteroids permit better determination of the families of the asteroids. With the improved ephemerides of the solar system, the millisecond pulsar observations can be more accurately reduced and the potential for a long-term time scale is improved. Also, the improved orbit determinations provide a capability of determining the dynamical time scale more accurately, and to test the equivalence between the atomic and dynamical time scales.

In time, these more accurate orbits may permit us to determine the changes taking place in the solar system, and the status of the evolution of the solar system. However, the capabilities in trying to trace back the solar system should be limited due to chaos.

More accurate observations will provide the astrometric observations of motions of stars, for studies of planets around other stars. This should provide information about other solar systems besides our own, which in turn should lead to knowledge about how solar systems are formed and the characteristics of other solar systems.

Finally, as we achieve more accurate observations, this should lead to the discovery of other phenomena. Systematic differences between the observations and the computed ephemerides have in the past, and should, in the future, indicate the causes of these discrepancies, and new phenomena.

5. Conclusion

Improved observational accuracies for solar system bodies will be much more difficult than for stars. However, if the improvements can be achieved, they will require improvements in the theoretical methods for computing orbits. This, in turn, provides benefits in improved accuracy and new knowledge based on the improvement in accuracy of both the observations and the computations.

GRAVITATIONAL LENSING BY STARS AND MACHOS AND THE ORBITAL MOTION OF THE EARTH

M. Hosokawa[1], K. Ohnishi[2], T. Fukushima[3] and M. Takeuti[4],
[1] *Communications Research Laboratory, Koganei, Tokyo 184, Japan. e-mail hosokawa@crl.go.jp*
[2] *Kansai Advanced Research Center, Communications Research Laboratory, Kobe 651-24, Japan. e-mail ohnishi@crl.go.jp*
[3] *National Astronomical Observatory, Mitaka, Tokyo 181, Japan. e-mail toshio@spacetime.mtk.nao.ac.jp*
[4] *Astronomical Institute, Tohoku University, Aoba-ku,Sendai,980 Japan. e-mail i4a0s4d@JPNTOHOK.ecip.tohok.ac.jp*

ABSTRACT. We showed that it is feasible to measure the mass of a single star by observing the variation of gravitational deflection caused by the orbital motion of the Earth. When the distance of a star is less than 60 pc and some appropriate sources are within 1 arcsec. in its background, not only the distance but also the mass of the star may be determined by measuring the deflection with an accuracy of 10 μ arcsec. In the case of photometric microlensing by a MACHO, the observation of astrometric gravitational deflection is also useful. By measuring the separation between the primary image and the secondary image, the ratio of mass to distance of the MACHO will be obtained. Further, the orbital motion of the Earth modifying the light curve of the source is discussed.

1. Introduction

Measurement of a stellar mass, especially that of a single star, is very difficult. In principle, gravitational light deflection by a star depends on its mass. Therefore this effect can be used to measure the mass of the star. Unfortunately, the deflection angle itself cannot be measured unless the configuration of the source, deflector and observer changes. That makes it difficult to apply this effect.

On the other hand, much attention has been paid to another aspect of the gravitational lensing, that is called the microlensing. The possibility of microlensing by galactic dark halo objects was discussed by Paczynski and Griest (Paczynski,1987; Griest 1991). Recently, many candidate events of this phenomena caused by MACHOs are found (Alcock et al. 1993; Aubourg et al. 1993; Udalski et al. 1993). In spite of these successful observations, this method has some limitations. The alignment must be so well that the event rate of this photometric phenomenon is very small, the order of a millionth per year per star. For the determination of the mass of the lensing matter, some assumptions are needed. Also it is impossible to know which star will be lensed.

Now, the direct observation of the variation of the gravitational deflection is becoming feasible thanks to the development of high accuracy astrometric observation methods. In the near future, an accuracy of 10 μas is expected to be achieved by some VLBI and optical interferometer (Lestrade et al. 1992; Sasao et al. 1993; Shao and Colavita 1992). The theory and an application of this effect is discussed in Section 2 (Hosokawa et al. 1993). The gravitational deflection will also be detectable in the case of microlensing phenomenon. This issue is discussed in section 3.

2. Parallactic Variation of Gravitational Deflection

E. Høg and P. K. Seidelmann (eds.),
Astronomical and Astrophysical Objectives of Sub-Milliarcsecond Optical Astrometry, 305–308.

2.1. THEORY

Consider that a light emitted from a point source S was deflected by the gravitational field of a foreground star P and reached to the Observer O. The distance between O and P, and that between O and S is denoted by D and Ds, respectively. Here we consider the case that the separation angle between P and S is about the order of one arcsecond, and the source S is sufficiently far from the deflector P. Configuration of the S, P and O is shown in Fig.1. The angle β is the true separation between P and S in the absence of gravitational lensing (the angular separation in the source plane) while θ is the apparent separation (that in the lens plane). The relation between β and θ is expressed as

$$\theta = \beta + \frac{4GM}{c^2 D\theta} , \tag{1}$$

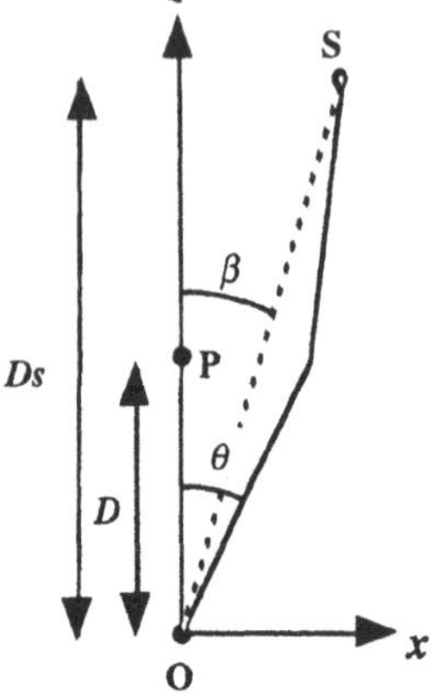

Fig. 1. Configuration of the gravitational lensing

where G is the gravitational constant, c is the speed of light and M is the mass of the deflector P (Schneider et al. 1992).

Next, we consider the case that the observer shifts its position by $\boldsymbol{r}$ (x, y, z). According to this shift, the position of the source relative to the deflector will change. Let us denote the change of the source's position in the source plane by $\Delta\beta$ and that in the lens plane by $\Delta\theta$, respectively. In the case $|r| << D\theta$, these quantities are obtained as follows by expanding the deflection angle in the first order.

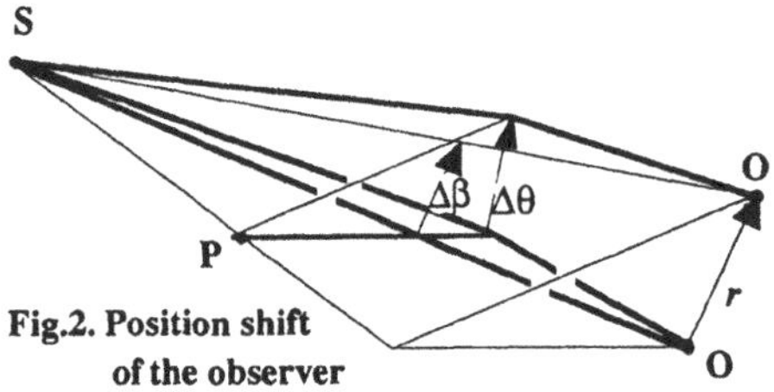

Fig.2. Position shift of the observer

$$\Delta\theta x = \Delta\beta x \left(1 - \frac{4GM}{c^2 D\theta^2} \right), \quad \Delta\theta y = \Delta\beta y \left(1 + \frac{4GM}{c^2 D\theta^2} \right). \tag{2}$$

Here, the position shift along the line of sight z will make no change in the angular position in this order. Note that the angular shift along the source direction is suppressed by the change of the impact parameter while that across the source direction is enhanced by β/θ owing to the deflection itself. So they are not parallel in general. Using this effect, we can determine the mass and the distance of the deflector simultaneously. If we take the mass of the deflector M as 10 M_{sun}, θ as 1 arcsecond, D as 100 pc and $|\boldsymbol{r}|$ as the order of one AU, then $\Delta\beta_x \sim \Delta\beta_y \sim 10$ mas and $4GM/c^2D\theta^2$ becomes the order of 1/1000. Hence, for a star with $D \sim 100$ pc and $M \sim 10$ M_{sun}, this effect will amount to 10 µas, the same value as the expected accuracy mentioned before. By solving Eqs (2), we obtain the following equations.

$$M = \frac{c^2\theta^2}{8G} \left(\frac{x}{\Delta\theta x} - \frac{y}{\Delta\theta y} \right), \quad D = \frac{1}{2} \left(\frac{x}{\Delta\theta x} + \frac{y}{\Delta\theta y} \right). \tag{3}$$

Then, we will discuss on two applications of these formulae to determine the mass of a star.

In the case that a pair of sources can be found close enough to the deflector, the change of the separation angle between two sources caused by the position shift of the observer is proportional to the mass of the deflector. Therefore in this case, we need not decompose the angular shift of the separation into two components. Let two sources S_1 and S_2 be close to a deflector P.

2.2. VARIATION DUE TO ANNUAL PARALLAX

In order to detect this variation of gravitational deflection, it can be considered to make use of the orbital motion of the Earth fqr the position shift of the observer. In this case, the parallactic ellipse of the deflector will suffer a shear from the gravitational deflection. According to the convention used in the present section, we choose the position of the deflector as a origin and express the parallactic ellipse as the trajectory of the motion of the source relative to the deflector. The case that the deflector P and the source S is in the same ecliptic latitude is illustrated in Fig. 3, where Π is the annual parallax of the deflector and $\varepsilon= 4GM/c^2D\theta^2$. Hence, if we assume an accuracy to be better than 10 μas, for a stars of $M > 10$ M_{sun} and $D < 100$ pc, or $M > 5$ M_{sun} and $D < 60$ pc, and that we have some appropriate sources within 1 arc second in its background, its distance and mass will be determined by detecting the shear of the parallactic ellipse caused by the gravitational deflection. This may be the most practical application of this effect.

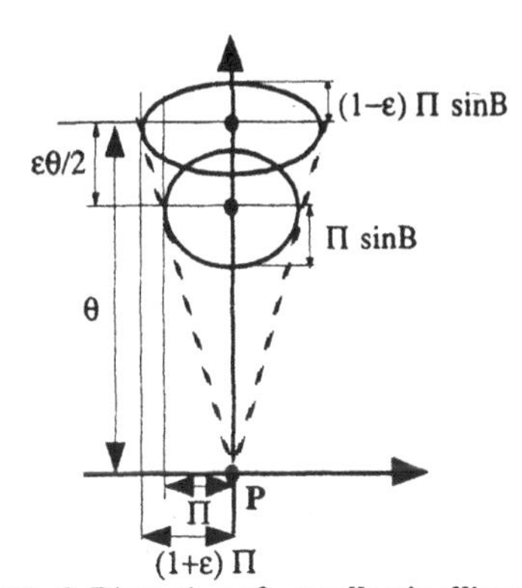

Fig.3. Distortion of parallactic ellipse

3. Microlensing and Gravitational Deflection

In the case of microlensing by the MACHOs, there are four unknown parameters. These are the mass M and the distance D of the MACHO, the relative proper motion between the background source and the MACHO μ, and the smallest impact parameter r_m. The source is assumed to be sufficiently far from the lensing matter, as in sec. 2. These quantities can be rewritten as the set D, Θ_0, A_{max} and t_e, where Θ_0 is the angular radius of Einstein Ring, A_{max} is the largest amplification of the luminosity and t_e is the characteristic time scale of the event. By observing the light curve of the microlensing event, however, we can obtain only two parameters; A_{max} and t_e. In order to obtain the rest of these parameters, the observation of the gravitational deflection will play an important role.

In the microlensing, the formulae on the luminosity amplifications and the deflection angles of the primary image and the secondary image of the source are well known (Paczynski,1987). As an example, the trajectories of the primary image and the secondary image in the case that the least impact parameter r_m is 0.2 $D\Theta_0$ is illustrated in Fig.4 (Hosokawa, et al. 1994; Høg, et al. 1994). Note that, if we take M as 0.1 M_{sun} and D as 10 kpc, we obtain the value of Θ_0 as 0.3 mas. Then, the separation between the primary image and the secondary image of the source will be larger than 0.6 mas even at the epoch of the closest approach. Though the luminosity of the source is very faint because of its distance, such as 20 mag. in the case of the stars in Large Magellanic Cloud, this angle itself is separable enough in the resolution of recent optical interferometers. As for the luminosity enhancement, that of the secondary image is amplified twice of the original one in the epoch of the closest approach (Fig.5). Though this amplification will

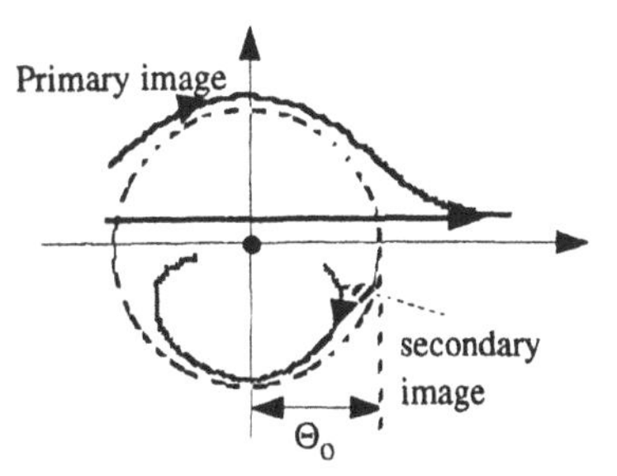

Fig.4. Trajectories of primary image and secondary image in microlensing

make it easier to observe these two images, it may be difficult for the interferometers of today and near future to observe a binary of the 19 mag. with the resolution of 0.1 mas. However, the observation facilities of the interferometer is being advanced rapidly (Shao, M., Colavita, M.M., 1992). If a powerful observation facility that enable us to resolve these images are completed, and the real time analysis of the photometric microlensing search, that could tell us which stars the candidates are, is realized in the future, then the separation between the primary image and the secondary image of the source may be measured. To resolve these two images has been considered by Gould (1992), who proposed that this high resolution observation could be applied to confirm the microlensing events caused by the heavy MACHOs (10^3 M_{sun}> M >10^3 M_{sun}). Now, many candidate events have been observed and the masses of the MACHOs are thought to be, under some assumptions, much lighter than in that range. So we should seek for the way to apply this high resolution observation to the MACHOs in this light mass range.

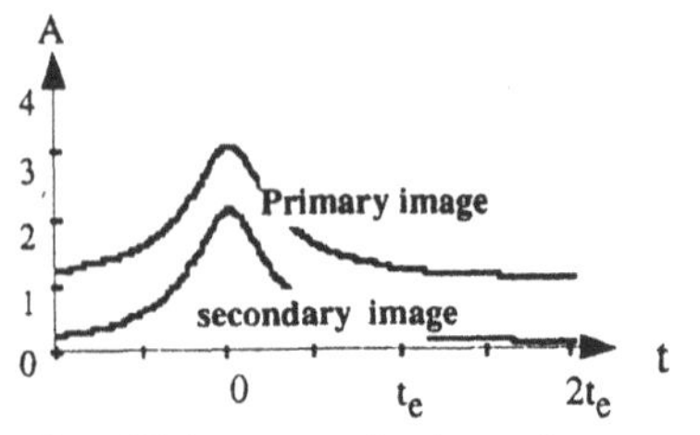

Fig.5. Light curves of primary image and secondary image in microlensing

Further, it should be noted that the orbital motion of the Earth modifies the light curve of the source. In this modification, the acceleration of the Earth is more useful than the variation of its position. In a typical duration of the events, about 30 days, the variation of the Earth's velocity amounts to 15 km/s . This effect causes not only the seasonal variation of event rate (Griest 1991) but also the asymmetry of the light curve. In the case that the acceleration of the Earth is almost parallel or anti-parallel to the proper motion of the MACHO relative to the source and that the event duration is short enough to regard the acceleration as almost constant during that period, this modification becomes maximum and is easily obtained. For example, the epoch of maximum brightness shifts 12 hours maximally from the center of profile when the event duration is 30 days and the transverse velocity of MACHO is 100 km/sec.

In conclusion, to detect the variation of gravitational light deflection is within reach. This can be used for the mass measurement of the stars and MACHOs. In the latter, to resolve the primary image and the secondary image is needed. Therefore it might be more difficult than the measuring the position shift as in the former. In both cases, the orbital motion of the Earth will be useful for making detectable variations in observable quantities.

References

Gould, A.,1992, ApJ, 392, 442.
Griest, K., 1991, ApJ, 366, 413.
Høg, E., Novikov,I.D., and Polnarev,A.G., 1994, A&A, to be published.
Hosokawa, M., Ohnishi, K., Fukushima, T. and Takeuti, M. , 1993, A&A 278, L27.
Hosokawa, M., Ohnishi, K., Fukushima, T. and Takeuti, M. , 1994, Proceedings of the Third Workshop on General Relativity and Gravitation, Tokyo, p.307
Lestrade, J., Phillips, R.B., Preston,R.A., Gabuzda, D.C., 1992,A&A 258,112.
Paczynski, B., 1987, ApJ, 304, 1.
Sasao, T., Kawano.N.,Hara.T., et al.,1993, Proceedings of iRiS '93, CRL, TOKYO, p.91.
Schneider, P., Ehlers, J., Falco, E.E., 1992, Gravitational Lenses, Springer - Verlag.
Shao, M., Colavita, M.M., 1992, A&A 262, 353.
Udalski, A., et al., 1993, ActaAstr. 43, 289.

IMPROVEMENTS TO REFERENCE SYSTEMS AND THEIR RELATIONSHIPS

P. K. SEIDELMANN
U.S. Naval Observatory
Washington, D.C.

Abstract.
Reference Systems include the reference frames and their relationships, time arguments, ephemerides, and the standard constants and algorithms.

The extragalactic, or radio, reference frame will be the basic frame. Achieving milli to microarcsecond accuracies at optical wavelengths will reduce the disparity between optical, radar, and radio reference frame determinations. Thus, the relationships and identifications of common sources should be much more accurate. Another significant change should be the ability to determine distances, and thus space motions on a three-dimensional basis, rather than the current two-dimensional basis of proper motions.

Improvements in ephemerides provide the opportunity to investigate the difference between atomic and dynamical time, the relationship between the dynamical and extragalactic reference frame and the values of precession and nutation.

Also, the relationships between the bright and faint optical catalogs, the infrared, and extragalactic reference frames should be better determined. Reference frames at other wavelengths will become determinable.

1. Introduction

We are on the threshold of the introduction of an extragalactic-based reference frame and a reference system consistent with the general theory of relativity. At the same time, new observational data are becoming available, and new techniques for observation are being considered. Thus, while in the past, arcseconds were the units for astrometry, now the milliarcsecond is

E. Høg and P. K. Seidelmann (eds.),
Astronomical and Astrophysical Objectives of Sub-Milliarcsecond Optical Astrometry, 309–314.

the standard unit, and the microarcsecond is being used. The technologies of interferometers and charge coupled devices (CCD), combined with space-based observing programs, have, and will, significantly change astrometry in the future. Infrared astrometry is in its infancy. Briefly, this paper will touch on all aspects of reference systems; reference frames, time, ephemerides, and the constants and algorithms that go with them.

2. Radio Reference Frame

The extragalactic reference frame being introduced has approximately 400 sources based on VLBI observations, achieving an accuracy of approximately a milliarcsecond. There are sufficient data available now to recognize that some of these sources may show apparent motions in the range of 25 microarcseconds/year which are mostly structural changes. Improvements in positional accuracy can be anticipated on the basis of space-based antennae combined with ground-based VLBI data and continued VLBI observations. In turn, we can expect that some sources will be resolved, or show motion, such that the defining sources will have to be revised.

3. Optical Reference Frame

The optical reference frame situation is indicated in Table 1. The FK5 is the fundamental catalog, today, with the Astrographic Catalog Reference Stars (ACRS) or the Positions and Proper Motions (PPM) catalogs providing supplementary reference catalogs. The Hubble Space Telescope Guide Star Catalog provides a source of positional information down to 17th magnitude.

A dramatic change in accuracies will be achieved when the Hipparcos and Tycho catalogs become available. The Navy Prototype Optical Interferometer (NPOI) is about to begin observations, and will produce significantly improved ground-based observations, providing a continuation of Hipparcos accuracy with improved proper motions. The Washington Fundamental Catalog, which is in preparation, will provide a final fundamental catalog.

There will be several faint star catalogs in the future. The Precise Measuring Microdensitometer (PMM) of the U.S. Naval Observatory will be measuring Palomar Sky Survey plates from both the first and second epochs, so they will provide a catalog of positions and proper motions for approximately 5 million stars in the 14-20 magnitude range. The Sloan Digital Sky Survey (SDSS) is in the design stage and should begin observing in 1996. This will cover the northern galactic area and survey both stars and galaxies down to 22nd magnitude. The astrometry is expected to provide 50 milliarcsecond accuracies for stars in the range of 8 to 22nd magnitude.

Table 1. Optical Reference Frame

	Current			Future		
mag	No. stars	Cat	Accuracy mas/mas yr^{-1}	No. stars	Cat	Accuracy mas/mas yr^{-1}
< 9	5000	FK5	20/2	1500	NPOI	3/1
				30,000	W.F.C.	20/1
< 11	300K	ACRS or PPM	80/4	100K	Hipparcos	2/2
				1M	TYCHO	30/5
7-16	15M	Guide Star	500-2000			
14-20				5M	PMM	150/7
8-22				100M	+SDSS	50
10-20					*Roemer+	0.01–0.1
< 15					*Newcomb	0.1
< 16					*GAIA	0.020
< 18					*Points	0.005

+ In development
* Concepts

There are several future space astrometric satellites in the planning stage. These include the Roemer Plus satellite, conceived by Dr. Høg, which would observe stars in the 10 to 20 magnitude range at the 100 microarcsecond accuracy level. There is the Global Astrometric Interferometer for Astrophysics (GAIA) which is designed to provide 20 microarcsecond accuracies for limiting magnitudes of 15 or 16. The proposed Points Optical Interferometer is designed to reach 5 microarcseconds with a limiting magnitude of approximately 18. A scaled-down version of Points to meet the current philosophy of "small, quick, cheap," is called the Newcomb Astrometric Satellite. This would reach 100 microarcsecond accuracy with a magnitude limit of approximately 15.

The added feature that goes with these improved accuracies is the distance scale measurements. So not only are the angular positions of the objects improved, distances are also improved, and a three-dimensional reference frame developed. In addition, the infrared reference frame is needed. For this purpose the USNO Catalog of Positions of Infrared Stellar Sources (CPIRSS), has been prepared by correlating the Infrared Astronomy Satellite Point Source Catalog as observed by (IRAS) with astrometric star catalogs. This catalog has about 34,000 stars with an accuracy of about 200 milliarcseconds and flux information for 12, 25, 60 and 100 microns from IRAS and an estimated value of the flux at 2.2μ. There is a planned observational effort called the 2MASS project, which plans to measure at the 2 micron wavelength two million stars with an accuracy of 2 arcseconds. This catalog can be matched with optical catalogs to achieve improved accuracies.

4. The Dynamical Reference Frame

Currently, the dynamical reference frame is defined by DE200/LE200, and the IAU 1976 Astronomical Constants. The uncertainties in this reference frame can be specified by the precession constant, with an uncertainty of 3 milliarcseconds per year, theory of nutation, which is uncertain at the 2 milliarcsecond level, and the equinox, with a 4 milliarcsecond uncertainty. With the adoption of the extragalactic reference frame as the principal reference frame, the question could be raised "Do we need a dynamical reference frame at all?" Observations can be made in terms of the extragalactic reference frame. However, it appears that we will still be observing with respect to the equator of the Earth, and affected by its precession and nutation. The remaining question, then, is whether to use the equinox, or a fiducial point, such as that called the non-rotating origin, or an arbitrary origin.

The solar system presents observational challenges. Lunar laser ranging, radar, and VLBI provide observational techniques at milliarcsecond accuracies. The millisecond pulsars provide significant tests of the ephemerides. The question is "can these observations be used to improve the ephemerides?" The extended objects with phase effects and defective illumination effects present a challenge to obtain milliarcsecond accuracy positional observations. Observing the satellites of the outer planets may be much more accurate than observing the primaries themselves. Alternatively, is there some new technique which can be used to obtain significantly more accurate observations of the bodies?

5. Terrestrial Reference Frame

The terrestrial reference frame is defined by Earth orientation data, determined from VLBI, lunar laser ranging, satellite laser ranging and Global Positioning System (GPS). There are many gravity models for general or specific purposes. The combination of GPS and GLONASS is significantly improving the relative terrestrial reference frame. Improvement of the absolute values of UT1 will be more difficult.

Similarly, the International Atomic Time has an accuracy today of about 10 nanoseconds with a prospective future accuracy of 1 nanosecond. However, this is based on a statistical combination of many independent standards, and, as such, is subject to possible low frequency systematic errors.

Can the millisecond pulsars provide a long-term independent time scale, or is there an alternative source for such a time scale? Does the atomic time scale agree with dynamical time, or is a relationship between these required? In addition, methods of time transfer can be significantly improved by means of GPS and two-way communication satellites.

6. Ephemerides

With new time scales, new constants, accurate planetary masses, new observational data, and the extragalactic reference frame, along with models that include the asteroids, the galaxy, the Kuiper belt, the Oort cloud, and the best possible formulation of the theory of relativity, new ephemerides for the solar system can be determined. Consistency between the radar and optically-based ephemerides can be achieved. Primary improvement should be in the outer solar system, with the exception of Pluto, which is going to require improvement based on a longer period of observational data. The determination of asteroid mass values is the current challenge.

7. Astronomical Constants and Methods

The astronomical constants and the methods used for reduction of observations will continue to be driven by the most accurate observations available. The introduction of an IAU computer database of values and standardized software packages should provide both international standardization plus the mechanism for future improvements by means of standard software replacements.

8. Reference Frame Relationships

With the accuracy of the extragalactic reference system at approximately the milliarcsecond level, the ties for other reference frames depend on how directly the transformations can be determined and on the number of source observations. Thus, where direct observations can be made, the ties can be at the accuracy of the observations. In other cases, a multi-step approach is necessary, due to the magnitude differences between the sources. These relationships will be improved in the future as more accurate catalogs of optical sources covering wider magnitude ranges become available.

9. Summary

In 1976, resolutions were adopted to change from the FK4 to the FK5 star catalogs, from B1950.0 to J2000.0, and to introduce new constants, ephemerides, and time-like arguments. The discrepancies between observations and the old system were one arcsecond and one arcsecond per century. The changes, which were introduced in 1984, have provided significant improvements in accuracy, and new knowledge from the improved astrometry. Also, more accurate observational techniques, such as CCD detectors, VLBI, and space astrometry, have developed.

Now, an extragalactic reference frame and astrometry at the milliarcsecond level are available. The changes must be made to start another cycle of improvement and new knowledge.

3. EXPECTED DEVELOPMENTS IN HIGH PRECISION ASTROMETRY

A NEW ERA OF GLOBAL ASTROMETRY.
II: A 10 MICROARCSECOND MISSION [1]

E. HØG
Copenhagen University Observatory - NBIfAFG
Østervoldgd. 3
DK-1350 Copenhagen K, Denmark

Abstract. A ground-based project for a small telescope about 20 cm aperture is proposed. It would be able to obtain 500 million astrometric observations per year of all stars between V=7 and 18 mag, based on the Hipparcos-Tycho reference net of one million stars. In addition, V and I magnitudes for all stars from V=7 to 17 mag would be obtained. — A new space mission is proposed, capable of obtaining 10 microarcsecond (μas) precision for parallaxes and proper motions of stars of $V = 11$ mag, and a precision of 0.5 millimagnitude (mmag) for intermediate-band photometry. The mission is here called Roemer Plus (or Roemer+) and the design is derived from Hipparcos and Roemer, but a larger telescope aperture and high-precision metrology is applied. At magnitude $V = 20$ a precision of 1.0 milliarcsecond (mas) would be achieved.

1. Introduction

The Hipparcos and Tycho catalogues (Lindegren 1994; Perryman 1994; Turon 1994; Høg 1994 - at this symposium) will be epoch-making for the understanding of the Galaxy and its stellar content. It is worth noting that these catalogues can also be regarded as a first, and necessary, step towards a new era of observational astronomy, where much higher astrometric accuracy can be obtained for fainter stars in even greater number, and where precision multi-colour photometry can be included. The Hipparcos mission results and the use of special CCDs are conditions for improvements by

[1]Part II of a paper. Part I was presented at a conference in Padova, Høg (1994)

E. Høg and P. K. Seidelmann (eds.),
Astronomical and Astrophysical Objectives of Sub-Milliarcsecond Optical Astrometry, 317–322.

many orders of magnitude, as discussed here for two proposed instruments, one on the ground and one in space.

The ground-based project can be implemented on an existing meridian circle. A reflector system is preferred instead of a refractor in order to increase the spectral band in good focus. The design of a reflecting telescope and of a CCD mosaic is given by Høg (1994), including performance data, assuming a 16 cm aperture. Fig. 1 shows a detail of the special CCD which might be manufactured piggyback on a wafer with other CCDs in order to lower the development cost. A meridian circle reflector with such CCDs would become a hundred times more efficient than if a standard meridian circle refractor were equipped with standard CCDs with square pixels. It would be able to obtain 500 million astrometric observations per year of all stars between V=7 and 18 mag. In addition, V and I magnitudes for all stars from V=7 to 17 mag would be obtained.

This ground-based survey could serve many purposes, one of them would be to produce an input catalogue for the following space project, Roemer+.

The design of a new astrometric satellite, called Roemer+, is discussed in Sect. 2. A precision of 10 microarcseconds (μas) for parallaxes and annual proper motions will be achieved for stars of V=11 mag from a 2.5 year mission, i.e. about six times smaller errors than for the basic Roemer satellite proposed for the M3 mission of ESA by Lindegren et al. (1993). The limiting magnitude will be fainter than V=20, and all stars to the limit can in principle be measured.

Precision intermediate-band photometry, i.e., standard errors about 0.003 mag or less, will be obtained for all stars in the interval $V = 6 - 16$ mag.

Sources of global astrometric data, including a possible future, are characterized in Table 2.

2. Ten microarcsecond astrometry by Roemer+

The proposed satellite is shown schematically in Fig.2. Two reflective telescopes of 70 cm aperture observe two separate fields on the sky. Due to the large size of the mirrors it will not be possible to use a beam-combiner technique, like in Hipparcos. The beam-combiner of that satellite joined two fields on the sky into a single focal plane image where the measurements were carried out. A beam-combiner much larger than the one of 29 cm diameter in Hipparcos would be very difficult to manufacture whereas the telescopes in Roemer+ could be manufactured to almost any size. Another advantage of having separate telescopes is that 'parasitic' stars and sky background from the other field of view will be absent. Each focal plane holds a mosaic of CCDs as shown in Fig. 3, similar to that in the basic Roemer satellite, as it is described by Høg (1994).

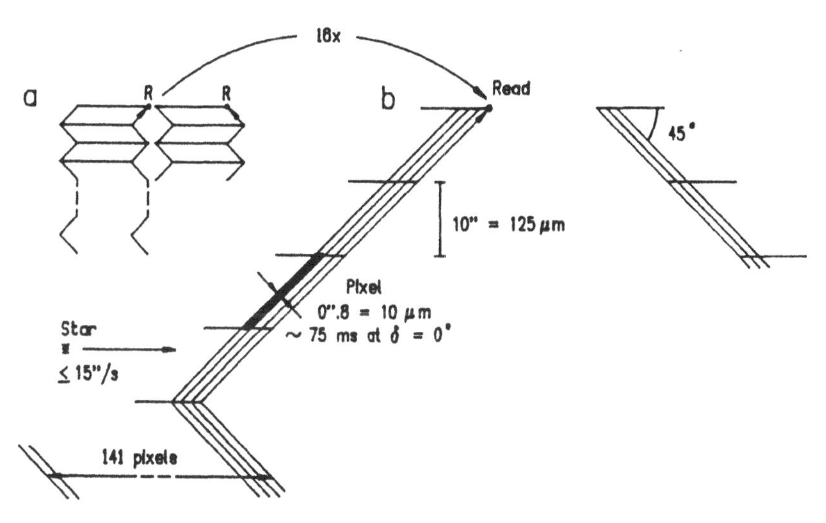

Figure 1. Special CCD for a ground-based telescope. The stars cross the CCDs horizontally from left to right. (a) Detail of two CCDs, further magnified in (b). Readout takes place at R each 75 ms for a star at equator, after 10 s integration time on each CCD.

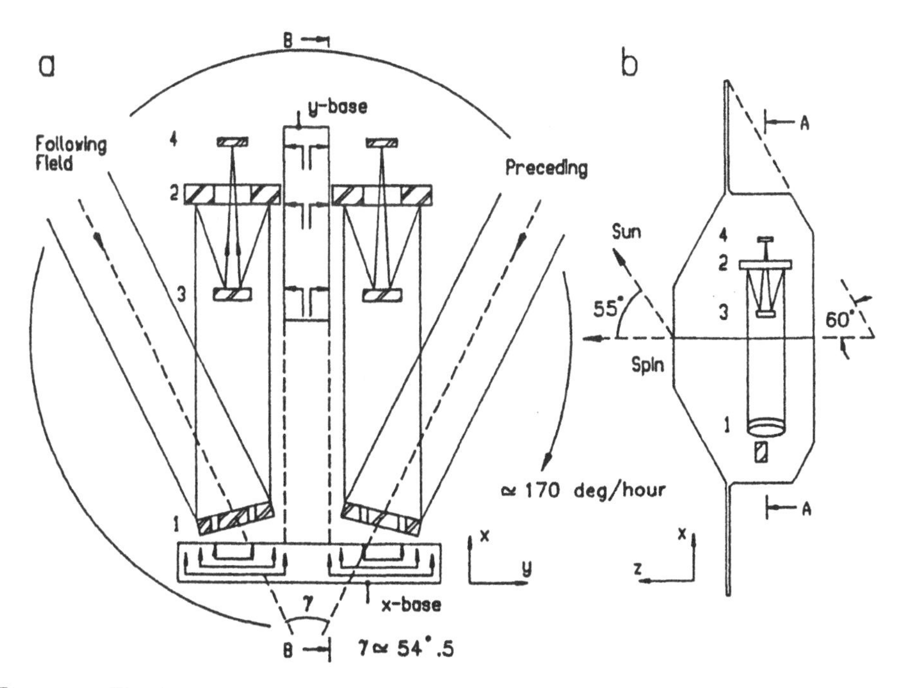

Figure 2. The Roemer+ satellite design. (a) Two Baker-Schmidt telescopes with tilted reflective corrector plates are pointed at the scanning great circle. The optical components are monitored by interferometric distance gauges with picometer precision, thus obtaining the variations of the basic angle as function of time. (b) Section of the rotationally symmetric satellite.

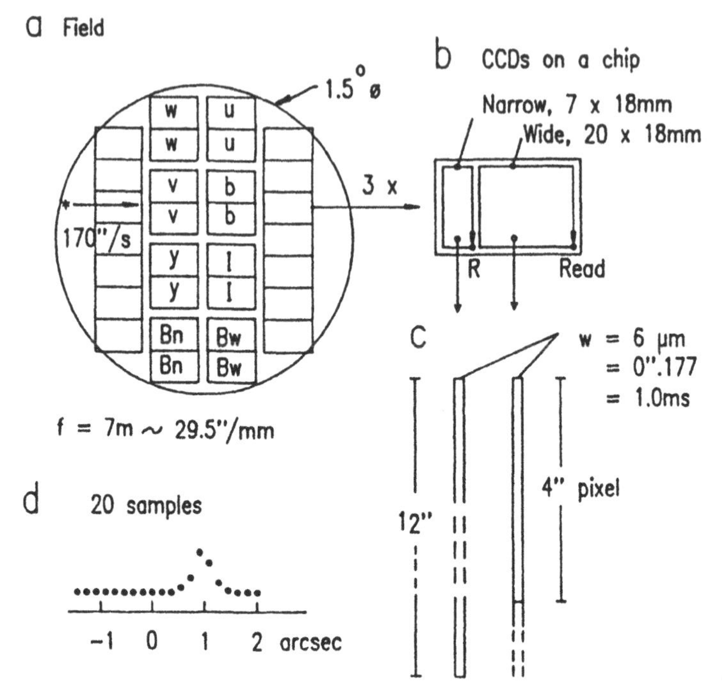

Figure 3. The CCD mosaic at the focal plane of the Roemer+ satellite. (a) The mosaic of identical CCD chips, those in the middle have colour filters as specified in Table 1. (b) A chip with a narrow and a wide CCD. (c) A pixel from each of these. (d) A sequence of observed samples of a star.

A Baker-Schmidt telescope as proposed here was studied for Hipparcos by Iorio Fili and Scandone (1979). The authors found that diffraction limited performance in a field of 1.5 degree or more could be achieved by a system with F-ratio=10.

In Roemer+ the basic angle is calibrated every few hours by means of a 360 degrees closure of the star observations. But variations at shorter time scales than a few hours must be monitored in order to determine the basic angle as function of time. An edge-to-edge tilt of a mirror by one picometer (pm $= 10^{-12}$ m) results in a change of the basic angle by one microarcsecond. The tilts of the six mirrors are monitored by six differential distance gauges mounted on an optical table, the x-base in Fig. 2. The arrows point in the direction of distance measurement. The distance of the four curved mirrors and the two CCD assemblies are monitored in the direction perpendicular to the optical axis by six distance gauges on the y-base, where ten times larger errors can be tolerated than for the tilts.

Distance gauges with a precision of a few picometer and stability over a few hours are required for tilt measurement. Such devices using inter-

TABLE 1. Predicted standard errors due to photon noise in astrometry and photometry for a G0-star from a 2.5 year Roemer+ mission, having two telescopes of 70 cm aperture. *Note:* The standard errors given here are $\sqrt{2}$ times smaller than those in the widely distributed preprint, where they were affected by a calculation error. Filter- and CCD characteristics are given at the bottom. A minus (-) at bright stars means non-linear response of the CCD, i.e. $> 3000e^{-}\mu m^{-2}$ in a pixel. At faint stars a minus means a signal-to-noise ratio ≤ 2.0 on a single CCD crossing. Unit: mas = milliarcsec.

	Astrometry		Photometry [millimagnitude]								
V	par.	p.m.	W	w	u	v	b	Bn	Bw	y	I
mag	mas	mas/year									
2	0.007	0.008	-	0.1	0.0	-	-	0.0	-	-	-
4	0.004	0.005	-	0.1	0.1	0.0	0.0	0.0	0.0	0.0	-
6	0.005	0.006	-	0.2	0.1	0.1	0.0	0.1	0.0	0.0	0.0
8	0.004	0.005	0.0	0.5	0.3	0.1	0.1	0.3	0.1	0.1	0.0
10	0.006	0.007	0.0	1.4	0.8	0.3	0.3	0.7	0.3	0.3	0.1
12	0.014	0.016	0.1	3.5	2.0	0.9	0.6	1.9	0.6	0.7	0.3
14	0.035	0.041	0.1	9.0	5.2	2.2	1.6	4.8	1.6	1.6	0.7
16	0.091	0.107	0.3	26.8	14.0	5.7	4.2	12.9	4.2	4.2	1.8
18	0.261	0.305	0.9	-	55.0	16.8	12.0	49.3	12.0	12.1	5.1
20	0.998	1.165	3.3	-	-	-	49.9	-	49.9	50.3	18.3
Central wavelength [nm]			-	320	350	411	467	486	486	547	800
Filter FWHM [nm]			-	20	30	25	25	3	25	25	140
Peak transmission			-	0.30	0.40	0.60	0.70	0.70	0.70	0.70	0.96
QE of CCD			-	0.80	0.85	0.93	0.97	0.97	0.97	0.95	0.77

ferometric techniques have been developed for the POINTS instrument by Noecker et al. (1993) as one step towards a space-qualifiable picometer distance gauge.

Values of the expected astrometric and photometric precision, defined as the standard error (s.e.) due to photon noise, are given in Table 1. The errors are six (6.0) times smaller than for the basic Roemer mission, given by Høg (1994). This is explained as follows. The s.e. should be inversely proportional to the square of the telescope diameter, D. One factor D comes from the number of photons collected and a second factor D from the improved optical resolution. The smaller F-ratio, compared to Roemer has also been taken into account, and CCD pixels of 6 μm width have been assumed, instead of 4 μm in Roemer. The total light collecting area is 8.0 times larger since there are two telescopes in Roemer+, each with 2.0 times larger D than Roemer.

Other sources of error than photon noise are expected to contribute

TABLE 2. Precision of global astrometric data, present and possible future. Typical V mag for which the precision is given. *Notes:* [1] The Tycho proper motions are obtained by means of first epoch positions from the Astrographic Catalogue, cf. Röser & Høg (1993). [2] Result of two years observations with the proposed reflective meridian circle. [3] A 2.5 years Roemer+ mission about 2010 has been assumed.

Source	N	Position at epochs:				Motion	Parallax
		1990	2000	2010	2020		
		mas	mas	mas	mas	mas/yr	mas
PPM North	200 000 stars	270	300	-	-	4	-
PPM South	200 000 stars	110	130	-	-	3	-
Hipparcos	120 000 stars	1.5	15	30	45	1.5	1.5
Tycho	1 million V~10.5	30	40	70	100	3 [1]	30
MC [2]	200 million V~17	-	50	50			
Roemer+ [3]	1 million V~11	-	-	0.01	0.1	0.01	0.01
	200 million V~17	-	-	0.20	2.0	0.20	0.20

relatively little, as shown by the theory and experience of Hipparcos, and in a study by Makarov, Høg and Lindegren (1994).

Acknowledgements: I am grateful for the information and inspiration received in discussions with U. Bastian, R. Florentin Nielsen, J. Geary, M. Lesser, L. Lindegren, and V.V. Makarov. This work was supported by the Danish Space Board.

References

Høg E., 1994, A new era of global astrometry and photometry from space and from ground, contribution at the *'G. Colombo' Memorial Conference: Ideas for Space Research after the year 2000*, Padova, 18,19 February 1994

Iorio Fili D., Scandone F., 1979, In *European Satellite Astrometry*, by C. Barbieri and P.L. Bernacca (eds.). Padova, Italia, p. 29

Lindegren, L. (ed.), Bastian, U., Gilmore, G., Halbwachs, J.L., Høg, E., Knude, J., Kovalevsky, J., Labeyrie, A., van Leeuwen, F., Pel, W., Schrijver, H., Stabell, R. and Thejll, P., 1993, *ROEMER, Proposal for the Third Medium Size ESA Mission (M3)*, Lund, Sweden

Makarov, V.V., Høg, E. and Lindegren, L., 1994, Accuracy of star abscissae in the ROEMER project. In preparation

Noecker M.C., Phillips J.D., Babcock R.W., Reasenberg R.D., 1993, Internal laser metrology for POINTS, In: *Proceedings of SPIE Conference*, **Vol. 1947**-22

Röser, S., Høg, E., 1993, 'Tycho Reference Catalogue: A Catalogue of Positions and Proper Motions of one Million Stars.' In: Workshop on Databases for Galactic Structure. Ed.: A.G. Davis Philip, B. Hauck and A.R. Upgren. Van Vleck Observatory Contr. No.13, 137. L. Davis Press, Schenectady, N.Y.

ON THE REGISTRATION SYSTEM OF THE AIST-PROJECT

M.S.CHUBEY, V.S.PASHKOV, I.M.KOPYLOV, T.R.KIRIAN AND V.V.NICKIFOROV
Pulkovo Observatory, 196140, St.-Petersburg, Russia
e-mail: mchubey@gaoran.spb.su

AND

S.V.MARKELOV AND V.P.RYADCHENKO
Special Astrophys. Observatory,
Nizhnij Arkhyz, Karachai-Circassian Republic, Russia, 357147
e-mail: markel@sao.stavropol.su

Abstract. The researches on the AIST-project are directed to the design of the focal plane assembly. The concept of the chip-mosaic and numeric simulation of the two-pixels mode of the signal sampling are presented.

1. Introduction

The work on the AIST-project was carried out in accordance with the concept described by Chubey *et al.* (1993). More recently the main researches have been devoted to the registration system on CCD chips.

The increase of the micrometer efficiency by 2-3 orders in the number of objects that can be registered and the increase of the astrometric and photometric accuracy because of the longer registration time - all this has changed the scope of task for the forthcoming second generation space astronomy experiment. Its scientific content evidently will be a close analogy with the "Carte du Ciel" with natural development towards electronic tools of registration and data processing. The data base which is planned to be gathered in the form of astrometric and photometric catalogues, will be of interest as the source material for fundamental investigations in astronomy. This opinion is supported by the AIST team being operating on the international cooperation. The goal of this paper is to present short information about the current stage of technical design of the project, beyond any crit-

E. Høg and P. K. Seidelmann (eds.),
Astronomical and Astrophysical Objectives of Sub-Milliarcsecond Optical Astrometry, 323–326.

icism of the concurrent projects, and mainly concerning the problem of the registration system.

2. Particular decisions in the project

Two telescopes on board in a stable arrangement allow to use three fields on the scanning great circle, two of which R_1 and S_1, Fig. 1, are folded in the focal plane F_1 by means of a beam combiner; the third is a full aperture single optical channel to the second focal plane F_2. This variant was discussed and described earlier (Høg and Chubey 1991). The advantages of this variant are:

- the ability to use the hardware and algorithmic separation of the folded fields R_1, S_1 with the simple subsequent separation of the frames R_1 and S_1 from their folded image $R_1 + S_1$, aided by the focal assembly F_2;
- photometric gain is evident: the round and twice as bright optical image of every object in the field F_2 doubles the "signal-to-noise" ratio;
- the choice of the angle between R_1 and R_2 is $\gamma_1 = 180^o$, implying absolutely symmetrical paths of the centers of areas R_1 and R_2 on the scanning motion.

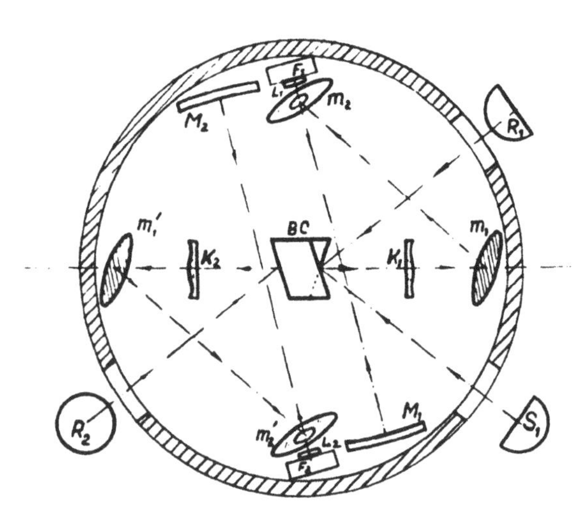

Figure 1. Option with a full aperture entrance pupil. Both focal plane assemblies F_1, F_2 include the field lenses L_1, L_2.

The definition of the attitude angles under the chosen dynamics of the space craft will be processed more precisely because of the near-rectangular distribution of the scanned directions: the base angle between R_1, S_1 is $\gamma_2 = 74^o$ in the discussed variant.

The main disadvantage of the double-telescope instrument is the rise of the information flow.

3. Micrometer based on CCD chips

The detailed research and design of the micrometer requires long time and great efforts. Only a short conclusion from our analysis of the problem is presented.

The use of micro channel plates chip for the micrometer is not supposed in our analysis because of the low QE, more narrow spectral band and the high level of noise on the cascade amplifier: the puls amplitude spectrum has the Farri's distribution (negative exponent), with the dispersion of the amplification coefficient equal to the square of this coefficient. This can not be improved significantly by hardware. The most important disadvantages are the loss of astrometric accuracy and the narrow dynamical range.

The CCD in fact is a television sensor, working in the super-small-frame regime with large possibilities to suppress the noises. Manufactured by ELECTRON Company (Russia) chips with $16 \times 16 \mu m$ pixels have a well-capacity up to $130000 e^-$ with a read-out noise of $5e^-$, covering a dynamic range of 26000, i.e. near to 11 magnitudes. The world level is somewhat better. We assume that the *time delay and integration, (TDI)* principle, or variant of it, must be the basic mode of operation.

The integration mode suggested by Høg and Lindegren (1993), based on the *TDI*, allows to analyze the photometric profile of the diffraction image of star. The splitting of the image in N parts applied to CCD could ensure a higher accuracy, but the large number of transfers demands a charge transfer efficiency of nearly 0.999 998 and leads to a decrease of the energy and restriction of the photometric possibilities. The accuracy estimation of the form modified by Yershov (1993), shows that the optimal number of splits for the AIST optics is $N = 3$.

An extended simulation was made for the *TDI* subclass (Pashkov & Chubey 1993) using *two-pixel integration* method. The exposure $T = 8.8\ ms$ followed by a synchronous shift forward were used when the star image crosses each pixel. A polychromatic model of the Point-Spread Function and the discriminating function were calculated for the high level noise condition. In Fig.2 the plots of summarized astrometric accuracies are shown. The Monte-Carlo simulation with 20 tests and 512 pixels path in the array column were used. Two levels of the readout noises $5e^-, 10e^-$ were used in the model and two Analog Digital Convertors (ADC) 8 and 12-bit encoding, for red and blue spectral bands, for three variants of chips (128, 256 and 512 pixels in column) - all these in the range of 8 - 18 magnitudes.

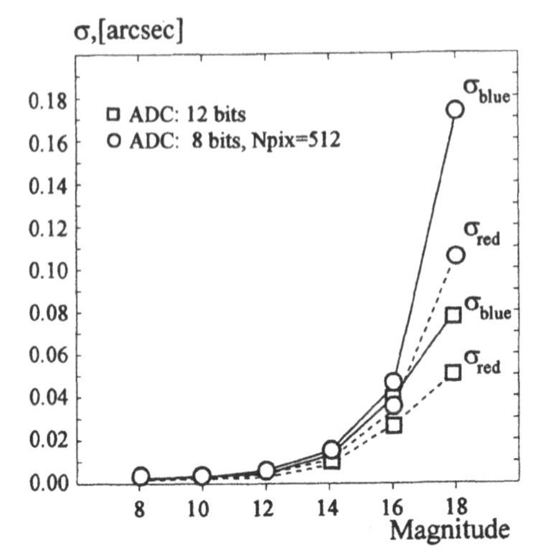

Figure 2. The one-array-readout error σ (mas) vs star magnitude.

4. The working version of a CCD-camera design

To perform the 7-colour photometry a larger number of CCDs must be used compared with the *Roemer* mission (Høg 1993). Two focal assemblies will be used. In each the CCD sensor must be a mosaic of 3 or 4×10 chips with the number of elements 512×512. Each CCD-array has a read-out system with virtual transfer of the frame and division of the horizontal register into two equal parts, each of which is read out on different sides. The small variations of the transfer speed, the usage of matrixes with nonperiodical structure and other methods from the *TDI* regime - all these are under careful testing for optimal final choice.

References

Chubey, M.S. *et al.*, (1993), Aist Project: Scientific and Technological Foundations, *Developments in Astrometry and Their Impact on Astrophysics and Geodynamics*, I.I.Mueller & B.Kolaczek (eds.), IAU Symp.156, pp. 415–420.

Høg, E., Chubey, M.S., (1991), Proposal for the Second Hipparcos, *Intern.Symp. "Etalon-91"*, Moscow, unpublished.

Høg, E., (1993), Astrometry and Photometry of 400 Million Stars Brighter than 18 Mag, *Developments in Astrometry and Their Impact on Astrophysics and Geodynamics*, I.I.Mueller & B.Kolaczek (eds.), IAU Symp.156, pp. 37–45.

Høg, E., Lindegren, L., (1993), A CCD Modulation Detector for a Second HIPPARCOS Mission, ibid, pp. 31–36.

Pashkov, V.S., Chubey, M.S., (1993), On the Accuracy Estimation in the AIST-Project, *Proceed. of the Russian Astrometric Conf.* held in St.-Petersburg, Oct. 4-8, 1993, in press.

Yershov, V., (1993), A Focal CCD Micrometer for the Astrometric Satellite, *Second Intern. Workshop on Posit. Astronomy and Celest. Mech.*, ed. A.Lopez Garcia, Valencia Univ., pp. 307–312.

PHOTOMETRIC FACILITIES OF THE AIST SPACE PROJECT

I.M.KOPYLOV , D.L.GORSHANOV AND M.S.CHUBEY
Pulkovo Observatory,196140 St Petersburg, Russia
e-mail: mchubey@gaoran.spb.su

Abstract. The arguments are given of the choice for the Vilnius seven color medium band photometric system as the most effective one for the *AIST* space project. A prediction is made of the limiting stellar magnitudes V for every Vilnius passband and the integral band T at $\sigma_{\Sigma} \leq 30mmag$ from a 3 year *AIST* mission.

1. Introduction

The main aim of the scientific program of the *AIST* project is to create a fundamental catalogue of positions, proper motions and parallaxes for faint stars at the epoch of mission with milliarcsecond precision, to carry out their high precision multicolor photometry and direct optical measurements A the precise quasar coordinates to fit a fundamental system. The last was the main reason for a choice of the limiting magnitudes to be of the order of $V_{lim} = 18mag$.

2. Instrumentation

We estimate for a two Schmidt telescopes instrument. The version is: one of two telescopes has a beam combiner, another one, with full pupil, has no beam combiner. As the standard detector units we use (Fig.1) a square CCD chips with 256 x 256 pixels ($4.1 \times 4.1mm$), the pixels being 16 $\mu m = 1.32''$ squares. The central diffraction spot for the full telescope pupil is $d_0 = 0.75''$. The adopted spectral sensitivity $QE(\lambda)$ of CCD has values 56,76,77 and 42% for the wavelength 300, 500, 700 and 900 nm respectively. At the angular velocity of $150''/s$ the star crosses the field of view within 34^s and the single CCD chip within 2.25^s. The main axis of rotation has a scan step $20'$ and small drift during one rotational period.

E. Høg and P. K. Seidelmann (eds.),
Astronomical and Astrophysical Objectives of Sub-Milliarcsecond Optical Astrometry, 327–330.

3. The Choice of the Photometric System

We have compared the parameters and qualities of several modern most usable multicolor photometric systems. The criteria have been used: (a) the of the selected passbands and their positions, fixing spectral type and luminosity, and domains of application of the system; (b) the bandwidths and their comparative abilities in achieving the necessary limiting magnitude of the whole system; (c) the system purity parameters (Jaschek and Frankel, 1986), i.e. the quality and ability of the photometric system to classify stars with different characteristics in the optimal way and to reveal and classify the various types of peculiar stars, for example, Ap, Am, carbon and metal-poor stars, subdwarfs, white dwarfs, some types of binaries etc.; (d) the availability of a sufficient number of standards, including both ordinary and peculiar stars observed in a given photometric system; (e) the possibility of transformation of the photometric data collected in the chosen system to other systems with a minimum loss of precision. These criteria were applied for the comparison of the $UBVRI, uvby + \beta_n + \beta_w$, Vilnius and Geneva photometric systems. The firm conclusion has been reached that the most effective one is the medium-band seven-color Vilnius system $WPXYZVS$ (Straižys 1977 and 1992).

After that, we compared main merits and problems of the basic versions of this system: Standard, VilGen, Interferometric and CCD-adapted (Straižys 1977; Straižys et al. 1992). As a final result, we have used the Interferometric version (Int) whose parameters are listed in Table 1. The contents of columns 1 - 4 are understandable. The 5th column contains the integrated filter transparency (the equivalent width W_λ).

4. Calculation Scheme

For the calculation of the expected limiting magnitudes V_{lim}, the magnitude error in 1^s exposure σ_1 and the error in the whole flight time σ_Σ we use the standard formula

$$\sigma = \frac{\sqrt{N_* + N_n + N_{CCD}}}{N_* \sqrt{t}}. \quad (1)$$

Energy distribution $E(\lambda_i)$ in the spectrum of the faint stars, of the faint stars background (SL) and of Zodiacal light (ZL) is similar to that of a star of spectral type G. With the adopted aperture, optics transparency $p(\lambda_i)$, $W(\lambda_i)$ and $QE(\lambda_i)$ we find for every band the value N_* in units $[e^-/s]$. For the calculation of N_n in formula (1) we have used the mean integrated sky brightness $\overline{m_i}(ZL + SL)$ per pixel given in column 8 of Table 1.

All the calculation for the band T (the last row of Table 1) was made taking into account only the losses of light in optics and the $QE(\lambda)$ of the

TABLE 1. The adopted version of the Vilnius system.

Band	λ_0 nm	$\Delta\lambda$ nm	τ_{max} %	W_λ nm	$\frac{W_\lambda(Int)}{W_\lambda(Std)}$	QE %	$\frac{m_i}{pix}$ mag	t_{total} s
1	2	3	4	5	6	7	8	9
W	350	52	65	35.5	2.10	62	21.7	670
P	375	20	58	12.2	1.11	65	21.7	1080
X	406	17	60	10.7	1.57	68	21.6	1005
Y	468	23	82	19.8	2.61	74	21.4	310
Z	518	18	74	14.0	1.47	77	21.1	380
V	547	26	85	23.2	3.14	79	20.9	285
S	656	32	85	28.6	7.94	80	20.5	205
T	625	587	100	548	–	80	20.2	105

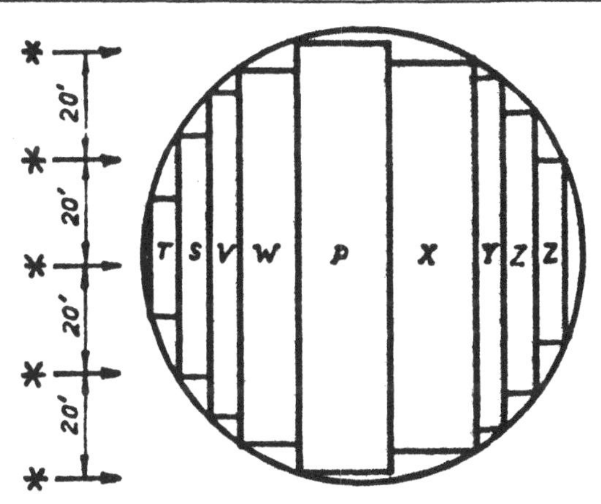

Figure 1. Relative areas in the field of view covered by 7 Vilnius filters (and by T-band), to reach the same V_{lim} magnitude of a G2 IV-type star.

CCD. The total noise N_{CCD} in the formula (1) was assumed to be $5e^-\ s^{-1}$. A "drift-scan" principle was used throughout calculations.

5. Analysis of the Results

Extensive tables were prepared, containing the relations between V and σ_1 for the bands of the Vilnius system and for the T-band. These tables demonstrate that because of the rather small W_λ values for the P and X bands and relatively low $E(\lambda_i)$ for the W band, the magnitudes V (or V_{lim}) at some adopted σ (or σ_{lim}) for these bands are smaller than $\overline{V}$ (or $\overline{V}_{lim}$) of the remaining four Vilnius bands by 1.6, 1.3 and 0.8mag respectively. The best possibility for an approximate equalizing of V_{lim} for different bands is to enlarge the relative widths and lengths of the chips within the field of view covered by the "poor" filters P, X and W.

The result of one such numerical equalization of V_{lim} (the number and size of CCD chips covered by the Vilnius filters) is presented in Fig. 1.

We have estimated also the approximate scanning times (column 9 of Table 1) for the Vilnius bands during the time of the $AIST$ mission (3 years) which would be necessary in order to reach nearly the same V_i.

In Table 2 we represent the dependences of the final errors for all the Vilnius and T bands. The upper row of asterisks marks $\sigma_\Sigma \leq 1mmag$, at $V \approx 11mag$, the lower one shows us the V_{lim} at $\sigma_\Sigma \leq 30mmag$.

TABLE 2. The final errors σ_Σ of the predicted stellar magnitudes V_i of G2IV-star. Total flight time is 3 years. Errors are given in $mmag$.

V	σ_W	σ_P	σ_X	σ_Y	σ_Z	σ_V	σ_S	σ_T
11	*0.92*	*1.07*	*0.94*	*0.92*	*1.00*	*0.98*	*0.92*	0.32
12	1.42	1.71	1.51	1.60	1.56	1.40	1.46	0.51
13	2.3	2.8	2.4	2.4	2.4	2.2	2.3	*0.80*
14	3.7	4.5	3.8	4.0	4.0	3.5	3.7	1.28
15	6.1	7.5	6.4	6.4	6.4	5.7	5.9	2.0
16	10.4	13.3	10.8	10.5	10.6	9.2	9.6	3.2
17	18	* 26 *	21	18	18	15.5	16	5.2
18	* 37 *	55	* 42 *	* 32 *	* 35 *	* 28 *	* 29 *	8.6
19	77	125	99	70	72	58	56	15
20	178		230	147	107	119	118	* 28 *

6. Conclusion

The calculation shows that the $AIST$ project would obtain medium band multicolor photometry of faint stars down to $18.0mag$ with a final error $\sigma_\Sigma \leq 30mmag$, photometry of bright stars down to $\approx 11mag$ with $\sigma_\Sigma \leq 1mmag$ and surface photometry of extended objects down to $\approx 18mag$ per pixel with $\sigma_\Sigma \leq 30mmag$. In addition, the T band allows to measure stars with the same precision but $2.0 - 2.5mag$ fainter.

References

Jaschek C., Frankel S., 1986, A&Ap, **158**, 174.
Straižys V., 1977, Multicolor Stellar Photometry, Mokslas, Vilnius.
Straižys V., 1992, Baltic Astronomy, **1**, 107.
Straižys V., Boyle R.P., Kuriliene G., 1992, Baltic Astronomy, **1**, 95.

NEWCOMB ASTROMETRIC SATELLITE

K. J. JOHNSTON AND P. K. SEIDELMANN
U.S. Naval Observatory
Washington, D.C. 20392-5420

AND

R. D. REASENBERG, R. BABCOCK AND J. D. PHILLIPS
Smithsonian Astrophysical Observatory
Cambridge, Massachusetts

Abstract.
Newcomb is a design concept for an astrometric optical interferometer satellite with a nominal single measurement accuracy of 100 microarcseconds. In a 30-month mission life, it will make scientifically interesting measurements of O stars, RR Lyrae and Cepheid distances, probe dark matter in our Galaxy via parallax measurements of K giants in the disk, establish a reference grid with internal consistency better than 50 microarcseconds, and lay the groundwork for the larger optical interferometers that are expected to produce a profusion of scientific results during the next century.

1. Concept

The idea is to achieve an initial optical interferometer in space by means of a small satellite with a simple design, which could be launched by a small rocket, such as the Pegasus, at a low cost to the mission. The spacecraft would hold a stack of simple optical interferometers pointing in certain fixed relative directions. The interferometers will provide high precision differential angle measurements with a nominal single measurement accuracy of 100 microarcseconds. The optics would be mounted on a three-axis gimbal with the spacecraft holding inertial orientation. The plan would be for an orbit of about 600 nautical miles in altitude with continuous sun illumination and a lifetime of two and a half years.

E. Høg and P. K. Seidelmann (eds.),
Astronomical and Astrophysical Objectives of Sub-Milliarcsecond Optical Astrometry, 331–334.

2. Justification

The Newcomb Astrometric Satellite would provide a precise inertial reference frame based on something over 1000 stars observed with an accuracy of 0.5 nanoradian, or 0.1 milliarcsecond, at the epoch of observation. The single instrument would provide all sky coverage. These results are compared to the current optical reference frame, which has an accuracy of 250 nanoradians, or 50 milliarcseconds, and worse when systematic errors are considered. The Hipparcos mission will provide 10 nanoradians accuracy, or 2 milliarcseconds, at the epoch of observation with an error increasing at the same value per year from the epoch of observation. Ground-based interferometer observations are expected to be limited to 10 nanoradians or 2 milliarcseconds accuracy.

Alternative future plans for an optical interferometer in space include POINTS and OSI which are not expected to be able to fly for 15 or 20 years. This is another reason for the Newcomb Astrometric Satellite, as it will provide a method for characterizing materials and techniques for optical interferometry in space, and be a testbed for future DoD or NASA interferometry missions such as AIM or TOPS. Also, this satellite mission will provide technology for calibrating inertial orientation systems and a means for future precise pointing for satellite systems.

3. Scientific Goals

Newcomb would provide a direct link between high precision optical astrometry and the radio reference frame (Fey, *et al.*, 1992). Observations of a few radio quasars can fix the rotation between the radio reference frame and the optical reference frame as developed by the Newcomb and Hipparcos observations.

Assuming that the design would develop a limiting magnitude around 15, there are a number of science studies that could be accomplished. Nineteen of the known Cepheids have parallaxes between 200 and 1000 microarcseconds and visual magnitudes of less than 10. So a 5% distance measurement would be useful in refining the cosmic distance scale. Absolute magnitudes of O stars are uncertain because none is close enough to allow a trigonometric parallax measurement from the ground. Since those stars are bright targets with magnitudes of 4 to 6 at 1 - 2 kiloparsecs, 25 - 500 microarcsecond parallax measurements are needed.

Age determinations of globular clusters depend, in part, on the calibration of the absolute magnitude of RR Lyrae stars as a function of period. The 20 brightest RR Lyrae stars have visual magnitudes in the range from 7 to 10. Parallax measurements are needed at the 1% level which corresponds to 40 microarcseconds for RR Lyrae itself. Distances to nearby subdwarfs in the magnitude range of 7.2 - 12 would calibrate the subdwarf luminosities which are used in fitting globular cluster main sequences. Parallaxes are needed with 30 microarcsecond precision. Parallaxes of bright, visual magnitude 10, K giants in the galactic disk, accurate at 50 microarcseconds, would probe the dark matter in our galaxy.

With 100 microarcsecond measurements and a sufficient mission duration, Newcomb could detect Jupiter-sized planets around nearby stars. The problem would be whether the 30-month lifetime envisioned for Newcomb would be long enough to detect the quadratic term in the motion of a star necessary to indicate the existence of a planet.

Gravity Probe B, an experiment which measures the general relativistic frame dragging due to the spinning Earth, needs a bright guide star, Rigel, with the proper motion known in an inertial frame to approximately 1 milliarcsecond per year or better. Newcomb could determine the proper motion of Rigel relative to a few bright quasars tied directly to the reference frame.

Newcomb will provide, for a limited number of stars, parallaxes and proper motions which are an order of magnitude more accurate than Hipparcos.

4. Satellite Characteristics

The satellite orbit would be approximately 600 nautical miles altitude, Sun synchronous, with continuous Sun illumination and 90° inclination. The payload weight would be in the range of 150 to 200 pounds, with a total satellite weight in the 500 - 600 pounds range. It could be launched from a Pegasus launch vehicle, or share an Argos mission. Approximate size would be 3.2 cubic feet. Electric power would be 28 volt DC with approximately 125 Watts average power. Attitude control based on the step bus would be $\pm 0.5°$ in inertial orientation. The object payload with gimbals would provide $\pm 0.001°$ to $\pm 0.0001°$. Thermal control would be a passive system for the spacecraft, with inertial attitude and sun soaker characteristics, and active thermal control for the critical optical elements.

5. Sensor Description

The Newcomb Astrometric Satellite would have a stacked set of 3, or 4, Michelson optical interferometers oriented at $0°, 40.9°$, and $60.5°$. Selected star pairs would be measured one by one. Free axis pointing of the optics payload would be used to acquire each star pair. The dispersed fringe pattern from the interferometer would be detected by a two-dimensional detector made up of 8000 x 32 elements, so the detector would work in a wide frequency band from 4000 to 8000 angstroms. The size of the detector and the small optics provide a field of view of several arcminutes. The detectors would do onboard image integration for the fringe detection. The stellar separations and instrument parameters would be determined from the data reduction process with the long-term 10 - 20 hour drifts in instruments calibrated by the observations. The interferometer would use a 30cm baseline with an aperture of 2 x 5 cm. The instrument field-of-view would be approximately 0.7° with a sensitivity of approximately 15th magnitude.

6. Technical Issues

The preliminary mission concept has been completed primarily based on the POINTS concept. The critical technical issues which have been identified are as follows:

Thermal control is necessary in order to maintain the stability of the optics payload.

The detectors necessary for the fringe detection must be identified.

The attitude control and payload pointing requirements must be achieved. Gimbals can provide 0.001° of arc while the payload pointing requires 0.0001°.

The optimum on-board fringe detection and data processing must be developed.

Mission planning must be accomplished to develop the observing mode and the catalog of target stars.

Finally, a detailed optical design trade-off study must be done to specify the optics sizes necessary to achieve a signal-to-noise ratio required for a successful mission.

7. Summary

In summary, the Newcomb Astrometric Satellite is a concept for a small, quick, inexpensive, initial optical interferometer in space. It could measure the angles between stars, to define a inertial reference frame with the precision of 0.1 milliarcsecond. By means of this observational capability, proper motions and parallaxes can be determined and other scientific investigations achieved.

References

Reasenberg, Robert D., Babcock, Robert W., Phillips, James D., Johnston, Kenneth J., and Simon, Richard S. (1993) "Newcomb, A POINTS Precursor Mission with Scientific Capacity" in Spaceborne Interferometry, Proceedings of SPIE Conference, 1947.

Fey, A., Russell, J. L., Ma, C., Johnston, K. J., Archinal, B.A., Carter, M.S., Holdenreid, E., and Yao, Z. (1992) *Astron J.* **104**, 891.

THE PROJECT OSI

M. SHAO

Abstract. The OSI project at JPL is a study of a series of concepts for space based astrometric interferometry. The various concepts are designed for slightly different types of astrometry. The original concept OSI was designed for wide angle/global astrometry by look at stars about 30 degrees apart. A derivative of OSI, called SONATA was designed for narrow angle astrometry (10 arcmin) at sub-microarcsec levels for exo-planet searches. This paper describes optical concepts of theses instruments with regard to metrology and systematic error. In addition to conceptual designs, system error budgets and covariance analysis the OSI project has a technology development program in place to investigate the most difficult problems.

Technology is being persued in three areas. One is laser metrology at the picometer level where sub-picometer accuracy for null metrology has been achieved and < 2 picometer accuracy has been achieved for relative metrology. A second area is the control of vibration on a large space structure. A 5 m truss structure has been built at JPL with an interferometer on the structure to test vibration isolation and control technologies. Tests with this structure will begin in winter of 1994. The key to vibration control for OSI will be the delay lines. The third technology project is a shuttle flight experiment, Stellar Interferometer Technology Experiment (SITE), currently in a phase A study, (part of NASA's INSTEP program). The goal is to fly the experiment on the Shuttle in 1997-1998.

E. Høg and P. K. Seidelmann (eds.),
Astronomical and Astrophysical Objectives of Sub-Milliarcsecond Optical Astrometry, 335.

POSSIBILITIES OF SUB-MILLIARCSECOND ASTROMETRY IN THE LOMONOSOV PROJECT

A.M. CHEREPASHCHUK, V.V. NESTEROV,
E.K. SHEFFER

Abstract. Scientific and technical problems of accuracy improvement of stellar position measurements in the *LOMONOSOV* project are discussed. In contrast to the *HIPPARCOS* project, in our project we use a steerable satellite with a three-axis spatial stabilization. A Cassegrain-type 1-meter mirror telescope will be mounted on board of the satellite. We discuss scientific goals in astrophysics, stellar astronomy and astrometry, for which the knowledge of stellar positions with sub-milliarcsecond accuracy is necessary.

E. Høg and P. K. Seidelmann (eds.),
Astronomical and Astrophysical Objectives of Sub-Milliarcsecond Optical Astrometry, 336.

A SMALL INTERFEROMETER IN SPACE FOR GLOBAL ASTROMETRY: THE GAIA CONCEPT

L. LINDEGREN
Lund Observatory
Box 43
S-22100 Lund
Sweden

AND

M.A.C. PERRYMAN
Astrophysics Division
European Space Agency, ESTEC
NL-2200 AG Noordwijk
The Netherlands

Abstract. We present a concept for a scanning interferometer for global (wide-angle) astrometry from space. The GAIA concept has been proposed for the European Space Agency's long-term scientific programme. It consists of three Fizeau-type interferometers with 2.5 m baselines, set at large and fixed angles to each other. Complete utilization of the instrument's resolution and sensitivity requires a new type of photon-counting detector, combining very high spatial and temporal resolution. An array of superconducting tunnel junctions (STJ) may ultimately provide this capability. Pending this development we describe a focal-plane configuration for GAIA using existing technology in the form of a modulating grid and CCD detectors. We estimate that 50 million stars brighter than $V = 15.5$ could be observed on the 10 to 20 microarcsec accuracy level. In addition, high-precision multi-colour, multi-epoch photometry is obtained for all objects.

1. Introduction

The highly successful Hipparcos mission has demonstrated that a small, dedicated astrometry satellite can perform wide-angle measurements over the whole sky with an accuracy ultimately determined by instrument reso-

E. Høg and P. K. Seidelmann (eds.),
Astronomical and Astrophysical Objectives of Sub-Milliarcsecond Optical Astrometry, 337–344.

lution and photon noise. The capacity to produce large quantities of high-quality data can be derived from a combination of a few simple ideas:

- use of two viewing directions, separated by a large angle, connecting all parts of the sky in a few steps;
- one-dimensional angular measurements along the great circle through the viewing directions;
- continuous scanning and observation of all programme stars as they pass over the fields of view, while maintaining a constant geometry with respect to the sun;
- determination of critical instrument parameters from the closure conditions on each complete rotation (few hours), making long-term instrument stability quite uncritical;
- positions, proper motions and parallaxes ideally solved in a single global solution together with all instrument and attitude parameters.

These ideas appear to be equally valid for a future space astrometry mission aiming at accurate wide-angle astrometry for very many stars. Indeed, they are the basis for at least two proposals submitted to the European Space Agency since 1993, the Roemer project (and subsequently Roemer+; see Høg, 1994) and GAIA (Lindegren et al., 1994). Both aim at sub-milliarcsec astrometry and multi-colour, multi-epoch photometry for many millions of stars.

The target accuracy for GAIA is 20 μas for the positions, parallaxes and annual proper motions of all compact optical objects brighter than V = 15–16, or some 50 million stars and a significant number of extra-galactic sources. In order to achieve this accuracy we propose to use a fixed configuration of small optical interferometers of a few metres baseline.

2. Scientific Objectives

A full survey of the scientific capabilities of an instrument like GAIA is clearly beyond the scope of this paper. The following few examples are mostly drawn from existing reviews (e.g., Kovalevsky & Turon, 1992, Ridgway, 1993, and the report of the ESA Interferometry Review Panel under the chairmanship of Dr. C. Dainty).

Stellar luminosities: Direct luminosity estimates are based exclusively on trigonometric parallaxes. At the 20 μas accuracy level the parallax method would reach to 5 kpc with a relative precision of 10%, or to 10 kpc with 20% precision. For the first time this would provide an extensive network of distance measurements throughout a significant part of our Galaxy, including the galactic centre, spiral arms, the halo, and the bulge. The volume would include numerous representatives of the rarer but evolutionary important classes of objects, such as the most massive stars, novae and nova-like variables, central stars of planetary nebulae, Cepheids, and RR Lyrae stars.

Planetary Detection: The detection of non-linear photocentric motions in the paths of nearby stars due to planetary companions has been extensively studied for example as part of NASA's TOPS (Towards Other Planetary Systems) programme. At 20 μas mission accuracy, GAIA would be able to detect Jupiter-mass planets with an orbital period smaller than the mission duration out to distances of 100 pc, while correspondingly smaller planets could be detected for very nearby stars. In principle all 50 million stars could be screened for possible signatures of planetary or brown dwarf companions, providing a complete census to well-defined and uniform detection limits.

Binary Systems: A rich variety of astrophysical phenomena in interacting binary systems could be addressed with very accurate astrometric data, including questions about accretion rates, precursors, mass distributions, and kinematic behaviour. For binary systems in general, GAIA would provide a vast statistical material for systems with angular separations greater than 1 mas, in many cases including orbital parameters and component masses.

Clusters: Studies of open clusters are important for numerous reasons, mostly because they represent a co-eval population of stars with well-defined initial chemical compositions. Some 20 open clusters are considered to lie within about 400 pc, sufficient to provide individual distances to better than 1% accuracy with GAIA. Ages of *globular clusters* have indicated a possible discrepancy with the age of the Universe derived from present estimates of H_0 and Ω. Direct distances are needed for an absolute determination of ages, and GAIA would provide sufficient accuracy to resolve the age conflict.

Galactic Dynamics: The huge number of stars and the impressive accuracy of proper motions and parallaxes would totally revolutionise dynamical studies of our Galaxy. Considerable advances would be possible in our understanding of the structure and motions within the spiral arms, the disc and the outer halo. Questions related to the existence and amount of dark matter could be addressed by direct measurements of the distribution and kinematics of stars in the solar neighbourhood and of the rotation curve beyond the Sun.

Proper Motions of the Magellanic Clouds and AGN's: Proper motions for large numbers of stars in LMC/SMC would clarify the dynamical behaviour of these systems, and in particular whether they are gravitationally bound to our Galaxy. Further out, the nuclei of active galaxies are sufficiently point-like that their relative proper motions may be measurable. An astrometric programme reaching 15–16 mag would include a number of

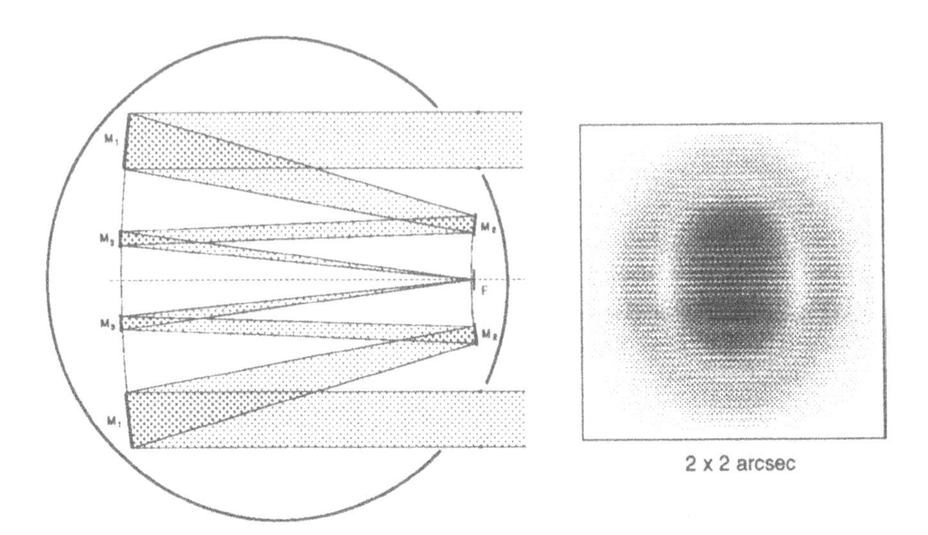

Figure 1. Optical layout of one of the Fizeau interferometers. The two pupils are 0.5 m in diameter and the baseline (distance between the pupil centres) is 2.5 m. The total width is 3.0 m and the focal length 11.5 m. A three-mirror design provides a diffraction-limited field of nearly 1°. The envelope (of 4.3 m diameter) indicates how the interferometer may fit into the Ariane 5 shroud. The focal surface (F) is conveniently located near the envelope to allow cooling of the detectors. To the right an example of the calculated diffraction image at field angle 0.31°, showing the Airy disk modulated by Young's fringes.

quasars allowing a direct tie of the stellar proper motions to the extragalactic (inertial) reference frame, which is critical for dynamical interpretation of the motions.

General Relativity: The gravitational light bending at right angle to the sun is about 4 mas. This quantity has been measured by Hipparcos to 1% accuracy. At the much higher precision offered by GAIA such measurements start to become interesting as a means to discriminate between General Relativity and alternative theories.

For all kinds of studies involving stellar distances, access to *absolute* parallaxes is essential. Direct determination of absolute parallaxes requires wide-angle astrometry, utilizing the different parallax factors in different parts of the sky. Wide-angle measurements are also necessary to build a distortion-free and rigid system of coordinates and proper motions over the whole sky, which is critical for all dynamical studies of large-scale motions. In order to take full advantage of the space environment it is thus essential that a future space mission aims at *global* astrometry rather than narrow-field measurements.

3. The GAIA Concept

A possible optical configuration for the GAIA concept consists of three mechanically connected Fizeau interferometers, each with two 50 cm apertures on a 2.5 m baseline (Fig. 1). Continuous scanning, as opposed to a pointing instrument, requires a large field ($\sim 1^\circ$ diameter) in order to accumulate sufficient observing time on each object, and hence to reach faint objects. All objects within the field of view can be observed simultaneously, resulting in extremely high efficiency in terms of the astrometric yield. The large field can only be realized with a Fizeau-type interferometer, where the two primary mirror elements (M_1 in Fig. 1) are parts of an imaginary single primary surface.

Preliminary studies indicate that this instrument could yield positions, parallaxes and annual motions to the level of 20 microarcsec at $V = 15$, and possibly down to a few microarcsec at $V = 10$. A mission length of 5 years should be targetted, especially in view of the very much improved ability to detect planetary companions and measure orbital binaries with periods of a few years. The limiting magnitude will be around $V = 16$, as set mainly by background light and confusion from other stars in the detector subfields (see below). No pre-selection of programme stars is required; the instrument would simply observe all objects above a given threshold.

3.1. DETECTORS

Exploitation of the full information at the fringe frequency by means of a detector (CCD) directly at the focal surface would require a prohibitively large number of pixels. This requirement, and the corresponding tolerances on the detector performances, can be relaxed dramatically with the inclusion of a modulating element, resulting in a data collection somewhat analogous to that employed with Hipparcos.

In this concept the focal surface is covered by a grid containing a large number of slits parallel to the interference fringes of the stellar images. The grid period must match the fringe spacing $\lambda_{\rm eff}/L$, where $L = 2.5$ m is the interferometer baseline, so that the transmitted light is modulated as the star images move across the grid (Fig. 2a). The phase of the detected modulated signal provides the positional information (star image relative to the grid), while the amplitude contains photometric information. A much improved efficiency can be obtained with a completely transparent 'phase grid' (Fig. 2b), in which the phase or delay of the light is modulated instead of the amplitude, e.g., by means of a corrugated grid surface.

An integrated 'light curve' of the modulation can be recorded by a CCD in which the electric charges generated by the photons are shifted back and forth synchronized with the modulation. Alternatively, arrays

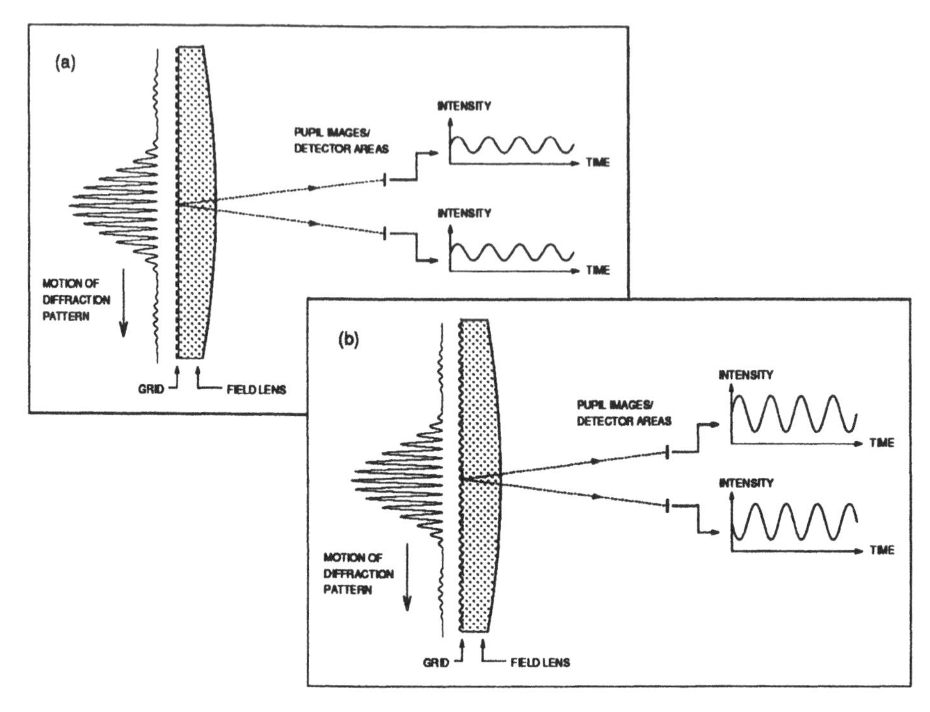

Figure 2. Interaction of the diffraction image with (a) an amplitude grid consisting of parallel slits having a period matching the interference fringes, and (b) a phase grid modulating the phase of the transmitted light. The phase grid gives more light on the detector and a stronger modulation, with the two pupil images modulated in anti-phase.

of photon-counting avalanche photodiodes (APD) could be used. These are characterized by high quantum efficiency, high time resolution, a large dynamic range, and a wide wavelength response, all of which are important qualities for the present application.

A detector option considered in parallel with these conventional detectors is based on superconducting tunnel junctions (STJ). Development work going on in ESA suggests the possibility of photon counting in the optical, using the superconducting phenomenon. Such a superconducting camera (SUPERCAM) would comprise individual pixels, as for the CCD, but with parallel rather than serial readout. The potential advantages of this detector are its very high quantum efficiency over a broad wavelength range, very high time resolution, low noise, and an intrinsic wavelength resolution of the order of 10 nm. Disadvantages are its low operating temperature of around 1 K and the high data rate generated by a large array.

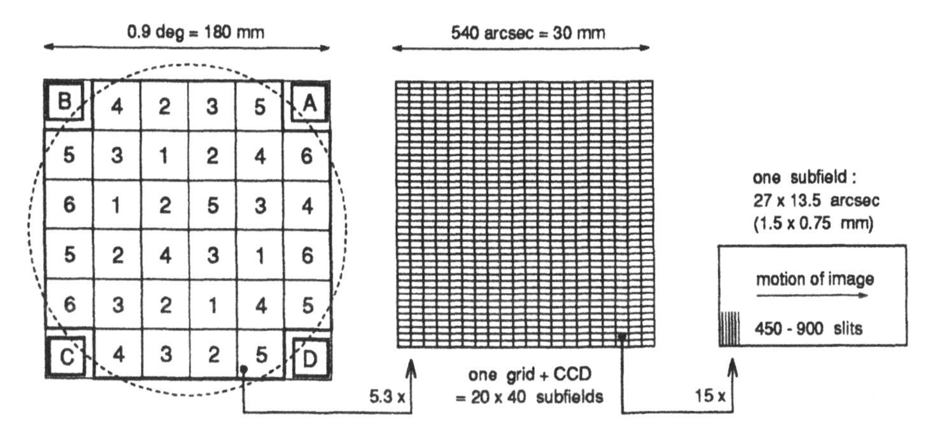

Figure 3. Focal-plane arrangement of grids and detectors. The one-degree field is divided in areas with different wavelength passbands (numbered 1 to 6 in the left diagram), each using its own CCD to record the light modulation. The areas are further divided into subfields of 27×13.5 arcsec^2. The field lens shown in Fig. 2 covers a single subfield. A–D are CCD detectors set directly in the focal surface and needed primarily for attitude determination (to ~ 2 mas) and active control of the mirrors. Additional CCDs around the central field may be used for lower-resolution astrometry and accurate photometry.

3.2. ARRANGEMENT OF GRIDS AND DETECTORS

With an observing programme of 50 million stars there are some 500–1000 objects in the field of view at any time. Ideally the modulated signal should be individually recorded for each of them. However, the light behind the grid is collected from the whole area covered by the field lens in Fig. 2. To avoid superposing the light curves of many stars, and also to reduce the background intensity level, the telescope field must be divided into a number of subfields, each equipped with its own small field lens and detector elements. Figure 3 indicates a possible focal plane arrangement. A subfield of 27×13.5 arcsec^2 gives an average background corresponding to $V \simeq 15.5$, increasing the photon statistical error by 50% at the faint limit. An array of 20×40 subfields defines an area of 9×9 arcmin^2, or 30×30 mm^2 in linear size, suitable for light registration by a single CCD.

All the subfields for a given CCD must use the same modulation frequency and hence the same grid period and the same effective wavelength. However, another CCD may use a different $\lambda_{\rm eff}$. The optimum relative bandwidth per CCD is given by the ratio of aperture diameter to interferometer baseline, or ~ 0.2. The visible wavelength range could be covered with six overlapping wavelength bands, adding multicolour photometry as a natural part of the mission. The use of six different grid periods would also solve the phase ambiguity problem for the reconstruction of point images from the detector signal.

3.3. WIDE-ANGLE MEASUREMENTS

Two identical interferometers, with optical axes set at a large angle (1 to 2 rad) to each other and perpendicular to the spin axis, are required to bridge the long arcs in a sufficiently short time to overcome intrinsic variations of the instruments. For redundancy we propose to have three interferometers stacked on top of each other, any two of them being sufficient for the nominal mission goals. Using the 360° closure condition on every complete great-circle scan means that the astrometric results are very insensitive to all instrumental variations on time scales longer than a few hours. A crucial question is the actual requirement in terms of the short-term stability, in particular of the 'basic angle' between the two interferometer axes. Monitoring of short-term variations of the relative geometries of the two instruments by internal metrology is probably necessary.

4. Conclusion

There is a very strong scientific case for global optical astrometry at the 20 μas accuracy level. Such measurements cannot be obtained from the ground due to the atmosphere and partial sky coverage, but are well suited for a space mission using a free-flying satellite of moderate size. We consider that a small space interferometer dedicated to global astrometry, such as GAIA, would be a suitable candidate for implementation within ESA's 'Horizon 2000 Plus' programme.

Acknowledgements

Optical studies for GAIA have been performed by Sacha Loiseau, Stuart Shacklan and Mario Lattanzi. We thank them and many other colleagues, including members of the Hipparcos project and of the various space interferometry study teams, whose support and ideas for a future astrometric mission have inspired this work.

References

Høg, E. (1994) A New Era of Global Astrometry. II: A 10 Microarcsecond Mission, this volume

Kovalevsky, J. and Turon, C. (1992), Scientific Requirements for Very Accurate Astrometry, *Advances in Space Research*, **Vol. 11**, pp. (2)5–(2)12

Lindegren, L., Perryman, M.A.C., Bastian, U., Dainty, J.C., Høg, E., van Leeuwen, F., Kovalevsky, J., Labeyrie, A., Loiseau, S., Mignard, F., Noordam, J.E., Le Poole, R.S., Thejll, P. and Vakili, F. (1994) GAIA — Global Astrometric Interferometer for Astrophysics, *Amplitude and Intensity Spatial Interferometry II*, ed. J.B. Breckinridge, SPIE, **Vol. 2200**, pp. 599–608

Ridgway, S.T. (1993) The scientific support for a space interferometry mission, *Spaceborne Interferometry*, ed. R.D. Reasenberg, SPIE, **Vol. 1947**, pp. 2–11

EUROPEAN SPACE INTERFEROMETRY

J.E. NOORDAM

Abstract. Optical interferometry is ensconced as an 'area of future interest' (a socalled Green Dream) in Horizon 2000, the long-term scientific plan of ESA. Over the years, there have been three large ESA workshops on Space interferometry, where many different concepts and designs were proposed, and several ESA committees have studied the possibilities. These committees were also involved, in an advisory role, in a modest technological research program (TRP) by ESTEC. In 1990, the Space Interferometry Study Team (SIST) recommended building an optical interferometer, consisting of 10-15 small telescopes attached to an 100m inflatable structure, as a scientifically interesting first step. The SIST even produced a workable design. It quickly became clear, however, that such an undertaking would cost much more than an ESA cornerstone mission, and was thus far too ambitious. Simultaneously, another ESA study team (LIST) came to the conclusion that the Moon, contrary to earlier beliefs, does not offer a particularly suitable environment for interferometry. At the Beaulieu workshop in 1992, it was decided to try to achieve cornerstone status for one or two smaller interferometry missions in Space: a 10m UV imaging interferometer, or an interferometric successor to the astrometry satellite Hipparchos. The latter seems to have a good chance at the moment, in the form of the GAIA proposal which has been selected for further study for the new 'post-Horizon 2000' program. GAIA may have some limited imaging capability, but a true imaging interferometer in Space will have to wait for a few decades yet.

E. Høg and P. K. Seidelmann (eds.),
Astronomical and Astrophysical Objectives of Sub-Milliarcsecond Optical Astrometry, 345.

ASTRONOMY FROM A LUNAR BASE

S. VOLONTE
European Space Agency
8–10 rue Mario Nikis
75738 Paris Cedex 15
France

ABSTRACT. The Moon is generally considered to be an ideal site for astronomy, offering excellent observing conditions and access to the entire electromagnetic and particle spectrum. A wide range of astronomical observations can be carried out from the Moon, but, as concluded in a recent ESA study (Mission to the Moon 1992), only a restricted number could be better implemented from a lunar site rather than from any other location. Very low frequency (VLF) astronomy, astrometry and interferometry fall into this category, as well as a transit telescope to map dark matter in the Universe. Whilst VLF and astrometric telescopes should be automatic, long baseline interferometers will probably require human intervention and will thus benefit from a manned lunar base.

1. Rationale for astronomy from the Moon

The Moon is regarded as an attractive site for astronomy, offering significant advantages over ground based observatories. It is very stable with only low levels of seismicity. Its attitude is very well known. The absence of atmosphere gives access to the entire electromagnetic and particle spectrum and, moreover, the far side is free from electromagnetic interferences.

Some of these advantages are also common to free flyers but the main benefits from the Moon are its mechanical and thermal stability and its large size giving the possibility of extendability and serviceability. There are some drawbacks such as micrometeorite impacts causing damage to instruments, pollution due to naturally levitated dust (or caused by vehicle or man made activities), large temperature variations between night and day, important thermal radiation from the lunar soil and no access to the full celestial sphere. But, a major problem is the considerably higher cost of Moon based instruments compared to orbiting instrumentation.

Nevertheless, in the context of a significant lunar programme (including manned activities) synergies could develop between lunar exploration missions and astronomical observations (e.g. sharing of power resources, communication, etc.), which would favour the development of almost any kind of astronomical application (Mumma and Smith 1990). Cost implications make it imperative that Moon based observations offer essential, if not unique, advantages over any other observational locations. Based on such considerations, a recent ESA study (Misson to the Moon 1992) has concentrated on three types of instrumentation for a lunar implementation; Radiotelescopes for Very Low Frequency (VLF) astronomy; Instruments for high precision astrometry, including Earth–Moon VLBI; Imaging interferometers for observation from UV to submillimeter wavelengths. More recently a lunar transit telescope has also been proposed as an attractive lunar instrument to search for dark matter in the Universe.

2. VLF Astronomy

VLF represents the extreme low energy part of the electromagnetic spectrum, corresponding to decametric (10–15MHz) and longer wavelengths. It is an almost unexplored window which would

E. Høg and P. K. Seidelmann (eds.),
Astronomical and Astrophysical Objectives of Sub-Milliarcsecond Optical Astrometry, 347–350.

allow to access the lowest practicable physical limit to study the radiation from as diverse objects as the Earth itself, the Sun, the giant planets, the spiral arms of own and other normal galaxies. Due to ionospheric absorption, VLF is practically unobservable from the ground except occasionally at a few favourable locations. Full exploration of this domain therefore awaits deployment in space and an ideal location would be the lunar far side free from electromagnetic interferences.

The resolution imposed by the angular size of discrete radio sources calls for large apertures achievable by interferometry. But interplanetary and interstellar scattering limit the ultimate angular resolutions from a few degrees to a few tens of arcsec translating into baselines from one to hundreds of kilometers. On the Moon, such VLF systems could consist of arrays of antenna deposited on the far side, connected to light, low–power receivers working at low data rate, allowing to carry out searches for extra–solar planets around nearby stars, as well as extending VLF astronomy towards extragalactic targets.

3. Optical Astrometry

The high angular resolution achievable by interferometry provides a most attractive capability for the future development of astrometry. In principle microarcsec accuracies can be achieved with baselines extending from 100 m to 1 km and faint magnitudes (m_v ~ 20) can be reached with 1 m–class telescopes. The scientific case for astrometry at this level is very strong indeed (Ridgway 1993) and the the Moon would appear as a suitable platform for this class of astrometric measurements. However, regarding global astrometry, interferometric measurement of large angles require either very large delay lines and movable telescopes, or re–orientation of the whole instrument. Since these motions must be controlled to an accuracy of a few nanometres and full sky visibility is needed, the Moon does not seem to be a convenient site for global astrometry which, indeed, is best achieved from free flyers.

For small angle measurements (small angle astrometry) of e.g. parallaxes and proper motions relative to background quasars, internal motions of clusters and nearby galaxies, search for planets, etc., the Moon, with its well known slow and smooth rotation could be useful. For instance, a fixed east–west interferometer with short delay lines could observe the transit of objects across a strip of e.g. 10 arcmin along the local meridian, providing one dimensional coordinate difference tied to the local lunar rotation rate. However, the sky coverage would be very limited and no astrophysically useful proper motions would be provided unless several different baseline orientations would be possible. Thus small angle astrometry from the Moon would be of restricted application even with very long baselines and several possible orientations.

4. Earth – Moon very long baseline interferometry (VLBI)

A single radio telescope located on the Moon, working in conjunction with ground telescopes would permit VLBI with a single baseline of some 400 000 km. Since the Moon's distance and position can be determined accurately to a few mm and milliarcsecs respectively, the Earth–Moon VLBI system would be particularly suited for very accurate astrometry. The dramatic increase in angular resolution would allow, for instance, to build a network of reference extragalactic sources to unprecendented accuracies (microarcsecs).

However such baselines limit detection to sources with high brigthness temperatures and interstellar scattering increases the apparent size of objects with wavelength, thus imposing the use of large high frequency antennae. Nevertheless, a precursor system could use a relatively low frequency single medium class antenna (~ 10 m) with a passively cooled receiver. The implementation of a more ambitious system would be dictated by transportation and construction issues.

5. Lunar Interferometry

The ultimate goal in modern astronomy is to reach diffraction limited imaging at all wavelengths for objects of small angular sizes. However, in the UV to submillimeter range the Earth's atmosphere either absorbs radiation (partially or totally) or its fluctuations restrict the spatial resolution (~ 1 arcsec in the visible). Techniques now exist (speckle interferometry, adaptive optics) which allow to approach the theoretical resolving power of ground based filled apertures. But the ultimate resolutions are reached with interferometers which combine coherently the signals of several telescopes. For ground based interferometers, the atmosphere limits the sensitivity because large, individual apertures are illuminated coherently only over small areas. In space, full sensitivity can be achieved at once but the apertures will be smaller. The Moon therefore seems to offer a significant potential as it combines most of the advantages of ground based and space borne observations.

The scientific case for lunar interferometry has been extensively made (Mission to the Moon 1992, ESA Interferometry Review Panel 1994). In the optical range, the long term goal is to provide for imaging at kilometric baselines with angular resolution better than 100 microarcsec. An imaging interferometer requires more than two telescopes to obtain good quality images and allow phase closure techniques to be applicable. The Moon's slow rotation rate imposes a sufficient number of telescopes in the system to obtain adequate uv plane coverage within a reasonable exposure time. In addition, long optical delay lines are required to compensate for the pathlength differences due to this rotation. This is indeed, the most difficult problem of a Moon interferometer because weak signals might need to travel over long distances close to the Moon's surface, possibly through a thin atmosphere of electrostatically levitated dust. The complexity of the system increases with the number of telescopes, which in turn, will remain limited, thus requiring the telescopes to be movable to provide the required spatial frequency coverage. In other words, a imaging optical interferometric array will be an extremely complex instrument and the need to deploy it on the Moon cannot be evaluated on scientific grounds alone. The choice will be dictated by technical and cost considerations related in particular to deployment and transportation issues.

In the infrared and submillimetre range, since the Earth atmosphere is almost completely opaque to this radiation, the issue here is the choice of the best location in space, the Moon being a possible site. The long term goal is to provide imaging at high spatial resolution in both the lines and continuum. This will be limited by sensitivity requirements imposing use of large antennas (a few meters in diameter). Sensitivity will also restrict the achievable angular resolution to about 0.1 arcsec even for the brightest sources thus limiting the baselines (a few tens to a few hundred meters) but favouring good imaging. So for this wavelength domain sensitivity requirements impose a large total collecting area necessitating large structures and probably deployable antennas. From the science standpoint, it is not obvious without further evidence, to conclude that a Moon based submillimetre interferometer would offer any definitive advantages over a free flying instrument. In both cases, the instrument would be very complex.

6. Lunar Transit Telescope

Fort and co–workers (Fort and Mellier 1994) have recently proposed to use the dense population of background galaxies as a reference grid at large distance to look at the statistical gravitational image distortions to detect the contours and masses of foreground structures. In other words, when the light emitted from distant sources passes near large deflecting masses, the beams are weakly bent thus introducing in the cross section a small gravitational ellipticity perpendicular to the projected gravity field.

On the ground, this effect is detected up to amplitudes of ~ 1–5 % allowing to study dense halos of dark matter around clusters and groups of galaxies. Extrapolation to the largest scale structures of the Universe requires to detect amplitudes of about 3.10^{-4} with a spatial resolution of about one degree. Even with the best seeing conditions, instrumental distortions of ground based observations cannot be reliably calibrated to this level due to a variety of reasons (minute mechanical flexures, tracking jitter, fast seeing changes and atmospheric refraction effects). The Moon, with its lower gravity and its stable, slow rotation rate would allow to overcome these problems with a fixed 1 m transit telescope (0.5 to 1 degree field of view) observing long strips of the sky with a large CCD mosaic. Preliminary simulations indicate that this technique would allow to reach the required accuracy. In addition, such a telescope could be used for deep surveys at high spatial resolution and it could also be seen as an automatic precursor telescope of more elaborate interferometric instruments.

Conclusions

Although almost any kind of astronomical observations could be carried out from the Moon, only a restricted number could make best use of the Moon environment. As outlined in this paper, some ambitious projects can and probably will, at some stage, be implemented on the Moon. The complex, long duration observations required by these lunar instruments will be more efficiently implemented for remote, automatic use. However, human intervention is expected to be requested in the deployment and maintenance of such complex facilities as e.g. large interferometric systems. Potential nuisances due to human presence and related activities (vehicles, etc.) will have to be limited as much as possible. In particular the lunar far side will need to be protected from electromagnetic pollution.

References

ESA Interferometry Review Panel, 1994, Report, in preparation
Fort B. and Mellier Y., 1994, Astron. & Astrophys. Review, in press
Mission to the Moon, 1992, ESA SP–1150
Mumma M.J., Smith H.J., Editors, 1990 : *Astrophysics from the Moon,* American Institute of Physics, Conference Proceedings 207, New York
Ridgway S., 1993, *The scientific support for a space interferometry mission; Spaceborne Interferometry,* SPIE Vol. 1947, Reasenberg R.D., Editor

POSTERS

1.1 DEVELOPMENTS IN GROUND-BASED ASTROMETRIC TECHNIQUES AND LARGE CATALOGUES

THE NEW CATALOGUE OF POSITIONS AND PROPER MOTIONS OF 5115 BRIGHT STARS COMPILED AS THE RESULT OF THE INTERNATIONAL "BRIGHT STARS" PROGRAM

O.A.MOLOTAJ, V.V.TEL'NYUK-ADAMCHUK, N.A.CHERNEGA, N.D.KANIVEC'
Astronomical Observatory of Kyiv Taras Shevchenko University
3 Observatorna str.
254053 Kyiv-53
Ukraine

ABSTRACT. At Kyiv University Observatory, in accordance with IAU resolution, the Combined Catalogue of Positions and Proper Motions of 5115 Bright Stars in the FK5 system for Epoch and Equinox J2000.0 (BSC, see reference) has been compiled using 20 source catalogues obtained within the framework of 'Bright Stars' International Program. The source catalogs are Bordeaux-66, Bucharest-65, Bucharest-68, Belgrade-79, Cape-68, Kharkiv-70n, Kharkiv-70s, Kyiv-66, Kyiv-73, El Leoncito-70, Moscow-76, Mykolaiv-65, Perth-70, Perth-75, Santiago-65, Santiago-67, Strasbourg-65, Tashkent-64, Tokyo-68, Washington-66. The BSC accumulates around 30,000 source catalogue positions observed within 1960-1980. The standard errors of BSC do not exceed 0.1 arcsec and 0.25 arcsec/cy for positions and proper motions, respectively. The systematic differences between BSC and FK5 in positions for the mean BSC epoch 1970 as a rule do not exceed several hundredths of arcsec. A comparison of the BSC with the PPM catalogue indicates that in the latter the accuracy of positions and proper motions of the ordinary bright stars is slightly worse than that of average values given by the authors of the PPM. It is important to note that the comparison of the BSC with PPM for various star subsets shows deviations of both the HPS bright star system and the rest of the PPM bright stars, on the one hand, from the stars of the FK5 Extension contained in the PPM on the other hand. To examine the BSC proper motion system the stellar astronomy constants and correction to precession have been determined using the BSC proper motions. The parameters obtained agree with the standard ones. But the Oort constants are slightly smaller than those of standard.

The BSC catalogue as a whole is acceptable for astrometry usage with respect to both accidental and systematic accuracy.

Reference

Molotaj, A.A., Tel'nyuk-Adamchuk, V.V., Chernega, N.A., Kanivec', N.D. (1992) 'Compiled Catalogue of Positions and Proper Motions of 5115 Bright Stars for Epoch and Equinox J2000.0'. Kiev, Zovnishtorgvydav Publisher. 132 p.

E. Høg and P. K. Seidelmann (eds.),
Astronomical and Astrophysical Objectives of Sub-Milliarcsecond Optical Astrometry, 353.

A PROGRAM OF INFRARED ASTROMETRY

A.S. KHARIN
Main Astronomical Observatory National
Academy of Sciences of Ukraine
Goloseevo, Kiev-22, 252650
Ukraine

There exist no special astrometric telescopes or devices suited to determine precise positions, proper motions and parallaxes in the infrared. Only astrophysical IR observations are carried out now in the range of 1-350 μm. But many astronomical problems concerning galactic structure and kinematics and others may be solved if astrometric instruments and methods are developed and implemented for IR observations. Development of new technology and infrared array receivers gives the hope that many astrometric problems in the infrared would be solved successfully.

The main contents of the proposed program consist of the three tasks:

1. Construction and manufacturing of IR astrometric instruments.
2. Creation and extension of an IR reference catalogue.
3. Connection of optical, IR and radio coordinate systems.

Besides of the above mentioned problems it is necessary to point out two important scientific problems that may be solved after precise IR astrometric instruments and methods are developed.

– Finding of close double stars or multiple systems that have different coordinates in infrared and visual or in IR and radio ranges. It is obvious that these stars must be excluded from all precise reference catalogues – optical, infrared and radio.

– Finding planetary systems of the nearest stars.

The proposed program must be coordinated with all other IR astronomical programs. Possibly some of these may be used also to fulfill a part of the above astrometric problems. For example the observations planned with the DENIS project could also be used for more precise position determinations of many stars in the I,J,K bands. A large (about 2 million stars) preliminary IR reference catalogue should be prepared.

E. Høg and P. K. Seidelmann (eds.),
Astronomical and Astrophysical Objectives of Sub-Milliarcsecond Optical Astrometry, 354.

JOINT PROGRAM FOR AUTOMATIC ANALYSIS OF ASTROGRAPHIC PLATES

A.LOPEZ GARCIA, J.L.VALDES NAVARRO, A.ORTIZ GIL AND J.M.MARTINEZ GONZALEZ
Valencia University Observatory
46010, Valencia, Spain
e-mail: obsast@vm.ci.uv.es

AND

V.N.YERSHOV, E.V.POLIAKOV, H.I.POTTER AND L.YAGUDIN
Pulkovo Observatory
196140 Saint Petersburg, Russia
e-mail: yersh@gao.spb.su

Pulkovo and Valencia observatories are developing measuring machines based on the "Ascorecord" microscope that allow to undertake automatic measuring programs of different kind of plates (Ortiz *et al.*, 1993). All aspects of astrometry work are covered now by our software: field maps and plate files creation from the PPM and the GSC; precise ephemeris calculation of minor planets and comets, including planetary perturbations; image identification and automatic plate measurement and reduction (Lopez *et al.*, 1994a). For these tasks, correction tables for GSC catalogue vs. PPM, systematic errors (Lopez *et al.*, 1994b) have been obtained. Observing programs using plates and CCD devices at telescope focus are planned (the search of new asteroids and comets, big planets satellites observation, new high density photographic catalogues and some other projects).

References

Lopez Garcia, A., Ortiz Gil, A., Martinez Gonzalez, J.M. and Poliakov, E. (1994a), Newsletter, No. 5. IAU WG on Wide-Field Imaging. (1994) UK.

Lopez, G.A., Martinez, G.J.M., Ortiz, G.A. and Yagudin, L.I. (1994b), Newsletter, N. 6. IAU WG on Wide-Field Imaging. (1994). UK. (In press).

Ortiz, G.A., Lopez, G.A., Martinez G.J.M. and Yershov, V. (1993) Automatic measurement of images in astrometric plates, *Proc. IAU Symp.160,*, UK (in press).

E. Høg and P. K. Seidelmann (eds.),
Astronomical and Astrophysical Objectives of Sub-Milliarcsecond Optical Astrometry, 355.

NEW MEASUREMENTS OF POSITIONS FOR THE ASTROGRAPHIC CATALOGUE STARS

G.A.IVANOV, V.S.KISLYUK, L.K.PAKULYAK, A.V.SERGEEV, T.P.SERGEEVA and A.I.YATSENKO

Main Astronomical Observatory, National Academy of Sciences of Ukraine, Golosiiv, 252650 Kiev-22, Ukraine, maouas@gluk.apc.org

Six observatories of the former USSR carry out photographic observations within the FON project (the fourfold coverage of the northern sky) since 1982. The same type wide-angle Zeiss astrographs (D=400 mm, F=2000 and 3000 mm) are used. At present Golosiiv Observatory almost finished the Kiev observational part of the project and the measurements of onefold coverage plates have been started. The special measuring machine PARSEC (Programming Automatic Radial-Scanning Coordinatometer) has recently been constructed for this purpose.

We intend to apply the measurements of the FON project plates to the determination of new positions and proper motions of all stars in the Astrographic Catalogue (AC) using the last one as the first epoch of observations. AC is applied also as input catalogue for PARSEC. About 500 plates have been measured at the moment and proper motions for more than 500000 stars in different areas of the northern sky have been obtained.

The accuracy of the FON positions and proper motions of stars from the measurements of onefold coverage plates is estimated to be 0.35 arcsec and 0.006 arcsec/yr, respectively.

Futher investigations will be carried out in cooperation with Sternberg Astronomical Institute (Moscow, Russia) and Astronomisches Rechen-Institut (Heidelberg, Germany).

The above-described reasearches became possible partly by Grant N V41000 from the International Science Foundation.

E. Høg and P. K. Seidelmann (eds.),
Astronomical and Astrophysical Objectives of Sub-Milliarcsecond Optical Astrometry, 356.

PROGRESS REPORT ON THE SOUTHERN PROPER MOTION PROGRAM

I. PLATAIS, T. M. GIRARD, V. KOZHURINA-PLATAIS
R. A. MENDEZ AND W. F. VAN ALTENA
Yale University Observatory
P.O. Box 208101, New Haven, CT 06520, U.S.A.
C. E. LOPEZ
Felix Aguilar Observatory
5413 Chimbas, San Juan, Argentina
AND
W. Z. MA
Beijing Normal University Observatory
Beijing 100875, China

We present the status of the Yale/San Juan Southern Proper Motion program (SPM) which is the southern hemisphere extension of the Lick Observatory Northern Proper Motion program with respect to faint galaxies (Platais *et al.*, 1993). To date, measurements and reductions in the South Galactic Pole region comprising $\approx$ 1000 square-degrees on the sky have been finished. At this stage of the SPM program particular attention has been paid to the plate model choice along with an assessment of and accounting for systematic errors. For our establishing of a secondary reference frame we have noticed the presence of a potentially dangerous effect, so-called field-independent coma which is caused by lens decentering. We acknowledge the superb Hipparcos preliminary positions without which such analysis would be virtually impossible. The SPM data at the SGP region have also been used to constrain a multi-component Galaxy model. First results of this analysis are presented.

References

Platais, I., Girard, T.M., van Altena, W.F., López, C.E. (1993) in *Workshop on Databases for Galactic Structure*, eds. A.G. Davis Philip, B. Hauck and A.R. Upgren, p. 153.

E. Høg and P. K. Seidelmann (eds.),
Astronomical and Astrophysical Objectives of Sub-Milliarcsecond Optical Astrometry, 357.

FIRST RESULTS OF THE BORDEAUX MERIDIAN CIRCLE EQUIPPED WITH A CCD DETECTOR

Y.RÉQUIÈME, G.MONTIGNAC, J.F.LE CAMPION, F.BOSQ and F.CHAUVET
Bordeaux Observatory, B.P.89, F-33270 Floirac, France
and
P.BENEVIDES-SOARES and R.TEIXEIRA
Instituto Astronomico e Geofisico USP, 04301-904 São Paulo, Brazil

The Bordeaux automatic meridian circle has been working for 10 years with a tracking photoelectric micrometer. This micrometer was recently replaced by a focal CCD camera in order to extend the differential position measurements to much fainter stars with improved accuracy. The main features of the Bordeaux CCD instrument are presently :

- 20 cm (f/12) meridian refractor : 87.5"/mm.
- Thomson 7895A thick CCD (MPP technology) 512*512, cooled at -40 deg.C. by Peltier thermoelectric elements and water circulation. Pixel size : 19 microns (1.65").
- Filter GG495 + BG38 : effective bandwidth 5200-6800 Å.
- Imaging in drift-scanning mode. Equivalent exposure time 56 secδ seconds.
- Real-time continuous display of the observed star field.

Preliminary tests on the sky carried out from March 1994 onwards give clear evidence of slow image motion due the atmosphere (Fig.1 : raw differences night to night).

Observing narrow strips with the instrument stationary and taking as reference the final Hipparcos and Tycho positions, the expected accuracy on the positions is 0.03"-0.06" for four-fold measurements in the magnitude range $9 < V < 15$ (Fig.2). In addition, magnitudes will be obtained with a mean error of 0.1-0.3 m.

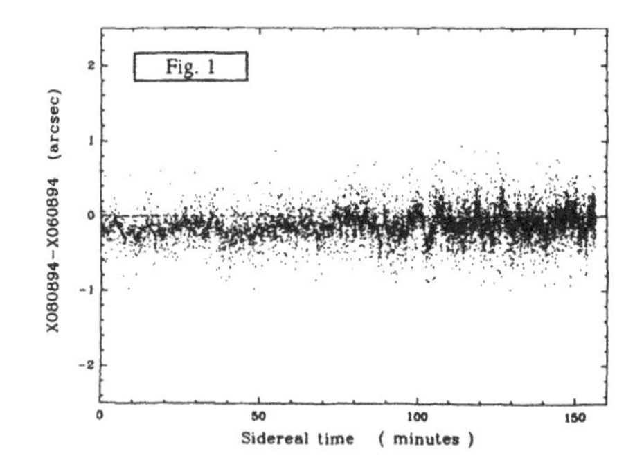

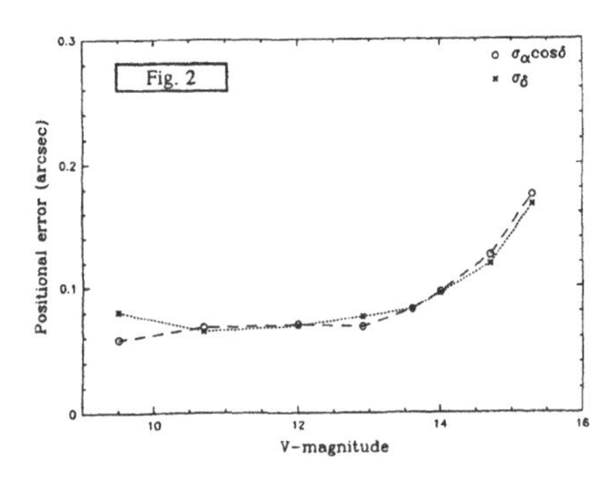

Scientific programmes are extension of the Hipparcos-Tycho frame up to V=15-16, link to the VLBI primary reference frame via faint radiostars, positions of satellites of planets and asteroids, survey of faint star proper motions in selected areas.

The first camera will be moved soon on the São Paulo meridian circle, a new one with a 1024*1024 CCD will be mounted on the Bordeaux meridian instrument.

ACKNOWLEDGEMENTS - This project was partially supported by the French CNRS (INSU) and the Aquitaine Region, and also by CNPq (Brazil) and CNRS through travel grants.

E. Høg and P. K. Seidelmann (eds.),
Astronomical and Astrophysical Objectives of Sub-Milliarcsecond Optical Astrometry, 358.

THE OBSERVATIONS OF GLIESE 623 AND SOME OTHER OBJECTS WITH SUSPECTED UNSEEN COMPONENTS

N.A.SHAKHT
Pulkovo Observatory, 196140, St Petersburg, Russia
e-mail: ddp@gaoran.spb.su

The star Gliese 623 ($RA = 16^h22.6^m$; $Decl = +48°28'$ [1950]), $\pi = 0.''138$; $m_v = 10.3$; Sp dM3, (Woolley *et al.*, 1970)) is observed at Pulkovo by means of the 26-inch Refractor. The series of 95 plates (about 600 individual positions) obtained during 1979 - 1994 has been studied. This set has shown perturbations in the motion of this star under the influence of a dark component which has been investigated earlier by different methods (see, for instance, Marcy & Moore, 1989). The following values of relative proper motion and parallax with their mean errors and preliminary elements of the photocentric orbit have been adopted on the base of our observations: $\mu_x = +1.''1483 \pm 0.''0010$; $\mu_y = -0.''4462 \pm 0.''0009$; $\pi_x = 0.''126 \pm 0.''019$; $\pi_y = 0.''136 \pm 0.''017$; $P = 3.76$ years; $e = 0.50$; $T_\circ = 1984.28$. With the use of modified Thiele - Innes constants the following elements of the orbit have been determined: $\alpha = 0.''052 \pm 0.''006$ (m.e.) and $i = 141°$; $\omega = 265°$; $\Omega = 126°$. We may suppose that the lower limit of the mass of the component is about 0.09 masses of the Sun. Our study has shown small periodic deviations in the distances between the components of δ Gem and ADS 11632, which may be evoked by substellar and stellar companions of low masses (Shakht, 1988).

References

Marcy G.W. and Moore D., (1989), The Extremely Low Mass Companion to Gliese 623, *Aph.J.*, **341**, pp. 961–967.

Shakht N.A. (1988), A Study of the Motion and a Determination of Kinematic Parameters of the Stars with Invisible Satellites Using Observations at Pulkovo *Izv.GAO*, **205**, pp. 5–14.

Woolley, R., *et al.*,, (1970), Catalogue of Stars within 25 parsecs of the Sun, *Royal Obs. Annals*, **5**, pp. 5–227.

E. Høg and P. K. Seidelmann (eds.),
Astronomical and Astrophysical Objectives of Sub-Milliarcsecond Optical Astrometry, 359.

A GIANT MERIDIAN CIRCLE - REFLECTOR

V.N.YERSHOV
Pulkovo Observatory,
196140 St.Petersburg, Russia
e-mail: yersh@gao.spb.su

A 1.5 m reflector is proposed for infrared and optical meridian observations in order to extend the fundamental coordinate system to faintest objects and to the K-infrared waveband. Classical meridian circles are unfit for the infrared observations because their lens objectives do not give good images in the infrared. But reflectors are almost never used as meridian circles due to uncertainties in their optical axis position. The main problem is that the secondary mirror is not connected with the micrometer and the circle reading system. In order to overcome this difficulty the author proposes to use an intermediary focal plane between the primary and the secondary mirrors where a luminous reference grid of wires might be placed. The Gregory optical scheme has such a focal plane, and its secondary mirror forms images of a star and the grid at the micrometer's detecting area. At the same time a special champher around the primary's central hole forms anautocollimated image of the grid near the grid itself. The micrometer measures the star image coordinates relative to two images of the reference grid. So, observations will not be affected by displacements of the secondary mirror and by those of the micrometer. The telescope's equivalent focal length has been chosen as 3 m, and the optical system has been transformed into an aplanatic Mersenne combined with an aplanatic focal reducer corrector (Popov, 1988). A new autocollimated circle reading system is chosen for the instrument (Yershov and Nemiro, 1994). The observations will be linked to the fixed optical axis of two long-focus collimators placed at the prime vertical plane.

References

Popov, G.M. (1988), Modern Astron. Optics, *"Science"*, Moscow, 146-147.
Yershov, V.N., Nemiro A.A. (1994), An Autocollimation Circle Reading System for the Infrared Merid. Instr., *IAU Symp.166*, The Haague, 1994

E. Høg and P. K. Seidelmann (eds.),
Astronomical and Astrophysical Objectives of Sub-Milliarcsecond Optical Astrometry, 360.

AN AUTOCOLLIMATION CIRCLE READING SYSTEM FOR THE INFRARED MERIDIAN INSTRUMENT

V.N.YERSHOV AND A.A.NEMIRO
Pulkovo Observatory,
196140 Saint Petersburg, Russia,
e-mail: yersh@gao.spb.su

A new autocollimation circle reading system is proposed for the reflector meridian circle (Nemiro and Streletsky, 1988). The instrument will be used for observations in the K-infrared waveband. Instead of the divided circle fixed to the instrument tube the new system has small spherical mirrors polished at the lateral surfaces of the primary mirror. The primary mirror is made from sitall and has an autocollimation system aimed at monitoring its optical axis position. The small spherical mirrors of the circle reading system link the circle readings with the primary's optical axis. The divided circles are fixed unmovable opposite to both lateral surfaces of the primary's optical block. Both surfaces have four spherical mirrors. The distance between the divided circles and the mirrors is equal to the mirrors' radii of curvature. The scales of each circle are illuminated from outside (where the measuring microscopes are placed). The mirrors form autocollimated images of the divisions at the plane of the divisions itself. Averaged coordinates of a division and its autocollimated image give the position of the mirror's optical center, and the semi-difference of the coordinates gives the angular position of the telescope. So, the measurements of the circle positions are differential ones, and any displacements of the microscope zero-points are not critical. The precision of measurements is estimated to be better then 0.05" (random) and 0.005" (systematical). The work was supported by the Russian Foundation of Fundamental Investigations (the project's code is 93-02-17095).

References

A.A.Nemiro, Yu.S.Streletsky, (1988), The Reflector Meridian Circle, *Izv.GAO*, No.205, Leningrad, 15-17.

E. Høg and P. K. Seidelmann (eds.),
Astronomical and Astrophysical Objectives of Sub-Milliarcsecond Optical Astrometry, 361.

STELLAR DIAMETER MEASUREMENTS WITH FIBER OPTIC DOUBLE FOURIER INTERFEROMETRY —EXPERIMENTAL STUDY

PEIQIAN ZHAO[1,2], V. COUDÉ DU FORESTO[1], J.-M. MARIOTTI[1], P. LENA[1] AND BIFANG ZHOU[2]
[1]*Observatoire de Paris-Meudon*
DESPA (CNRS, URA 264)
F-92195 Meudon, France
[2]*Nanjing Astronomical Instruments Research Center*
Chinese Academy of Sciences
Nanjing, 210042, China

Abstract. Long baseline optical interferometry has been successfully employed to measure the diameters of stars. In this technique, bandwidth smearing can affect the measurement accuracy. These bandwidth smearing effects can be, to some extent, eliminated by dividing the whole observing spectral band into sub-bands and calculating the star's diameter based on the visibilities and spatial frequencies at the corresponding sub-bands. In the visible range, dividing the whole spectral band can be implemented by introducing a spectrograph, while in the IR domain, this operation can be performed efficiently with the technique of double Fourier interferometry (DFI) without losing the advantage of multiplexing. In particular, the use of IR single-mode fiber optics for DFI will make the interferometer extremely compact, light, insensitive to surrounding conditions, etc. We established an IR single-mode fiber optic double Fourier interferometer in the laboratory, in which the optical path difference modulations are generated by stretching fiber arms and the beam combination is carried out with a fiber optic directional coupler. In this paper, we report on experiments and experimental results from measurements of the diameter of an artificial star with the technique of fiber optic DFI.

E. Høg and P. K. Seidelmann (eds.),
Astronomical and Astrophysical Objectives of Sub-Milliarcsecond Optical Astrometry, 362.

STELLAR IMAGE LOCATION WITH MAXIMUM CORRELATION PROCEDURE

ZHU ZI
Shaanxi Astronomical Observatory
P.O.Box 18, Lintong, China

ABSTRACT The location method and accuracy analysis has become a very important subject in astrometry since the photoelectric technique was widely applied. The location methods of the stellar images observed with the slit micrometer are discussed in this paper, and a new estimation method called the maximum correlation procedure is presented. This procedure has already been used to process and analyze the observational data. Because of the ideal mathematical performance of the correlation function, the procedure provides us a good way to determine image locations. By means of simulated observations according to the theoretical analysis, we find that the locating accuracy and the limiting magnitude for this method is much higher than those of traditional locating method. For the observational data of about one thousand stars obtained with the Photoelectric Astrolabe at Yunnan Astronomical Observatory, the positional accuracy of a single star using the maximum correlation is improved by 0.045 arcsec over that with the median. The maximum correlation procedure has also increased the data reduction ability for faint star observations with very low signal to noise ratios.

References

Helmer, L., and Morrison, L.V., 1985, Vistas in Astron., **28**, 505.
Hoeg, E., 1970, Astron. Astrophys., **19**, 27.
Li Dongming et al., 1983, Astrolabes and Their Application in Astrometry(in Chinese), Sciences Press, 22.
Lindegren, L., 1978, IAU Coll., **48**, 197.
Yoshizawa, M. et al., 1985, Astron. Astrophys., **147**, 227.
Yoshizawa, M., and Suzuki, S., 1989, Publ. Natl. Astron. Obs. Japan, **1**, 127.
Zhu Zi, 1989, Publ. Shaanxi Astron. Obs., **12**, 43.

E. Høg and P. K. Seidelmann (eds.),
Astronomical and Astrophysical Objectives of Sub-Milliarcsecond Optical Astrometry, 363.

GROUND-BASED ASTROMETRY WITH SHORT-FOCUS CCD-ASTROGRAPHS

I.S. GUSEVA
Pulkovo Observatory
196140 St Petersburg
Russia

Abstract. The main complication to use CCDs for astrometry is their small linear size. Nevertheless, a wide range of traditional astrometric problems can be solved by use of short-focus astrographs equipped with available CCDs. They are: creation of large astrometric and photometric catalogues, link of optical catalogues to an extragalactic reference frame, observations of the solar system bodies, etc. It can be shown that a telescope with focal length of 1-2 meters, equipped with a single CCD can produce a huge amount of simultaneous positional and photometric observations of high precision. It will be especially effective when the Tycho catalogue will be completed, because the problem of precise reference system in a small field of view (about 1 square degree) will be solved with extremely high accuracy. The majority of stars up to 18-20 magnitude can be observed with such an instrument. The problem of resolving "close" stars arises only for some very dense areas on the sky, such as globular clusters etc. – The results of CCD-observations made at Pulkovo with a small short-focus astrograph (D=100 mm, F=712 mm, FOV=45 $\times$ 67 arcmin) in 1993-1994 show that the precision of single observation is not worse than: in right ascension – 0.15 to 0.35 arcsec, in declination – 0.10 to 0.25 arcsec (the size of pixel was 18 $\times$ 24 μm^2), and in magnitude – 0.05 to 0.15 mag in a wide range of magnitudes from the brightest objects to V=16. From a series of 5–10 frames it is possible to obtain positions and magnitudes of a large number (up to 500–600 in one frame) of stars with a precision quite comparable with that of normal astrographs and other "large" telescopes.

E. Høg and P. K. Seidelmann (eds.),
Astronomical and Astrophysical Objectives of Sub-Milliarcsecond Optical Astrometry, 364.

IMPROVEMENT OF STAR POSITIONS BY A NEW AXIAL MERIDIAN CIRCLE WITH NEGLIGIBLE SYSTEMATIC ERRORS

G.I. PINIGIN, A.V. SHULGA, P.N. FEDOROV, A.N. KOVALCHUK

AND

A.E. MAZHAEV, A.G. PETROV, YU.I. PROTSYUK
Mykolayiv Astronomical Observatory
Observatorna, 1. Mykolayiv, 327030 Ukraine

Abstract. The Mykolayiv axial meridian circle (AMC) consists of a horisontal telescope with a pentagon prism of 'CITALL' in front of the objective of D=180mm and F=2480mm in the meridian, and a fixed aligned vacuum collimator of D=180mm and F=12360mm.

Investigations of the AMC parameters show that: horizontal flexure is negligible and about −0.037"; collimation is stable and can be described by the formula $C = C' + a \cdot t$, where t[C] is the temperature, a = 0.026" and $C' \cdot$ const. The variation of collimator inclination is of the order 0.09" per $1°C$. Test observations of FK5 stars show variations of the AMC intrumental system of not more than 0.1".

A CCD micrometer is being developed. Test observations with a pilot micrometer with a CCD of 288 x 256 pixels have started. It appears that the AMC magnitude limit (with cooling) will be about 14 and the positional accuracy 0.1 arcsec.

Finally, the expected capabilities of the fully automatic Mykolayiv AMC permit its efficient use for extention of the stellar reference frame to fainter magnitudes, for linking the optical/radio reference frames. In particular, for future observations a programme of $12 - 14^m$ reference stars around extragalactic radio sources is being prepared.

E. Høg and P. K. Seidelmann (eds.),
Astronomical and Astrophysical Objectives of Sub-Milliarcsecond Optical Astrometry, 365.

POTENTIAL OF MCP BASED PHOTODETECTING SYSTEMS FOR HIGH ACCURACY ASTRONOMICAL OBJECT'S OBSERVATION

I.D. RODIONOV, M.YU. KNIZHNIKOV AND E.B. SOLOMATIN
Reagent Research and Development Center
125190, Post Box 192, Moscow, Russia, tel. 7(095) 155 4526
e-mail: rid@reagent.msk.su

An ideal receiver of radiation should record the arrival time and coordinates of each incoming photon. Available CCD arrays (accumulation mode) and microchannel plate detectors (photon counting mode) are mere approximations to the ideal configuration.

Reagent R&D Center has developed with its own technology a range of MCP based photodetecting systems (PDS) which give simultaneous recording of the coordinates and time of arrival of individual photons. A counting rate of 10^6 events per second, and a spatial resolution of 100 lines per mm have been achieved owing to a special geometry of the readout system – so called "coded collectors". The coordinate reconstruction algorithms involve a rough evaluation of the position of the electron avalanche and an accurate calculation of its center of gravity. In comparison with the MAMA detector, the present system uses more sophisticated algorithms but a simpler electronics.

The information on a point object (star) is a track in the (x, y, t) space both in a guiding mode and in a scanning over the celestial sphere. An analysis of the time coordinate enables an algorithmic evaluation and the elimination of an a priori unknown blur of the image that can occur, for example, due to vibration of the instrument.

A special software package and a supporting mathematical model have been developed to tailor the PDS to the recording of fast processes (high-speed spectroscopy and photometry) and also to high-accuracy astrometric experiments and to the recording of faint objects. The new approach to the processing of time-coordinate information enables an efficient use of MCP based PDS as detectors in space astrometric experiments.

E. Høg and P. K. Seidelmann (eds.),
Astronomical and Astrophysical Objectives of Sub-Milliarcsecond Optical Astrometry, 366.

SCHMIDT PLATE ASTROMETRY: SUBPLATE SOFTWARE

N. ZARATE, B. BUCCIARELLI AND L.G. TAFF.
Space Telescope Science Institute
Baltimore, MD 21218

1. Introduction

A portable computer program called SUBPLATE was developed to implement the sub-plate overlap technique proposed by Taff (1989 *AJ* 358,359). The main point of the technique is to divide the plate into smaller pieces and then to overlap them, each with a minimum number of reference stars to solve for a linear solution.

2. The Sub-plate Software

Because of the number of input parameters that the user needs to feed to the SUBPLATE program, a text file that can be manipulated with any editor is a requirement. A good compromise was achieved by choosing Fortran Namelist as the text file to edit, presenting a minimum number of constrains that the user can easily follow.

There are 3 namelist files. The first has the input and output information. The second one corresponds to plate and observatory information The third is a catalogue description. A description is a set of up to 6 parameters per column stating column name, starting position, fortran format, units and, scale and offset factor. This method emphasizes flexibility in accepting any kind of input catalogue format.

The program allows as input a comparison catalogue to check for a good fitting, by calculating residuals between the computed solutions and the comparison values. The user can substitute this comparison file with a list of (x,y) positions to obtain the corresponding equatorial coordinates.

This program eventually will be part of the IRAF system to make use of the Tables facilities for a better handling of the input catalogues; also for easier interaction with the STSDAS astrometric software package.

E. Høg and P. K. Seidelmann (eds.),
Astronomical and Astrophysical Objectives of Sub-Milliarcsecond Optical Astrometry, 367.

ASTROMETRIC PLATE REDUCTION WITH ORTHOGONAL FUNCTIONS AND MILLIARCSECONDS ACCURACY IN DEEP PROPER MOTION SURVEYS

DEVENDRA OJHA AND OLIVIER BIENAYMÉ
Observatoire de Besançon, BP 1615, F-25010 Besançon, France

We have been doing a sample survey in UBV photometry and proper motions as part of an investigation of galactic structure and evolution. The 3 fields in the direction of galactic anticentre (l = 167°, b = 47°), centre (l = 3°, b = 47°) and antirotation (l = 278°, b = 47°) have been surveyed. The high astrometric quality of the MAMA machine (CAI, Paris) gives access to micronic accuracy (leading to $<$ 2 mas per year) on proper motions with a 35 years time base. The kinematical distribution of F and G–type stars have been probed to distances up to 2.5 kpc above the galactic plane. We have derived the constrain on the structural parameters of the thin and thick disk components of the Galaxy (Ojha et al. 1994abc) :

• The scale lengths of the thin and thick disks are found to be 2.6±0.1 and 3.3±0.5 kpc, respectively. The density laws for stars with $3.5 \leq M_V \leq 5$ as a function of distance above the plane follow a single exponential with scale height of $\sim$ 260 pc (thin disk) and a second exponential with scale height of $\sim$ 800 pc (thick disk) with a local normalization of 5–6 % of the disk.

• The thin disk population was found with $(\langle U+W \rangle, \langle V \rangle) = (1\pm4, -14\pm2)$ km/s and velocity dispersions $(\sigma_{U+W}, \sigma_V) = (35\pm2, 30\pm1)$ km/s. The thick disk population was found to have a rotational velocity of V_{rot} = 177 km/s and velocity dispersions $(\sigma_U, \sigma_V, \sigma_W) = (67,51,42)$ km/s. No dependence with r and z distances was found in the asymmetric drift measurements of the thick disk population.

References

Ojha D.K., Bienaymé O., Robin A.C., Mohan V., 1994a, A&A 284, 810
Ojha D.K., Bienaymé O., Robin A.C., Mohan V., 1994b, A&A 290, 771
Ojha D.K., 1994c, Ph.D. thesis, Observatoire de Strasbourg, France (in preparation)

E. Høg and P. K. Seidelmann (eds.),
Astronomical and Astrophysical Objectives of Sub-Milliarcsecond Optical Astrometry, 368.

ACCURATE PROPER MOTIONS AS PROBES OF DISTANCE AND DYNAMICAL STRUCTURE: THE CASE OF M13

M. WYBO AND H. DEJONGHE
Sterrenkundig Observatorium, Krijgslaan 281, Gent, Belgium

We explore the proper motion distributions of anisotropic Plummer models, and show that accurate measurements on the proper motions and radial velocity of individual stars can lead to information on the distance of a stellar system. We consider a one-parameter family of models, which all have the same Plummer law in the mass density, but have different anisotropic 2-integral distribution functions $F(E, L)$, and hence different orbital structure (Dejonghe 1986). The input data for M13 are the photometry and the radial velocities as measured by Lupton, Gunn & Griffin (1987, AJ, 91, 1114), and the proper motion data as measured by Cudworth (1979, AJ, 84, 774). We represent the distribution function as a power series expansion in Fricke-type form. The Quadratic Programming (QP)-algorithm (Dejonghe 1989, Ap.J., 343, 113) determines the coefficients in the expansion by minimising a χ^2-type variable, subject to the constraints that the distribution function must be positive in phase space. From the photometry we calculate the spatial mass density (the well-known Abel inversion) and the potential. The total mass is accounted for by applying a suitable scaling on the kinematical data. All models have a constant M/L. We use surface luminosities and the 3 (anisotropic) pressures, which depend on the three velocity dispersions. By re-iterating the procedure for various distances (and masses), the best distance estimate can be obtained.

Conclusions: Proper motions provide valuable information on orbital structure, independently of the distance. Simple inspection of the distribution of proper motion suffices to detect whether tangential orbits or radial orbits are prevalent. With a QP-algorithm we determined the distance to M13, and found agreement with previous estimates. The relative error on the distance is of the same order as the relative error on the proper motions, wherein both the accuracy in measurements and small number statistics have a contribution.

E. Høg and P. K. Seidelmann (eds.),
Astronomical and Astrophysical Objectives of Sub-Milliarcsecond Optical Astrometry, 369.

A METHOD FOR STAR IMAGE SEPARATION FROM THE BACKGROUND IN AUTOMATIC COORDINATE MEASUREMENTS

K.N.ZAMARASHKIN
Pulkovo Observatory, 196140 St Petersburg, Russia
e-mail: ddp@gaoran.spb.su

A method for star image separation from the background is suggested. Applicability of the method is shown on the basis of automatical coordinate measurements of the double star images on astroplates with the use of the "Fantazia-2" Automatic Measuring Complex (AMC) (Kiritchuk *et al.*, 1984). The method provides a minimal dispersion of distances between components of a binary if the statistics of the centre of mass (CM) of discretized image is used for measurements. It have been shown that the CM cannot be considered as an accurate characteristic of the star image position. Nevertheless, the CM showed to be a valuable tool for determination of some local astroclimate parameters (isoplanetic square size). It was assumed that the process of star image formation could be easily checked with the help of the CM statistics if considered level by level in the sense of optical density . Positions of the center of mass of each component in the binary were determined as a function of the threshold and then the cross-correlations between the trajectories were calculated. Strong correlation between images of components of the double star ADS 3353 (angular distance $r = 4''$) have been discovered, so the components of a binary should not be processed alone in this case. Additional experiments with the double stars ADS 8397 (angular distance between components $14''$) and ADS 1090 ($20''$) have shown that the correlation between the components of the binary became less, but still exists, while the distance is increasing.

References

Kiritchuk, V.S. et al., (1984), The Automatic Measuring Complex based on 'Zenit-2', *Avtometria*, **4**, pp. 97-101.

E. Høg and P. K. Seidelmann (eds.),
Astronomical and Astrophysical Objectives of Sub-Milliarcsecond Optical Astrometry, 370.

THE ATMOSPHERIC LIMITATION ON THE PRECISION OF GROUND-BASED ASTROMETRY

I.S. GUSEVA
Pulkovo Observatory
196140 St Petersburg
Russia

Abstract. Anomalous refraction remains to be the most critical problem in the meridian astrometry measuring large angles on the sky. I study slow quasi-periodical variations of refraction caused by the processes in the middle and upper atmosphere, such as gravity waves, etc., which can not be detected and calibrated out by use of any on-ground meteorological measurements. For this study, very old observations at large zenith distances of 80 to 90 degrees made by V. Fuss at Pulkovo Observatory in 1867–1869 [1] were used. The Deeming's method [2] of spectral analysis of data was applied to examine the characteristic variations of refraction in a wide range of periods. Very powerful quasi-periodical processes with periods of 7-8, 11-14, 18-22, 36-44 minutes and with amplitudes of 0.3 to 0.5 arcsec in the zenith were found when short sets of observations (1-5 days) were considered. They increase random errors of astrometric observations with meridian circles, transit instruments, astrolabes, etc. The periods of very slow variations – 152, 122, 93, 82.5, 73, 61 and 50 days, – are close to the well known periods discovered in other astronomical phenomena, for instance, in solar activity and in Earth rotation. I note also, that some of the long-period variations of refraction may cause quasi-systematic errors in astrometric measurements and catalogues.

References

1. Fuss, V. (1872) Mem.Acad.Imp.Sci.de St.-Petersbourg, T.XVIII, N3.
2. Deeming, T.J. (1975) Astr. Space Sci., 36.

E. Høg and P. K. Seidelmann (eds.),
Astronomical and Astrophysical Objectives of Sub-Milliarcsecond Optical Astrometry, 371.

PRECISION VS. ACCURACY IN STAR CATALOGS

L. G. TAFF, J. E. MORRISON, AND R. L. SMART
STScI, 3700 San Martin Drive, Baltimore, MD 21218 USA

As better precision is achieved and more sophisticated reduction methods are created previously invisible biases surface. This has been especially true in astrometric Schmidt plate work. The problem of their amelioration is not fully solved and precision per se is meaningless in the presence of poor accuracy of comparable amplitude. Continuing to benignly neglect this issue puts us in the position of standing on only one statistical leg. New techniques have been designed to further minimize systematic errors. Of especial interest to star catalog analysis is the method of infinitely overlapping circles (Taff, Bucciarelli & Lattanzi, ApJ **361**, 667, 1990; Taff, Bucciarelli & Lattanzi, ApJ **392**, 746 1992; Bucciarelli, Taff & Lattanzi, J. Stat. Comp. and Sim. **48**, 29 1993). With it almost complete success has occurred with regard to the removal of systematic errors which creep into compilation catalogs as a result of inadequate treatment of catalog-to-catalog systematic errors; they can essentially be eliminated *a priori* or *a posteriori* (Bucciarelli, Lattanzi & Taff, in press in ApJ 1994; Taff & Bucciarelli, in press in ApJ 1994). What infinitely overlapping circles does can be briefly described as follows: Let X (x) be the measured (true) value of a standard coordinate, S(x,y) (ε) be the systematic (random) error in x at this point, let w_∞ be the infinitely overlapping circle weight, σ be the standard deviation of the random error in x, N be the total number of stars in this circle which has radius R, and x_0, y_0 be the coordinates of the center of this circle. Then, the relation between the measured value X and the true value x is given by

$$X = x + S(x,y) + \varepsilon.$$

Expanding S in a Taylor series centered at (x_0,y_0) yields---$\mathbf{r} = (x,y)$---

$$X = x + S(x_0,y_0) + \nabla S(x_0,y_0)\bullet(\mathbf{r} - \mathbf{r}_0)$$

$$+ \tfrac{1}{2}(\mathbf{r} - \mathbf{r}_0)\bullet\nabla\bullet\nabla S(x_0,y_0)\bullet(\mathbf{r} - \mathbf{r}_0) + \text{higher order terms} + \varepsilon.$$

Now, multiply this expression by the infinitely overlapping weight and integrate over the circle. Using the assumption that the stars are uniformly distributed over the circle and the intrinsic symmetry of the circle itself yields

$$\int (X - x) w_\infty \, dA = S(x_0,y_0) + \int \varepsilon w_\infty \, dA + \tfrac{1}{2}\nabla^2 S(x_0,y_0)\langle R^2\rangle + \text{quartic terms}$$

$$\sim S(x_0,y_0) + \sigma/N^{1/2} + \tfrac{1}{2}\nabla^2 S(x_0,y_0)\langle R^2\rangle + \text{quartic terms}.$$

Thus, the bias at the center of the circle is recovered, the random errors are diminished, and the next term is only the Laplacian of the systematic errors.

Research supported by NASA Contract NAS5-32496.

E. Høg and P. K. Seidelmann (eds.),
Astronomical and Astrophysical Objectives of Sub-Milliarcsecond Optical Astrometry, 372.

1.2 SPACE MISSIONS

A TEST OF THE ASTROMETRIC QUALITY OF THE SOUTHERN GUIDE STAR CATALOG 1.2

B. Bucciarelli (OATo, Italy),

C. R. Sturch, B. M. Lasker, M. G. Lattanzi (STScI, USA),

T. M. Girard, I. Platais, W. F. Van Altena (Yale University)

We used preliminary positions of the Yale Southern Proper Motion (SPM) catalogue (Platais, Girard et al., Astronomy from Large Database II, eds. Heck & Murtag, 1992) in a region of 5 fields around the South Galactic Pole to assess the astrometric accuracy of the mask solution (Taff, Lattanzi and Bucciarelli, ApJ **361**, 667,1990), which will be used (in combination with the subplate method) for the construction of the Guide Star Catalogue (GSC) version 1.2. Another semi-external check is done by direct comparison of GSC positions of stars lying onto overlapping plate areas. Results in tables 1 and 2 show that the average rms of the GSC–SPM differences is quite satisfactory ($\sim$ 0.33 *arcsec* **per coordinate**), while an error degradation (up to $\sim$ 1 *arcsec* **positional**) can occurr within a limited area at the plate corners, its signature varying from plate to plate. This can be cured on a plate-by-plate basis by the use of a filtering technique, e.g., as provided by the Collocation method (Bucciarelli, Lattanzi and Taff, ApJ Suppl. **84**, 91, 1993), wherever a suitable reference star density is available.

Table 1: GSC–SPM *rms* for the 5 fields

	S409	S349	S410	S471	S472
No. of stars	1342	934	1219	989	1237
$\sigma_{\Delta\alpha cos\delta}$	0.″30	0.″44	0.″30	0.″34	0.″32
$\sigma_{\Delta\delta}$	0.″29	0.″35	0.″35	0.″33	0.″22

Table 2: GSC–GSC *rms* between **S409** and the four contiguous overlapping plates

	S409/S349	S409/S410	S409/S471	S409/S472
No. of stars	1106	1023	3119	1128
$\sigma_{\Delta\alpha cos\delta}$	0.″23	0.″36	0.″24	0.″20
$\sigma_{\Delta\delta}$	0.″28	0.″26	0.″28	0.″39

E. Høg and P. K. Seidelmann (eds.),
Astronomical and Astrophysical Objectives of Sub-Milliarcsecond Optical Astrometry, 375.

RANDOM ERRORS OF TYCHO ASTROMETRY [1]

V.V. MAKAROV AND E. HØG
Copenhagen University Observatory - NBIfAFG
Østervoldgade 3, DK-1350 Copenhagen K, Denmark

Abstract. The paper focuses on the problem of estimation of Tycho astrometry random errors, caused by photon noise. A theoretical model for the random errors was developed on the basis of the Maximum Likelihood estimator. The model provides a satisfactory agreement with experimental data in the wide range of star magnitudes from 4 to 10 for both vertical and inclined slit transits. It is confirmed theoretically, that for fainter stars, which constitute a half of the Tycho star sample, this model is not valid. Moreover, it is demonstrated that a reliable astrometry could hardly be achieved for these faint stars, were it not for a kind of Bayesian approach which is in fact implemented in the processing. This approach uses implicitly *a priori* astrometric information on positions of the stars. A major drawback of the method is that it introduces some bias in astrometry estimation, of presently unknown size. Nonetheless, no transfer of systematic errors from the input catalogue is expected, for the used prior information comes from Recognition, hence it is based purely on the satellite's own observations. – The inadequacy of the pure ML theory leads us to correct the model empirically, in order to provide reliable formal errors for the astrometric parameters in the final catalogue, for all Tycho magnitudes. In this way, a rms normalized residual is used for each star individually as a scale factor or correction to the formal covariances. Corrected in this way, formal errors are compared directly with external errors, calculated from absolute differences between the Tycho provisional parallaxes in a 30 months solution and the Hipparcos parallaxes of some 100 000 common stars. Analysis of a standard external error shows that the corrected formal errors are robust, and probably even overestimated.

[1]Based on observations made with the ESA Hipparcos satellite

E. Høg and P. K. Seidelmann (eds.),
Astronomical and Astrophysical Objectives of Sub-Milliarcsecond Optical Astrometry, 376.

1.3 EXTRAGALACTIC REFERENCE FRAME

A COMPILED CATALOG OF OPTICAL POSITIONS OF EXTRAGALACTIC RADIO SOURCES

W. J. Jin[1], J. L. Li[2], T. Q. Xu[1], S. H. Wang[1]
[1]Shanghai Observatory, 200030 Shanghai, CHINA
[2]Yunnan Observatory, 650011 Kunming, CHINA

ABSTRACT. Based on 27 individual catalogues, which have been published since 60s' and consist of the positions of the counterparts of extragalatic radio sources, we have compiled a combined catalogue by means of Aoki (Aoki, S. et al. 1983) method for transferring the reference frame from FK4 to FK5 and the referred epoch from B1950.0 to J2000.0 and IERS method (Feissel, M. et al. 1988), which is used for compiling the combined catalog of radio sources. Of course, in the compiling procedure the systematic differences of Peth 70, AGK3RN, and AGK3 (Schwan, H. 1985) have been considered. The precisions of optical positions in the combined optical catalog are as follows:

Type	Number	$E_\alpha \cos\delta$	E_δ
P	56	±0."123	±0."093
S	62	0."265	0."241
C	392	0."603	0."374

where P indicates primary sources
S indicates second sources
C indicates complementary sources

Comparison between the compiled optical catalog and RSC(IERS)93 C 01 (Feissel, M. et al. 1993) was made by using the primary sources. The rotation angles between two frames are as follows:

A_1=0."16 ± 0."03, A_2=-0."03 ± 0."03, A_3=0."11 ± 0."02

And local relative deformation of the combined optical catalog is not obvious within the precisions of optical observations.

REFERENCES

Aoki,S. et al. (1983). A. Ap. **128**, 263
Feissel, M. et al. (1988). BIH Annual Report for 1987, D-113
Feissel, M. et al. (1993). IERS Annual Report for 1992, 11-19
Schwan, H. (1985). A .Ap. **149**, 50

E. Høg and P. K. Seidelmann (eds.),
Astronomical and Astrophysical Objectives of Sub-Milliarcsecond Optical Astrometry, 379.

HIPPARCOS EXTRAGALACTIC LINK

Preliminary Bonn, Potsdam and Kiev solutions

P. BROSCHE[1], M. GEFFERT[1], S. HIRTE[2], N. KHARCHENKO[3], V. KISLYUK[3], M. ODENKIRCHEN[1], S. RYBKA[3], E. SCHILBACH[2], R.-D. SCHOLZ[2], H.-J. TUCHOLKE[1] AND A. YATSENKO[3]
[1] *Sternwarte der Universität Bonn, Germany*
[2] *WIP Astronomie, Universität Potsdam, Germany*
[3] *Main Astronomical Observatory, Kiev, Ukraine*

Hipparcos proper motions contain an unknown angular velocity ω relative to a non-rotating system. The basic equations for its derivation are:

$$\begin{aligned} \Delta\mu_\alpha \cos\delta &= -\omega_1 \cos\alpha \sin\delta - \omega_2 \sin\alpha \sin\delta + \omega_3 \cos\delta \\ \Delta\mu_\delta &= +\omega_1 \sin\alpha - \omega_2 \cos\alpha \end{aligned} \quad (1)$$

where $\Delta\mu_\alpha$ and $\Delta\mu_\delta$ are absolute minus Hipparcos proper motions.

	Bonn	Potsdam	Kiev
photographic plates	astrograph	Schmidt	astrograph
m link fields	8	10	183
n link stars	33	104	1015
galaxies per field	1 to 5	300 to 2000	3 to 5
base line [years]	70 to 90	20 to 40	20 to 40
random p.m. error per star [mas/yr]	0.5 to 1.5	3 to 5	5 to 12
syst. abs. p.m. error per field [mas/yr]	1.0 to 1.5	~ 2	~ 4
rms of solution of (1) [mas/yr]	5	8	14
$\omega_1 \pm \sigma(\omega_1)$ [mas/yr]	$+1.2 \pm 1.0$	$+0.8 \pm 1.0$	-1.5 ± 0.7
$\omega_2 \pm \sigma(\omega_2)$ [mas/yr]	$+3.2 \pm 0.7$	-0.7 ± 1.0	-2.0 ± 0.5
$\omega_3 \pm \sigma(\omega_3)$ [mas/yr]	$+0.0 \pm 1.1$	$+0.5 \pm 1.0$	$+1.2 \pm 0.5$

The Table describes three different absolute proper motion programmes and shows preliminary link results with H 30 data. The number of Bonn and Potsdam link fields will be increased (to 15 and 50, respectively) so that the influence of possible systematic effects - not represented by the formal errors $\sigma(\omega_i)$ - can be further reduced. We expect to provide an accuracy of the final link of the Hipparcos proper motions of better than 1 mas/yr, competitive with other link programmes (Lick/Yale, VLBI, HST).

E. Høg and P. K. Seidelmann (eds.),
Astronomical and Astrophysical Objectives of Sub-Milliarcsecond Optical Astrometry, 380.

RORF – A RADIO OPTICAL REFERENCE FRAME

N.ZACHARIAS, A.L.FEY, J.L.RUSSELL, K.J.JOHNSTON
USRA / NRL / ARC / USNO, Washington D.C.

The radio observations are based on more than one million pairs of group delay and phase delay rate observations from *all* applicable dual frequency Mark–III VLBI data from 1979 until the end of 1993.

A subset of 436 sources, nearly uniformly distributed over the entire sky, has been selected to define a celestial inertial frame which is presented to the IAU Working Group on the Radio/Optical Reference Frame.

The precision of most sources is below 1 mas per coordinate. A comparison to the JPL94R01 catalog has been made. The $\Delta\delta$ vs. δ plot shows an offset of up to 1 mas for southern declinations, while a comparison with the IERSC01 catalog shows no systematic differences larger than 0.5 mas. Details are given in Johnston et al.(1994).

Because of the faintness of the optical counterparts ($V \approx 17...21$) of the compact extragalactic radio sources a multi–step approach is required to obtain optical positions with respect to the optical reference system based on bright stars (FK5, IRS, Hipparcos).

The systematic errors as function of magnitude are controled by use of objective gratings with the astrographs and different exposure times with the other instruments, including wide field CCD observations. Depending on the available data a standard error of 10 to 50 mas can be obtained for the internal precision of the mean position of a source. About 95% of the astrograph work and 30% of the source observations are completed. A precision of about 1 mas for the radio–optical link is feasable with this technique. A precision of about 2.5 mas is estimated with currently available data.

The radio catalog is available via anonymous ftp at maia.usno.navy.mil (192.5.41.22) or via Mosaic with the URL "file://maia.usno.navy.mil/rorf". The complete poster paper is available from nz@pyxis.usno.navy.mil.

Johnston,K.J., et al. (1994) "A Radio Reference Frame", submitted to *AJ*

E. Høg and P. K. Seidelmann (eds.),
Astronomical and Astrophysical Objectives of Sub-Milliarcsecond Optical Astrometry, 381.

ANALYSIS OF THE FK5 SYSTEM IN THE EQUATORIAL ZONE

YU.B. KOLESNIK
Institute of Astronomy of the Russian Academy of Sciences
48 Piatnitskaya St.
109017 Moscow
Russia

ABSTRACT. 15 catalogues produced in the eighties and 12 catalogues of the sixties-seventies have been used to assess the consistency of the FK5 system with observations in the declination zone from -30° to 30°. Positions of the FK4-based catalogues have been transformed at the equinox and equator J2000.0.Classical δ-dependent and α-dependent systematic differences (Cat-FK5) have been formed for individual instrumental systems of the catalogues by a method close to the classical Numerical Method. The weighted mean instrumental systems for the two subsets of catalogues centered at the epochs 1970 (MIS 60-70) and 1987 (MIS 80) and for all types of systematic differences have been constructed. The mean errors of the total systematic differences in α and δ have been estimated as 14 mas and 21 mas, respectively, for the catalogues of the 60-70ies, and 10 mas in both α and δ for the catalogues of the 80ies.

It has been found that the mutual consistency of individual instrumental systems of catalogues of the 80ies with respect to δ-dependent systematic differences is superior by the factor 1.5 comparing with the catalogues of 60-70ies, while the consistency of both catalogue selections with respect to α-dependent systematic differences is comparable. Random accuracy of the FK5 positions and proper motions at the epochs under analysis has been assessed as close to expected from the formal considerations.Actual systematic discrepancies of the FK5 with observations at the respective epochs have been detected. For systematic differences $\Delta\alpha_\delta \cos\delta$ and $\Delta\delta_\delta$, the absolute deviations of the MIS 80 are, in general, within 40 mas, those of the MIS 60-70 are within 30 mas. For systematic differences $\Delta\alpha_\alpha \cos\delta$ and $\Delta\delta_\alpha$, the absolute deviations reach 30-40 mas for both MIS. For total systematic differences, local deformations of the FK5 system in the equatorial zone in both right ascension and declination has been found exceeding expected ones from the formal errors of the FK5 system by a factor about 1.5 for the MIS 60-70, and by a factor about 2 for the MIS 80. Consistency in area distribution between both MIS for the total systematic differences$\Delta\alpha\cos\delta$ has been detected. Quick degradation of the FK5 system with time due to optimistic estimation of the errors of its proper motion system is supposed to be one of the main causes of its discrepancies with observations.The results in declination are recognized to be less reliable due to larger inconsistency of the individual instrumental systems.

Before the next space astrometric mission will be realized, ground-based observations will continue to be the only available check of an external systematic accuracy of the HIPPARCOS catalogue. Evidently, random and, possibly, systematic accuracy of each individual catalogue observed from the Earth surface would be inferior to that of the HIPPARCOS catalogue. Taken as an ensemble, however, a certain selection of catalogues might give a rather definite idea about the actual distortions of the HIPPARCOS system. This study shows to which level of accuracy such ensembles of different selections of catalogues might check the HIPPARCOS system in the equatorial zone. The analysis of the FK5 gives also an idea about levels of random and systematic discrepancies which may be expected in the equatorial zone when the HIPPARCOS catalogue will be compared with the FK5 at different epochs.

E. Høg and P. K. Seidelmann (eds.),
Astronomical and Astrophysical Objectives of Sub-Milliarcsecond Optical Astrometry, 382.

CONFOR PROGRAM: DETERMINATION OF RELATIVE ORIENTATION PARAMETERS BETWEEN VLBI AND FK5 REFERENCE FRAMES

(1) I.I. KUMKOVA, (2) V.V. TEL'NYUK-ADAMCHUK, (2) Yu.G. BABENKO, (2)O.Ya. VERTYPOLOKH

(1)--Institute of Applied Astronomy, Russian Academy of Sciences
8 Zhdanovskaya str.,
197042 St. Peterburg, Russia

(2)--Astronomical Observatory of Kyiv Taras Shevchenko University
3 Observatorna str.,
254053 Kyiv, Ukraine

ABSTRACT. The primary goal of the CONFOR Program (**CON**nection of **F**rames in **O**ptics and **R**adio) is to study the connection between VLBI and traditional optical reference frames as well as to study regional features of the FK5 (Gubanov, Kumkova & Tel'nyuk-Adamchuk, 1990). The main concept of this program is the usage of fixed systems of reference stars in field containing the extragalactic radio sources (ERS) and as a result the creation of a reliable base for astrometric reduction of photographic plates with ERS images. Fulfillment of the Program is embracing several stages.

1. Star lists have been prepared of 2575 both intermediate reference stars and radio stars for meridian observations and about 7 thousand stars of 12-14 mag. for astrographic observations in 238 fields with extragalactic compact radio sources.
2. Observations have been organized of these stars with meridian circles and astrographs in several Eastern observatories (Kyiv, Odesa, Bucharest, Mykolaiv, Kharkiv). The meridian observations are finished in Kyiv and Odesa. Positions of radio stars were determined from Kyiv meridian observations (two epochs: 1984 and 1990).
3. Databases of both the optical and radio interferometric observations have been compiled of radio stars, ERS as well as optical observations of intermediate stars in the fields centered in ERS.
4. Observations were carried out with several astrographs (Kyiv, Bucharest, Abastumany, Tautenburg, Kitab, Sanglok) and positions were determined of 57 ERS.
5. Determination was carried out of the angles of relative orientation between VLBI and FK5 reference frames as well as estimation of regional features of the FK5. For this purpose both collected and original radio source coordinates were used.

This work was supported by grant of Academy of Sciences of former Soviet Union and ISF grant U52000.

Reference

Gubanov, V.S., Kumkova, I.I., and Tel'nyuk-Adamchuk, V.V. (1990) 'CONFOR: a new program for determining the connection between radio and optical reference frames', in J.H. Lieske and V.K. Abalakin (eds.), Inertial Coordinate System on the Sky, Kluwer Academic Publishers, Dordrecht, pp. 75-76.

E. Høg and P. K. Seidelmann (eds.),
Astronomical and Astrophysical Objectives of Sub-Milliarcsecond Optical Astrometry, 383.

PLACING THE GUIDE STAR CATALOG ON THE HIPPARCOS SYSTEM

U. BASTIAN[1], B. BUCCIARELLI[2,3], J. HAYES[2], B. M. LASKER[2], J. E. MORRISON[2], S. RÖSER[1], R. L. SMART[2], C. R. STURCH[4], AND L. G. TAFF[2]

[1] Astronomisches Rechen-Institut, Mönchhofstr. 12-14, 69120 Heidelberg, Germany
[2] STScI, 3700 San Martin Drive, Baltimore, MD 21218 USA
[3] On leave from Torino Observatory, Torino, Italy
[4] Computer Sciences Corp., Astronomy Programs, STScI, Baltimore, MD 21218 USA

The HIPPARCOS Catalogue, linked to the system of extra-galactic radio sources, will soon be *the* reference system in optical astronomy. The Guide Star Catalog (GSC; Lasker *et al.* AJ **99**, 2019ff, 1990), with its 18 million objects, already serves most astronomers for the planning of small field observations, optical identifications, and the reduction of CCD images. The utility of the GSC will be greatly enhanced if it is transformed to the HIPPARCOS system and its remaining systematic errors minimized. To do so we will use the forthcoming TYCHO Reference Catalogue (Röser & Høg, in Workshop on Databases for Galactic Structure, eds. Davis Philip, Hauck & Upgren, pp. 137, 1993). The reductions will use a combination of the mask and sub-plate methods (Taff, AJ **98**, 1912, 1989; Taff, Lattanzi & Bucciarelli ApJ **361**, 667, 1990), the co-location method (Bucciarelli, Lattanzi & Taff, ApJ Suppl. **84**, 91, 1993), the new "fixed number of stars filter" of Röser, Bastian & Kuzmin (in press in the proceedings of IAU 148, 1994), and the infinitely overlapping circles (IOC) version of the latter (Taff, Bucciarelli & Lattanzi, ApJ **358**, 359, 1990; Bucciarelli, Taff & Lattanzi, J. Stat. Comp. Sim. **48**, 29, 1993). We present in the table below test results using the PPM as the reference catalog (typically only 4 sub-plates per plate; 25 stars for the Square Filter; 16 stars per circle for the IOC method). The first line presents the estimates for $\sigma_\alpha \cos\delta$ and σ_δ, both in arcsec, relative the to PPM Suppl (Röser, Bastian & Kuzmin, A&A Suppl. **105**, 301, 1994); the second line contains them based on plate overlaps (all for the -30° declination zone). When comparing the numbers note that in the first line the GSC and PPM Suppl (0."12) errors are contributing whereas in the second line the GSC errors are present twice (in the rms sense).

	GSC 1.0	MASK	SUB-PLATE	SQUARE FILTER	IOC
PPMSuppl	0.72,0.60	0.34,0.35	0.57,0.56	0.35,0.34	0.33,0.30
Overlaps	0.87,0.62	0.80,0.60	1.10,0.49	0.64,0.44	0.57,0.31

Further work is required to select the optimum method (which may vary with the plate). However, it is clear that any modern small-area reduction procedure, based on a high quality, dense reference catalog, produces results far superior to global Schmidt plate modeling. Research supported by NASA Contract NAS5-32496 and the STScI Collaborative Visitors Program.

E. Høg and P. K. Seidelmann (eds.),
Astronomical and Astrophysical Objectives of Sub-Milliarcsecond Optical Astrometry, 384.

ARC LENGTH DIFFERENCE METHOD IN THE SELECTION OF PRIMARY SOURCES OF COMBINED EXTRAGALACTIC RADIO SOURCE CATALOGUES

J. L. LI[1] and W. J. JIN[2]
[1] *Yunnan Observatory, Kunming 650011, P. R. China*
[2] *Shanghai Observatory, Shanghai 200030, P. R. China*

ABSTRACT. The coordinates of common extragalactic radio sources will be different if the orientations of individual frames are different. If coordinates in each of the frames are in good consistence, i.e. small or non local relative deformations exist, common arcs in frames will be equal to each other (within precision) whether the orientations of the frames coincide or not. In addition, if relative deformations exist between frames, these deformations will be reflected partly on the difference of the lengths of common arcs. It is inferred that we can select candidates for primary objects by comparing the lengths of common arcs in individual frames. We call this as the Arc Length Difference (ALD) selection method.

Considering that it is unrealistic to distinguish between positions with or without deformations beforehand (Arias et al., 1988), and that an arc length difference gives us nothing definite about the consistency of a specified source in frames because of the variable values and the statistical zero mean of the differences (Li & Jin, 1994), we take the Mean value of all the Absolute Arc Length Differences (MAALD) related to each specified source as the criterion for the recognition of deformation. The large the deformation for a particular source is, the larger (statistically) the absolute value of the arc length difference or the MAALD related to this source will be. Taking all the MAALDs of all the common objects of two frames as a random series, and deleting large MAALDs in this series by means of statistics, we can select candidates for primary objects which have better internal and external consistencies in frames.

Here we make emphasis on the following two aspects of the ALD method. First, if we take all the absolute values of the arc length differences as a random series and perform statistical deletion on this series, there will come out the situation that among all the absolute values related to a specified source some will be deleted and others will be kept because of the variable value nature of the arc length difference. Therefore the characteristics of consistency of the source will be concealed and we will be faced to the ambiguous case to decide whether the source should be deleted or not. The effect of the variable value nature of the arc length difference can be avoided and the characteristics of consistency of a source is highlighted by using the mean of all the absolute values related to this source. Second, since the value of MAALD is the mean of all the absolute arc length differences related to a source, a specified arc length difference will enter into two MAALDs, so that all the MAALDs are then interweaved with one another. After one of the MAALDs is deleted, the rest will be changed accordingly. So we emphasize a progressive process to avoid a large number of sources being deleted in a single step.

Our tests show that the ALD selection method is stable and reliable. It can emphasize the characteristics of each source on the premise of the consistency of all the sources and the selection results of this method can reflect appropriately the total number of sources, the position precision, the sky coverage and so on of individual frames.

Reference

Arias, E.F., Feissel, M, Lestrade, J.-F., 1988, In: BIH Annual Report for 1987, D-113
Li, J.L. & Jin, W.J., 1994, Annals of Shanghai Observatory, Academia Sinica, No.15, 98

E. Høg and P. K. Seidelmann (eds.),
Astronomical and Astrophysical Objectives of Sub-Milliarcsecond Optical Astrometry, 385.

2.1 STELLAR ASTROPHYSICS

ANALYSIS OF STELLAR OBSERVATIONS: SYSTEMATIC EFFECTS, RADIO STARS, PROPER MOTIONS

V.A.F. MARTIN AND N.V.LEISTER
Instituto Astronômico e Geofísico-Universidade de São Paulo

An observational stellar program with a prismatic astrolabe was started at "Observatório Abrahão de Moraes" - OAM - São Paulo, Brazil, in 1974

The primary goal of the observational programme is to produce a general astrolabe catalogue at OAM involving 800 stars, approximately, observed with the same instrument for 20 years. The observations are obtained at two different zenith distances (30° and 45°), so that it is possible to observe absolute declination and compute systematic effects, as well as obtain fundamental reference system corrections [1].

The basic problems concern the optical-radio system connection and the deterioration to which the HIPPARCOS system is subject. So, there is a requirement of exact and systematic observations with optical instruments from the ground.

In such case, the radio stars included in this observational programme are aimed at determining the local systematic effects between the optical and radio reference systems in the inertial reference context [2].

The possibility of proper motions determination is due to the fact that the observational programme has lasted a long time. So, not only the elapsed period is well suited for the analysis but the number of stars is large and guarantees the efficiency of the method [3]. The residuals of each star are related with: a) RA and Dec corrections; b) group corrections; c) azimuth and parallactic angle of the star; d) equinox, equator and zenith distance corrections and e) proper motion in RA and Dec.

References

[1] Martin, V.A.F., Clauzet, L.B.F., Benevides-Soares, P., Leister, N.V. (1994) "Absolute declinations from astrolabe data", *A&A* (submitted)
[2] Kovalevsky, J. (1990) "Astrométrie Moderne", *Lecture Notes in Physics*, **358**
[3] Kovalevsky, J. (1991) "Objectives of Ground-Based Astrometry After HIPPARCOS", *Astrophysics and Space Science*, **177**, pp.457-464

E. Høg and P. K. Seidelmann (eds.),
Astronomical and Astrophysical Objectives of Sub-Milliarcsecond Optical Astrometry, 389.

CCD PHOTOMETRY OF HIPPARCOS STELLAR SYSTEMS

E. OBLAK AND M. CHARETON
Observatoire de Besançon, France †

With the introduction of CCD detectors, it now appears feasible to obtain accurate photometric data for each of the components of close visual double stars with an angular separations between 1 and 12".

A programme of systematic and homogeneous acquisition of precise colour indices for several thousand components of double and multiple systems has been carried out by a European Network of Laboratories (see footnote) and supported by an ESO key programme.

The principle scientific objective is to provide the missing photometric data needed to supplement the high quality of Hipparcos astrometric data. VRI CCD photometry has been carried out in observatories in both hemispheres (La Palma, Calar Alto, OHP). The relative position of the components is obtain as a by-product thanks to CCD astrometric observations developed by the group in a few numbers of selected fields in right ascension. Since neither the crowded fields, nor the isolated star (aperture) software packages are convenient, specific reduction softwares had to be developed for PSF-fitting of overlapping star profiles.

Both the photometry and astrometry of visual double stars are being obtained at accuracy of 0.01 level in angular separation and photometry. The preliminary comparison based on the results of the observational campaign and the Hipparcos solutions permits the correction of the grid step ambiguity of the satellite and corrects for color term effects giving, after correction, more consistent results from the two methods.

A photometric database of stellar systems is compiled and will be generalized to all categories of stellar systems.

A parallel work consists of examining and correcting the data on stellar systems of the 'Centre de Données astronomiques' of Strasbourg (CDS).

† with the collaboration of A.N. Argue, P. Brosche, J. Cuypers, M. Geffert, M. Grenon, J.L. Halbwachs, M. Irwin, G. Jasniewicz, E. Martin, J.L. Mermilliod, F. Mignard, W. Seggewiss, D. Sinachopoulos, E. Van Dessel, S. Zola.

E. Høg and P. K. Seidelmann (eds.),
Astronomical and Astrophysical Objectives of Sub-Milliarcsecond Optical Astrometry, 390.

SOME PHYSICAL PARAMETERS AND UVBYβ PHOTOMETRY OF TRAPEZIUM TYPE MULTIPLE SYSTEMS

G.N.SALUKVADZE AND G.SH.JAVAKHISHVILI
Abastumani Astrophysical Observatory
383762 Abastumani, Republic of Georgia

Abstract.

The presented paper deals with the results of electrophotometric observations of 59 components of 19 trapezia in Strömgren and Crawford six-colour photometric system. The multiple systems, selected from the Abastumani Catalogue of Trapezia (Salukvadze, 1978), are: ABAO 2, 8,34, 48, 51, 62, 75, 94, 245, 312, 313, 316, 324, 348, 356, 363, 387, 396.

Observations were made on the 125-cm mirror telescope with the use of a one-channel photometer, based on photon counting, with diaphragms 10" and 20". Reduction was done on the Observatory computer with a procedure described by Salukvadze and Javakhishvili (1989).

We calculated the indices [m1],[c1] and [u-b] as in (Strömgren 1967, Philip and Egret 1980). The unreddened indices (b-y), m1 and c1 were calculated by the formulae of Crawford (1975).

Semi-empirical calibrations for effective temperature, bolometric correction and mass for early-type stars, using Strömgren photometric indices c0 and beta, are given by Balona (1984). In order to determine absolute magnitudes we used the calibration from Balona and Shobbrook (1984).

References

Salukvadze, G.N: 1978, Bull. Abastumani Astrophys. Obs. 49, 39.
Salukvadze, G.N., Javakhishvili,G.Sh.: 1989, Bull. Abastumani Astrophys. Obs. 66, 45.
Strömgren, B.: 1967, Proc. AAS-NASA Symposium, 461.
Philip, D.A.G., Egret D.: 1980, Astron. Astrophys. Suppl. Ser.,40, 199.
Crawford, D.L.: 1975, Astron.J., 80, 955.
Balona, L.A.: 1984, Mon. Not. R. Astr. Soc., 211, 973.
Balona, L.A., Shobbrook, R.R., 1984, Mon. Not. R. Astr. Soc., 211, 375.

E. Høg and P. K. Seidelmann (eds.),
Astronomical and Astrophysical Objectives of Sub-Milliarcsecond Optical Astrometry, 391.

SPECKLE INTERFEROMETRY OF DOUBLE STARS FROM THE SOUTHERN HEMISPHERE

E. HORCH, W. F. VAN ALTENA AND T. M. GIRARD
Yale University Observatory
P.O. Box 208101, New Haven, CT 06520, U.S.A.

C. E. LÓPEZ
Felix Aguilar Observatory
San Juan, Argentina

AND

O. FRANZ
Lowell Observatory
Flagstaff, AZ, U.S.A.

We have started a new program of double star observations in the southern hemisphere which utilizes the technique of speckle interferometry. Observations are made using the Stanford University speckle interferometer on the 76-cm reflector at the Cesco Observatory at El Leoncito, Argentina (jointly run by Universidad Nacional de San Juan and Yale Southern Observatory), although we will also have access to larger aperture telescopes. The Stanford system, formerly used at Lick Observatory, is on long term loan to us from Dr. Gethyn Timothy and features a multi-anode microchannel array (MAMA) detector as the imaging device. This new program of double star research will help alleviate the continuing problem of fewer speckle observations in the southern hemisphere. In combination with other data such as the eyepiece interferometer measures of Finsen and Hipparcos parallaxes, it should also eventually contribute to a better understanding of the lower portion of the main sequence mass-luminosity relation.

In 1994 July, we installed the instrumentation at the Cesco Observatory 76-cm telescope and observed 66 double stars. Preliminary results from these data indicate that we can produce speckle image reconstructions with nearly diffraction-limited resolution. In the near future, we will have observing time both at the 76-cm telescope and the 2.15-m CASLEO telescope which is also located at El Leoncito.

E. Høg and P. K. Seidelmann (eds.),
Astronomical and Astrophysical Objectives of Sub-Milliarcsecond Optical Astrometry, 392.

MACHO PHOTOMETRY AND ASTROMETRY

E. HØG AND I.D. NOVIKOV
Copenhagen University Observatory - NBIfAFG
Østervoldgd. 3, DK-1350 Copenhagen K, Denmark

AND

A.G. POLNAREV
Astronomy Unit, Queen Mary and Westfield College
University of London, Mile End Road, E1 4NS London, UK

Abstract. MACHOs (Massive Astrophysical Compact Halo Objects) have been discovered by their relativistic amplification of light from distant stars as they crossed very near to the line-of-sight. The very few events were detected from more than a billion photometric measurements of millions of stars in the LMC. – A mathematical theory of analysis of astrometric and photometric measurements of microlensing events is presented. It is shown that three photometric measurements and three one-dimensional astrometric measurements during an event are, in principle, sufficient to determine the – precisely – six observable physical parameters of the MACHO, including the proper motion and distance of the dark body, provided the position, motion and distance of the undeflected star has been determined from observations outside the event. The practical possibility of such observations is discussed by comparison with the proposed ROEMER satellite, an instrument concept with the potential for obtaining the many billions of photometric and astrometric observations of the required quality. At least 300 photometric amplifications of light from stars brighter than $V = 16.5$ mag could de detected during a ROEMER mission, but the prospects for significant astrometric observations are meagre.

References

Høg E., Novikov I.D., Polnarev A.G. 1994, MACHO Photometry and Astrometry, Astron.& Astrophys. (in press)

E. Høg and P. K. Seidelmann (eds.),
Astronomical and Astrophysical Objectives of Sub-Milliarcsecond Optical Astrometry, 393.

THE INTERSEASONAL TREND IN PHOTOGRAPHIC DOUBLE STARS OBSERVATIONS

O.V.KIYAEVA
Pulkovo Observatory,
196140 St Petersburg, Russia
e-mail: kov@gaoran.spb.su

Photographic observations obtained in Pulkovo and USNO for four visual double stars (ADS 48, 7251, 11632, 12815) have been investigated.

The Pulkovo data were taken from our catalogue (Kisselev *et al.*, 1988) sent to Strasburg. The USNO data were taken from Josties et al. 1969-78, but only uniform automatic measurements were used.

Analysis of (O-C) relative to the orbital motion shows one year period in both series due to systematic seasonal variations in the conditions of observations. The value of the interseasonal trend (a systematic residual of coordinates) reach up to $0.015''$ in X and $0.030''$ in Y for Pulkovo and $0.007''$ in X and $0.015''$ in Y for USNO. These values are almost of the same range as the accidental errors of observations, but they may be comparable with amplitudes of oscillations caused by a probable invisible satellite and thus they should be taken into account.

Comparisons of Pulkovo and USNO observations have been carried out for each star under the study. The relative motions obtained from both series are in close agreement with each other for all stars. This fact is essential for the orbit determinations.

References

Kisselev, A.A. *et al.* (1988) The catalogue of relative positions and motions for 200 visual double stars observed with 26-inch refractor at Pulkovo in 1960–1986. *Nauka, Leningrad.*

Josties, F.J. *et al.* (1969-78) Photographic Measures of Double Stars. *Publ. Naval Obs.***Vol. no. 18, 22, 24**,pp. 7–104, 7–88, 7–63.

E. Høg and P. K. Seidelmann (eds.),
Astronomical and Astrophysical Objectives of Sub-Milliarcsecond Optical Astrometry, 394.

THE CATALOGUE OF THE COMPONENTS OF DOUBLE AND MULTIPLE STARS (C C D M) - First edition.

J.DOMMANGET & O.NYS
Royal Observatory, Belgium
3, avenue Circulaire
1180 - Bruxelles

The Hipparcos mission required the realisation of an Input Catalogue giving the positions of 100.000 stars (single or components of double and multiple systems) to an accuracy better than 1"5. At the start of this work (1981) no specific catalogue of double and multiple stars provided these data. The only general data base on double stars available to us, giving positions to ±1', was the Index (1961,0) updated at the USNO by C.E.Worley till 1976,5 and of which a copy was communicated by P.Muller of the Observatoire de Meudon. It has then been decided to reformat this Catalogue in such a way as to allow the introduction of all necessary information for the mission. This permitted a correct cross-identification with the Hipparcos Input Catalogue (of finally 118.000 stars). It was later called: the *Catalogue of the Components of Double and Multiple stars* (CCDM). Since then, it has been developed and its aim remains to furnish the best accurate locations and descriptions of the double and multiple systems on the sky for all double and multiple star research.

In comparison with the INDEX (1976,5), its completion consisted of 4.700 additional Durchmusterung, 16.200 AGK2/3, 13.100 SAO, 25.500 HD, 3.900 BDS (if the ADS are not given) and 14.700 Hipparcos identificators as well as 45.000 accurate individual positions of components or systemic photocenters. Presently more than half of the 64.000 systems (135.000 records) of this catalogue are correctly located on the sky within an error of generally ±1". This part of the catalogue (34.031 systems) has been made available to the astronomical community: its introduction is published in *Communications de l'Obs. R. de Belg.* (Série A, n°115) and the catalogue it-self is available at the Centre de Données Astronomiques at Strasbourg (France). The remaining part (29.432 systems) for wich accurate positions are still missing, is available on request to the authors.

A further edition will be extensively completed in all data fields and, beside all known double and multiple stars observed by Hipparcos, will contain all those discovered by the satellite.

E. Høg and P. K. Seidelmann (eds.),
Astronomical and Astrophysical Objectives of Sub-Milliarcsecond Optical Astrometry, 395.

A DYNAMICAL INVESTIGATION OF NINE WIDE VISUAL DOUBLE STARS IN THE NEIGHBOURHOOD OF THE SUN

A.A.KISSELEV AND L.G.ROMANENKO
Pulkovo Observatory,
196140 St Petersburg, Russia,
e-mail: ddp@gaoran.spb.su

The dynamical states of nine wide visual double stars (ADS 7251, 10329, 10386, 10759 [psi Dra], 11061 [40&41 Dra], 11632, 12169, 12815 [16 Cyg]) are considered. The 20-30 – year series of photographic observations obtained with the Pulkovo 26-inch refractor (Kisselev *et al.*, 1988) supported by the data of relative radial velocities of the components and parallaxes are used. On this base the vectors of relative spatial positions and velocities of the components are determined or estimated with confidence. Families of orbits, satisfying observational data are determined by assuming stability of motions in the systems. The orbits of three nearby binaries ADS 7251, 11632 and 61 Cyg were calculated earlier by classical methods and belong to the family mentioned above. Thus the authenticity of our analysis is proved (Kisselev, Kiyaeva, 1980). It is shown that elliptical motion in the couples ADS 10759 and 11061 could be explained only if an additional (hidden) mass exists. This mass consists of one solar mass for the first couple and two solar masses for the second one. The dynamical orbital elements of the binaries belonging to each family may differ greatly, but the geometrical elements are fairly stable. The orientation of the orbit planes of binaries with respect to the Galaxy plane is determined. In one half of the cases the orbital planes are approximately orthogonal to that of our Galaxy.

References

Kisselev, A.A., et al. (1988) *The catalogue of relative positions and motions for 200 visual double stars observed with 26-inch refractor at Pulkovo in 1960-1986*. Nauka, Leningrad.

Kisselev, A.A., Kiyaeva, O.V., (1980), Applications of the Apparent Motion Parameters (AMP) Method for Determination of the Visual Double Star Orbits, *Soviet. Astron.*, **57**, pp. 1227–1241.

E. Høg and P. K. Seidelmann (eds.),
Astronomical and Astrophysical Objectives of Sub-Milliarcsecond Optical Astrometry, 396.

STAND-ALONE PROGRAM PACKAGE FOR IMPROVEMENT OF PARAMETERS OF CELESTIAL BODIES MOTION

T. V. IVANOVA, F. A. NOVIKOV and E. Yu. PARIISKAYA

Institute of Theoretical Astronomy of the Russian Academy of Sciences, St.Petersburg
E-mail ita@iipah.spb.su

A stand-alone Program Package for Improvement (PPI) of motion parameters of celestial bodies from observations is described. The PPI is oriented for solving conditional equations system by the method of least squares and is implemented for IBM PC in MS-DOS in the Turbo-Pascal programming language v.5.0. There are two versions of the system:
A) using the standard conventional memory 640 Kb and
B) using the external memory.

About 250 parameters can be improved by means of the version A simultaneously. In the version B the maximum number of the improving parameters is much greater and is limited only by the amount of the external memory available.

The most important feature of the PPI is a special improvement language with wide range of facilities for the presentation of input and output data and modes of improvement process. This language allows to set an iterative process of improvement on the $n\sigma$ criteria (where n is a real number and σ is a standard deviation) and to repeat the improvement process with different subsets of unknowns as well. In the case of non-intersecting subsets of unknowns it gives a possibility of independent improvement of parameters with different physical meaning in one task (for example, global and local parameters for each series of observations).

The PPI package can treat simultaneously observations of any kind (optical, radiointerferometric, laser etc.) using scale factors for different observed data. The PPI allows to improve parameters for unlimited number of observations due to its ability of accumulation of the normal system in several consequent tasks. The observations themselves should be provided as files together with information on conditional equations coefficients. It is possible to modify and set new coefficients by means of usual mathematical formulas directly in the task not changing the source data. The PPI package allows to reject observations depending on various conditions. Furthermore each group of observations might be supplied with its own weight function.

The package discussed is the further development of our package described in Ivanova and Novikov (1990).

References

[1] Ivanova T.V. and Novikov F.A. (1990) *Program Package for Improvement of Parameters of Equations of Condition for IBM PC*, Preprint of the Institute of Applied Astronomy of the USSR Academy of Sciences (Leningrad), **21** (in Russian).

E. Høg and P. K. Seidelmann (eds.),
Astronomical and Astrophysical Objectives of Sub-Milliarcsecond Optical Astrometry, 397.

VELOCITIES OF ESCAPERS FROM UNSTABLE HIERARCHICAL TRIPLE STARS

L.KISELEVA[1],J.ANOSOVA[2], J.COLIN[3]
[1] *Institute of Astronomy, Madingley Rd., Cambridge CB3 0HA, UK*
[2] *National Astronomical Observatory, Tokyo 181, Japan*
[3] *Observatoire de Bordeaux, Floirac, France*

We consider unstable hierarchical triple stars which are disrupted by the straightforward ejection and escape of the distant companion. The main aim of the present work is the study of velocities of escaping stars from these unstable systems, and the dependence of these velocities on the masses of the stars and on the distances of the escapers from the centers of mass of the remaining binary and of the triple system as a whole. Velocities of escapers are estimated for actual star systems. It is shown that in the triple star λ Tau, when it becomes unstable as a result of mass transfer in its close semi-detached binary, the velocity of the escaping distant companion with mass $M \approx 0.7 M_{\odot}$ can be about 100 km/s. A possible application of these results to the problem of anomalous high-velocity stars in the Galaxy is discussed.

E. Høg and P. K. Seidelmann (eds.),
Astronomical and Astrophysical Objectives of Sub-Milliarcsecond Optical Astrometry, 398.

Near Infrared High Angular Resolution Observations of Stars and Circumstellar Regions by Lunar Occultations

T.Chandrasekhar, N.M.Ashok and Sam Ragland
IR Astronomy Area, Physical Research Laboratory
Ahmedabad 380 009, India

The high angular resolution technique of lunar occultations enables one dimensional source structure in the direction of occultation to be extracted from the observed fringe pattern after detailed analysis taking into account the frequency response of the detection system, the optical filter bandwidth and the telescope size. A program of observing lunar occultations in the near infrared from 1.2m telescope at Gurushikhar,Mt Abu, India (72°47′E, 24°39′N, 1680m), is currently being pursued. Several occultations have been successfully observed in K band (2.2μm) including a day time event. The instrument used was a InSb based infrared high speed photometer, the details of which are given in a earlier paper (Ashok N.M., Chandrasekhar T. and Sam Ragland, 1994, Experimental Astronomy, 4, 177).

Table 1 lists the angular size values in milliarcsecond (mas) derived from the occultation measurements. In some cases the values represent upper limits only.

TABLE 1: Results from occultation light curves

Sl. No.	Source	Date	Angular Diameter (mas)		Remarks
			measured	[1]From F_v Vs $[V\text{-}K]_o$	
1	IRC 10013	26 Dec 90	[2]4.2±0.5	3.0	Possibily extended
2	IRC 30094	22 Feb 91	[2]3.0±0.5	2.5	Large Polarisation in R - Circumstellar Shell?
3	IRC 20190	22 Feb 91	2.6±0.5	2.1	
4	IRC 20200	25 Mar 91	3.3±0.3	3.4	3.1mas in K (Ridgway, 1982)
5	IRC 00198	17 Mar 92	[2]2.9±0.3	1.8	<2mas in V (Beavers, 1978)
6	IRC 20169	4 Feb 93	[2]1.8±0.3	2.2	1.8mas in K (Ridgway, 1982)

[1]Modified Barnes-Evans relation
[2]Most likely an upperlimit

Comparison of angular size derived from our analysis with those from a modified Barnes-Evans relationship yields a reasonable level of agreement within errors. Extension of the occultation program to longer wavelengths and fainter stars at the 1.2m Gurushikar telescope are in progress.

Useful comments and suggestions from A. Richichi are also gratefully acknowledged. This work was supported by the Department of Space, Government of India.

E. Høg and P. K. Seidelmann (eds.),
Astronomical and Astrophysical Objectives of Sub-Milliarcsecond Optical Astrometry, 399.

DYNAMICS OF THE ALPHA CEN SYSTEM.

J. ANOSOVA[1], V. ORLOV[2]
1. National Astronomical Observatory, Tokyo 181, Japan;
2. Astronomical Institute, St. Petersburg University,
Bibliotechnaya pl. 2, 198904 St. Petersburg, Peterhof, Russia

The triple star system αCen-AB and *Proxima Cen* - the component C - is the nearest to the Sun. The study of its dynamics has shown that this system is probable non-chance. The motion of the component C (Proxima) with respect to the centre of mass of the pair AB is hyperbolic with the probability $P = 1.0$. We observe, therefore, a slow passage of C close to the pair AB. We propose the hypothesis that this system is a part of the stellar moving group. We list the probable members of this group amongst the nearby stars. Amongst them we have the binaries *Gliese* 140.1 *and* 676, the triple system ADS 10288 (Gliese 649.1), and the six single stars. The probability to find by chance these stars inside the velocity space cube with the side of 20 km/s around αCen is equal to about 2%.

E. Høg and P. K. Seidelmann (eds.),
Astronomical and Astrophysical Objectives of Sub-Milliarcsecond Optical Astrometry, 400.

2.3 EXTRAGALACTIC ASTROMETRY

2.4 REFERENCE FRAMES AND SOLAR SYSTEM

DARK MATTER IN TRIPLE GALAXIES

LUDMILA KISELEVA
Institute of Astronomy, University of Cambridge,
Madingley Road, Cambridge CB3 0HA, England

On the basis of numerical simulation of the dynamics of triplets of galaxies, virial estimation of the individual masses of triplets is shown to be unreliable because of their strong nonsteadiness and projection effects: the spread in estimates due to these two factors reaches 2 orders of magnitude. However, the mass of a typical small galaxy group can be estimated statistically, from data on a whole homogeneous ensemble of groups. We propose two different methods of such statistical estimation. Triplets of galaxies offer a good opportunity to measure the amount of dark matter in them, especially because one can use the extensive data set on triplets by Karachentsev *et al.* (1989). The mass estimates we obtain for the typical group from the Karachentsev's list have more than 5 times excess compared to the visible mass, when a standard mass-to-light ratio is assumed. The typical masses of loose triple galaxies selected from Huchra & Geller (1982) and Maia *et al.* (1989) catalogues of galaxy groups are also estimated as $\approx 21 M_L$ and $\approx 100 M_L$ correspondingly. The influence of dark matter distributed in the common envelope on the dynamical properties and the merging rate in small galaxy groups are also considered. It is found that the significant dark matter makes motion of galaxies in groups more stocastic, increases the number of close double approaches between galaxies and increases slightly the merging rate. At the same time, the dark matter significantly decreases the number of long-lived temporary binary subsystems formed inside triplets.

References

Huchra, J.P., Geller, M.J. (1982) *Astrophys. J.*, **257**, 423
Karachentsev, I.D., Karachentseva, V.E., Lebedev, V.S. (1989) *Izv. Spets. Astrofiz. Obs. Akad. Nauk USSR* **27**, 67
Maia, M.A.G., da Costa, L.N., Latham, D.W. (1989) *Astrophys.J.Suppl.Ser.* **69**, 809

E. Høg and P. K. Seidelmann (eds.),
Astronomical and Astrophysical Objectives of Sub-Milliarcsecond Optical Astrometry, 403.

CURRENT FUNDAMENTAL SYSTEM ERRORS AND POST-HIPPARCOS UTILIZATION OF GROUND-BASED OPTICAL INSTRUMENTS

G.A. GONCHAROV
Pulkovo Observatory
196140, St.Petersburg, Russia

Several modern ground-based optical instruments have indicated similar systematic differences of their catalogues with respect to the FK5. Thus, the FK5 apparently needs a correction of about $60 \cdot \sin(6\delta)$ mas for northern hemisphere declinations, other sine corrections of order 20 mas and some equator correction. These errors are the result of systematically erroneous proper motions due to position errors, particularly in old observations.

A typical old instrument, the Pulkovo Struve-Ertel vertical circle, one of the main participants in the FK5, is compared for instrumental errors and their impact on positions with a modern instrument, the Pulkovo photographic vertical circle. Investigations of their divided circles, reading systems, flexures, levels and seasonal change of instrumental systems indicate that neglect and bad account of the errors in earlier observations could lead to the errors of the fundamental system. It is shown that all the instrumental errors up to 50 mas can be investigated if all environmental and electrical values are continuously registered.

Thus, modern instruments can make observations free from any errors of the fundamental system. They can link the system of Hipparcos/Tycho stars with extragalactic objects by observing all celestial objects up to the 18th magnitude and provide within a few years both coordinates and multicolor photometry of the objects with positional accuracy of about 50 mas. The resulting catalogue as an extragalactic reference frame can be used for the ROEMER or another astrometric satellite (see E. Høg, 1994, Contribution at the 'G.Colombo' memorial conference, Padova).

Acknowledgement: the author thanks the American Astronomical Society for financial support.

E. Høg and P. K. Seidelmann (eds.),
Astronomical and Astrophysical Objectives of Sub-Milliarcsecond Optical Astrometry, 404.

CCD ASTROMETRY OF NEAR–EARTH ASTEROIDS

A.K.B. MONET
U.S. Naval Observatory Flagstaff Station
P.O. Box 1149, Flagstaff, AZ, 86002, USA

Automated observations of near–Earth asteroids have been underway at the USNO Flagstaff Station since February 1994. Observations are made with an 8-inch transit telescope utilizing a CRAF/Cassini 1024×1024 CCD. The scale of 1.2 arcsec/pixel provides a field 20.5 arcmin on a side. Pointing and imaging are done automatically using nightly schedules of transit times and declinations of the asteroids to be observed. Observations are made in scan mode, allowing an integration time of 80 seconds at the celestial equator and a limiting magnitude of about 17.5. Flats, collimation and scale frames, and photometric standard fields can all be included in the nightly schedules. The telescope operates unattended during the night, but is monitored remotely by observers at another USNO telescope.

The 40–50 asteroids to be observed each night are selected from a list of hundreds of candidates proposed by collaborators at USNO, Edward Bowell of Lowell Observatory, and Donald Yeomans at the Jet Propulsion Lab. The USNO candidates comprise objects of special historical or dynamical interest. Those proposed by Yeomans are generally future targets for spacecraft encounters. By far the largest number of candidates is provided by Bowell and includes Earth–crossing asteroids, planet–crossers, objects on the MPC critical list, and other asteroids whose orbits need improvement.

Asteroid coordinates are computed using a transformation derived by a least-squares fit of pixel coordinates of field stars to an astrographic projection of their catalog positions. The accuracy of the resulting coordinates is thus highly dependent on the quality of the catalog employed. At present the Guide Star Catalog is used, yielding a typical accuracy of ±0.3 arcsec. Magnitudes are estimated by linear least-squares fit of the star images' intensities to their catalog magnitudes. Using the GSC magnitudes leads to an uncertainty of ±0.5 magnitudes. Accuracies should improve significantly with the use of the Hipparcos catalog, and the number of observations will increase due to planned hardware and software improvements.

E. Høg and P. K. Seidelmann (eds.),
Astronomical and Astrophysical Objectives of Sub-Milliarcsecond Optical Astrometry, 405.

4. CONCLUSIONS

CONCLUDING REMARKS

JEAN KOVALEVSKY
Observatoire de la Côte d'Azur/CERGA,
Av. Copernic 06130 Grasse, France

We are now at the end of a long symposium which lasted five (but not all full) days. The objective was first to review the best present achievements of astrometry and to present projects for a much more advanced astrometry in terms of precision attained versus the number of stars and limiting magnitudes. The word sub-milliarcsecond astrometry was intentionaly vague. It has rightly been often understood as designating a range of precisions of the order of a few tens microarcseconds.

The possibility of getting such precisions – and in some cases for tens of millions – stars was presented and at least eight space projects were described, from USA, Japan, Russia and Western Europe. These projects are based upon different designs, but have at least one common feature: the use of CCD receivers either by a direct view of images formed in the focal plane or by the analysis of some interferometric pattern. They belong essentially to two classes of astrometry. One – like Hipparcos/Tycho – are survey instruments intended to scan the entire sky. The other is designed – like the HST astrometric capabilities – to analyse single objects or at least identified small fields. The possible applications of these instruments are of course quite different.

But nowadays – independent of financial or political arguments which have unfortunately often the highest weight – no project can be approved unless there is a strong scientific case behind it. The various sessions of this symposium showed how many applications of very precise astrometry exist in all domains of astronomy and astrophysics. We have heard some examples, they will be published in the proceedings of this symposium and it is not my intention to present them again, something that I would do much worse than the speakers. What I would like to do, is only to highlight some conclusions that I believe come out of these presentations.

First of all, let me say that the presentations of techniques used and of instruments in existence, or in construction, for ground-based observations

E. Høg and P. K. Seidelmann (eds.),
Astronomical and Astrophysical Objectives of Sub-Milliarcsecond Optical Astrometry, 409–432.

were very useful. They are previews of future astrometry in space as well as on the ground. Clearly, sub-milliarcsecond astrometry is not reserved to space. Actually it already exists on ground. VLBI is the best example, and it was striking to learn that already now, apparent motions of some radio-sources have been obtained with a formal error of 2.5 μas/year, so that such motions were revealed for many sources. It is even more striking to hear that the continuation of this program could ultimately reach conclusions that have been obtained by the satellite COBE. In the same class of presently achieved sub-milliarcsecond precision or their equivalent are the timing of pulsars and the monitoring of Earth's rotation. In all these cases, the interpretation needs careful analyses in the framework of General Relativity.

But even in the optical domain, the achievements of the Mark III or the French I2T interferometers show that sub-milliarcsecond accuracies are already available in very small field astrometry and imaging of stellar discs. For similar accuracies in astrometry, the future USNO optical astrometric interferometer and its extension for the measurement of stellar diameters and binary systems will introduce the Earth based observations in the middle of the sub-milliarcsecond domain. I believe that these progresses are fundamental for astrometry which cannot be based only on prospective but rare space programs. Earth based astrometry must strongly engage in these techniques - or find new ones - which will allow this kind of precision. Another very important job for ground-based astrometry is to extend, although with less precision, the existing catalogues to fainter objects, first up to magnitude 15-17 and later more. The GSC2 is such a program that should be strongly supported. A recommendation in this direction has been prepared and will be submitted to you for approval (see resolution B7 in annex).

But, as was shown yesterday, this is not enough. The preparation of survey-type astrometric missions or the astronomical and astrophysical exploitation of such missions necessitate the knowledge of many other stellar parameters. Radial-velocities have been mentioned several times, but also good and well calibrated photometry, good a priori positions, spectral types and classes, as would for instance be provided by a survey analogous to the Sloan survey, but dedicated to stars rather than to extragalactic objects. Automation and computing facilities could make such a program quite feasible.

We have been delighted to see the first astrometric results of HST and the progress report on Hipparcos and Tycho which shows that this mission is giving results significantly beyond the nominal 2 milliarcsecond objective. The Hipparcos catalogue will become, once linked to the extragalactic reference frame, the optical extension of the new IAU reference frame. A working group sponsored by five of the commissions which organized the present symposium, has made the first step towards the definition of this extragalactic reference frame. Rather than present it to the five commission meetings, it seemed to us that the result could be presented to you and that we shall ask you, under the form of a recommendation, to allow the work to continue in order to be completed for the next General Assembly (see recommendation B6).

Let us now turn towards the scientific applications of very accurate astrometry. The situation is different in the various domains of astronomy, so let us consider them one by one.

Let us first consider stellar astrophysics. In order to model either the atmosphere or the interior, one needs to know all, or the majority, of the following physical parameters: mass, luminosity, radius, effective temperature, surface gravity, and helium and metal contents. Clearly, astrometry cannot provide all of that. Of course, there exist relations between some of these parameters, but there still remains the necessity of highly accurate photometric and spectroscopic observations. The new astroseismology will also add information. It also results from some of the presentations, that the theories themselves, particularly the theory of stellar interiors and evolution, are far from being in a satisfactory stage and much work has to be done to catch up with the improving precision of astrometry.

Concerning the stellar interiors, the evolution of stars, and consequently the age determinations, the theoretical problems are still big – in particular the evaluation of the overshooting parameter which is the position of the transition between radiative and convection zones within a star – and the knowledge of the primordial helium content is fundamental. The discussion of this quantity, as a function of metal content, passes through the main-sequence fitting as a function of metallicity. This means that one has to place main-sequence stars exactly in the HR diagram and this requires the improvement of luminosities, and hence of the parallaxes.

Another parameter of importance is the radius of the star. Very few radii are known and, if they are, it is with an insufficient accuracy, so that they do not provide the additional parameter which would simplify the a priori theoretical constraints used in the models. Here, we need at the same time an accuracy of 1% parallax determinations and 1% in apparent radius determinations, which call for 10 μas interferometric precision – quite reasonable goals for new interferometers and sub-milliarcsecond global astrometry.

Calls for significantly better parallaxes than those provided by Hipparcos were heard on behalf of studies of various classes of low luminosity stars. They are necessary in determining surface gravities and checking spectroscopic estimates of other atmospheric parameters. This includes studies of the later stages of stellar evolution and the evolution of population II stars. Clearly these are not all the needs. It is just as true for population I and for many other stellar types or evolutionary stages which were not presented here. I am sure that one would have heard the same type of arguments.

The problem of masses is different. Most of the direct determinations with a significantly good accuracy (of the order of 1%) are obtained for eclipsing binaries. Very accurate orbits of spectroscopic detached binaries should soon permit to obtain a similar precision provided that accurate radial velocities of the components are obtained. There is a challenge to double star observers to fill up the mass-luminosity relation as a function of metallicity and to extend it to stars with masses smaller than the solar mass. At present, stellar masses are generally evaluated from equations involving different stellar parameters, and hence are model dependent which may present important biases.

So, for stellar astrophysics, accurate astrometry is one, but not the only domain which controls progress. Hipparcos will provide enough good data to trigger more studies and observations that would lead to advances in this domain. The prospect of significantly better accuracies for a large number of stars should be accompanied by similar prospects in getting other parameters. This means, in particular, more very high resolution spectroscopy and obtaining bolometric, rather than current magnitudes. Astrophysical needs have been the major reason that led to the decision to make Hipparcos. Now the prospects of sub-milliarcsecond astrometry, together, I believe, with the on-coming astroseismology, should be a trigger to new developments in stellar physics.

The outcome of the papers dealing with dynamics of clusters, the Galaxy, and the Magellanic clouds, is much simpler. Proper motions and parallaxes are used directly without complicated relations with other parameters. Only the need of more radial velocities was stressed. Here, the more precise the better could be the motto. If one wishes to study a significantly large number of open clusters, 10 microarcsecond, and often 1 microarcsecond, precision are sought for parallaxes. This kind of accuracy would allow the determination of a great variety of HR diagrams to calibrate variations in luminosity as a function of age and metallicity and get detailed stellar evolution tracks. Several clusters will be spatially resolved so that, with proper motions and radial velocities, three dimensional dynamical studies can be made. In dynamical models of the Galaxy, one should be able to distinguish between cylindrical and spherical models and test the extension, the shape and the mass of the halo. The problem of the existence and distribution of the missing mass is at stake. The most accurate data towards the centre of the Galaxy should provide clues about the existence of a bar and other irregularities, such as the dynamical effect of clumps. They also should determine the motions of the globular clusters, possibly over the whole life of the Galaxy. Similar precisions of a few microarcseconds per year are needed for studying the motions of and within the Magellanic clouds. All these studies also imply that this accuracy is obtained for stars of magnitude 15-16.

The problem of the cosmic distance scale is still open, since various determinations of the Hubble constant are still spread in the 50 to 100 km/s/Mpc interval. If Hipparcos could, by its observations of Cepheids and some RRLyrae, reduce the uncertainty to 10-20%, better accuracies would certainly permit a calibration of these stars as a function of chemical composition and greatly reduce the uncertainty of this very important cosmological parameter.

Observations of the effects of gravitational lensing are also an objective of very accurate astrometry. Although it is difficult to predict what are the stars to observe, extensive astrometric surveys could identify optical couples of stars which may produce observable light deviations.

In the domains of solar system research and reference frames, the increase of accuracy of astrometric observations poses a number of new problems. Millisecond pulsar timings, as well as direct observations of minor planets, will permit drastic improvements in the determination of the solar system reference frame and help to compare the dynamical reference system to the extragalactic reference system, bringing an important result for the understanding of the various reference frames in General Relativity theory.

In conclusion, in any astronomical domain in which some parameter is subject to astrometric observations, there is a possibility of progress, provided that some level of accuracy is obtained. In comparison with the present situation illustrated by HST, Hipparcos and the existing optical interferometers, a gain of two orders of magnitude in precision and number of stars will produce remarkable new science. This is the main lesson of this symposium, a lesson publicized by resolution B8.

The speakers, the authors of posters and all the attendance of this symposium are thanked for their contributions, for all the efforts they have made to prepare their presentations, to discuss the matters of interest or to participate to the discussions. I would like to thank particularly Prof. Høg and Dr Seidelmann who are now taking the burden of publishing the proceedings of this meeting, the members of the Scientific Organizing Committee who helped me to set up this meeting, and the Dutch National and Local Organizing committees for providing these very nice meeting facilities.

GENERAL DISCUSSION

A general discussion was opened by P.K. Seidelmann who prepared a number of questions to which some answers were given. They are summarised below :

1 - *Can we get better accuracy from the ground ?*
Good prospects to reach a few microarcseconds exist in very small field interferometry. Progress in conventional large field and global astrometric techniques is probably marginal.

2 - *Future for photographic astrometry*
Despite difficulties in getting plates and their cost, photography has the advantage of long term availability for new measurements and reduction. It also is the best tool for relative astrometry for wide fields of a few degree range. It also permits more accurate colour calibration.
On the other hand, CCD astrometry with its high quantum efficiency and dynamic range, especially with the scan technique and mosaics of CCD chips, permit a survey of faint magnitudes, allowing a densification of existing catalogues up to magnitude 20 or so. Precisions of 30-40 mas are possible.
In conclusion, it seems that both techniques have a future at least until the next generation of global space astrometry projects.

3 - *Future for meridian circles*
They need to have a new mission. They could be the intermediaries between the Hipparcos/Tycho catalogues and the CCD densification. New design may be necessary.

4 - *Is it more important to go fainter than more accurate ?*
Both are very important and should be sought. Actually we need both at the same time and also additional wavelengths.

5 - *Future star catalogue requirements*
There is a great need, in addition to better and more extensive astrometric catalogues, for large surveys of different astrophysical quantities.

6 - *Maintenance plan for the extragalactic reference frame*
It should include not only densification and improvement of accuracies, but also the study and analysis of the source structures (imaging to better than 1 mas resolution). One should also test the quality of Hipparcos proper motions of radio stars, especially in the Southern hemisphere. This problem is addressed in the conclusions of the IAU working group on reference frames.

RECOMMENDATIONS

The symposium has adopted three recommendations which were in due course approved as resolutions by the 22nd General Assembly of the IAU. We give here these resolutions as adopted by the General Assembly after some slight amendments.

Resolution B6 on the working group on reference frames

The General Assembly of the IAU upon the advice of the participants of symposium 166

considering
that an IAU working group on reference frames consisting of members of Commissions 4, 8, 19, 24 and 31, the International Earth Rotation Service (IERS) and other pertinent experts has been formed to produce a list of candidate extragalactic radio-sources for defining the new conventional reference frame and secondary sources that may later be added or replace some of the primary sources,

noting
that a list of sources which defines the conventional reference frame together with a list of candidate sources which may, at some future date, be added or replace the defining sources has been made

adopts
this list of defining sources as the first stage in the definition of the new reference frame, and

requests
that the working group on reference frames be continued and its membership be reviewed by Commissions 4, 8, 19, 24 and 31 and the IERS to

1. specify the positions of the radio sources on the list,
2. determine the relationship of this frame to an optical frame defined by stars, and
3. recommend to the XXIIIrd General Assembly (1997) that a way be found to organize the work for the maintenance and evolution of this frame and its extension to other frames at other wavelengths.

The list, consisting of 606 objects is given in the annex.

Resolution B7 on the second generation of STScI Guide Star Catalog

The General Assembly of the IAU upon the advice of the participants of symposium 166 on *Astronomical and astrophysical objectives of sub-milliarcsecond optical astrometry*, meeting at the occasion of the 22nd IAU General Assembly in The Hague, 15-19 August 1994 :

taking into account
the immense importance to the entire astronomical community of the STScI's Guide Star Catalog (GSC);

taking into account
the expected characteristics of the proposed GSC-II project;

taking into account
foreseen implications of the availability of the GSC-II for countless applications in ground-based and space-based astrometry over the next decades;

taking into account
the anticipated distribution of compressed second generation plate scans to the astronomical community; and

taking into account
the scientific technical competence at the STScI, the availability of the plate material and digitizing facilities, and the team's willingness to undertake the GSC-II project

urges
the Executive Committee of the IAU to approach NASA and other relevant national and international funding agencies to do their utmost to ensure the necessary funding for timely completion of the second generation plate scanning and the construction of the GSC-II at STScI, and urges the international community to engage in broadening the support and in pursing derivative collaborative projects.

Resolution B8 on the need to develop optical sub-milliarcsecond astrometry

The General Assembly of the IAU

considering

that the symposium 166 has discussed the many aspects of solar system, galactic and extragalactic astronomy and astrophysics requiring high accuracy optical astrometry,

1. emphasizes the strong need for sub-milliarcsecond accuracy astrometric data for very large numbers of stars,

2. notes that satellite options have been proposed, orders of magnitude more accurate and productive than the very successful Hipparcos/Tycho mission.

3. urges the space agencies to study the possibilities of sub-milliarcsecond optical astrometry with the aim to develop optimal projects as soon as possible, taking advantage of the present high level of expertise and dedication.

ANNEX TO RESOLUTION B6

List of extragalactic objects identified as sources which define the new conventional celestial reference frame together with candidate sources which may, at some future date, be added or replace the defining sources

- d : defining sources
- c : additional sources
- o : optical objects

TABLE 1. Sources

	Name			R.A.			Dec.	Alias
d	0003-066	0	6	13.89	-6	23	35.3	PKS 0003-066
d	0007+106	0	10	31.01	10	58	29.5	IIIZW2, PKS 0007+106
d	0007+171	0	10	33.99	17	24	18.8	4C+17.04
d	0008-264	0	11	1.25	-26	12	33.4	PKS 0008-264
d	0010+405	0	13	31.13	40	51	37.1	B3 0010+406
d	0013-005	0	16	11.09	0	-15	12.5	PKS 0013-005
d	0014+813	0	17	8.48	81	35	8.1	S5 0014+81
d	0016+731	0	19	45.79	73	27	30.0	S5 0016+73
d	0019+058	0	22	32.44	6	8	4.3	PKS 0019+058

	Name		R.A.				Dec.	Alias
d	0026+346	0	29	14.24	34	56	32.2	OB343, S4 0026+34
d	0039+230	0	42	4.55	23	20	1.1	PKS 0039+230
d	0047-579	0	49	59.47	-57	38	27.3	PKS 0047-579
d	0048-097	0	50	41.32	-9	29	5.2	PKS 0048-097
d	0056-572	0	58	46.58	-56	59	11.5	PKS 0056-572
d	0056-001	0	59	5.51	0	6	51.6	4C-00.06
d	0059+581	1	2	45.76	58	24	11.1	
d	0104-408	1	6	45.11	-40	34	20.0	
d	0106+013	1	8	38.77	1	35	0.3	4C+01.02
d	0109+224	1	12	5.82	22	44	38.8	
d	0111+021	1	13	43.14	2	22	17.3	
d	0112-017	1	15	17.10	-1	27	4.6	PKS 0112-014
d	0113-118	1	16	12.52	-11	36	15.4	PKS 0113-118
d	0119+115	1	21	41.59	11	49	50.4	PKS 0119+115
d	0119+041	1	21	56.86	4	22	24.7	IRAS F01177+
d	0123+257	1	26	42.79	25	59	1.3	
d	0131-522	1	33	5.76	-52	0	4.0	PKS 0131-522
d	0133+476	1	36	58.59	47	51	29.1	
d	0135-247	1	37	38.35	-24	30	53.9	
d	0134+329	1	37	41.30	33	9	35.1	3C48, 4C+39.25
d	0146+056	1	49	22.37	5	55	53.6	PKS 0146+056
d	0148+274	1	51	27.15	27	44	41.8	
d	0149+218	1	52	18.06	22	7	7.7	PKS 0149+218
d	0150-334	1	53	10.12	-33	10	25.9	PKS 0150-334
d	0153+744	1	57	34.96	74	42	43.2	
d	0159+723	2	3	33.38	72	32	53.7	
d	0201+113	2	3	46.66	11	34	45.4	PKS 0201+113
d	0202+149	2	4	50.41	15	14	11.0	4C+15.05
d	0202-172	2	4	57.67	-17	1	19.8	PKS 0202-172
d	0202+319	2	5	4.93	32	12	30.1	B2 0202+31
d	0208-512	2	10	46.20	-51	1	1.9	PKS 0208-512
d	0212+735	2	17	30.81	73	49	32.6	S5 0212+73
d	0215+015	2	17	48.95	1	44	49.7	
d	0219+428	2	22	39.61	43	2	7.8	
d	0220-349	2	22	56.40	-34	41	28.7	PKS 0220-349
d	0221+067	2	24	28.43	6	59	23.3	
d	0224+671	2	28	50.05	67	21	3.0	4C+67.05
d	0230-790	2	29	34.95	-78	47	45.6	PKS 0230-790
d	0229+131	2	31	45.89	13	22	54.7	4C+13.14
d	0234+285	2	37	52.41	28	48	9.0	4C+28.07

	Name		R.A.			Dec.		Alias
d	0235+164	2	38	38.93	16	36	59.3	PKS 0235+164
d	0237+040	2	39	51.26	4	16	21.4	PKS 0237+040
d	0238-084	2	41	4.80	-8	15	20.8	NGC1052, PKS 0238-084
d	0239+108	2	42	29.17	11	1	0.7	PKS 0239+108
d	0248+430	2	51	34.54	43	15	15.8	S4 0248+43
d	0252-549	2	53	29.18	-54	41	51.4	PKS 0252-549
d	0256+075	2	59	27.08	7	47	39.6	
d	0259+121	3	2	30.55	12	18	56.7	
d	0300+470	3	3	35.24	47	16	16.3	OE400, 4C+47.08
d	0302-623	3	3	50.63	-62	11	25.6	PKS 0302-623
d	0302+625	3	6	42.66	62	43	2.0	
d	0306+102	3	9	3.62	10	29	16.3	OE110
d	0308-611	3	9	56.10	-60	58	39.1	PKS 0308-611
d	0312-770	3	11	55.25	-76	51	50.9	PKS 0312-770
d	0309+411	3	13	1.96	41	20	1.2	
d	0319+121	3	21	53.10	12	21	13.9	PKS 0319+121
d	0326+279	3	29	57.67	27	56	15.5	0326+277
d	0332-403	3	34	13.65	-40	8	25.4	PKS 0332-403
d	0333+321	3	36	30.11	32	18	29.3	NRAO140, 4C+32.14
d	0336-019	3	39	30.94	-1	46	35.8	CTA26, PKS 0336-019
d	0338-214	3	40	35.61	-21	19	31.2	PKS 0338-214
d	0341+158	3	44	23.17	15	59	43.4	
d	0342+147	3	45	6.42	14	53	49.6	
d	0400+258	4	3	5.59	26	0	1.5	PKS 0400+258
d	0402-362	4	3	53.75	-36	5	1.9	PKS 0402-362
d	0405+305	4	8	20.38	30	32	30.5	
d	0406-127	4	9	5.77	-12	38	48.1	
d	0406+121	4	9	22.01	12	17	39.8	PKS 0406+121
d	0414-189	4	16	36.54	-18	51	8.3	PKS 0414-189
d	0420-014	4	23	15.80	-1	20	33.1	PKS 0420-014
d	0420+417	4	23	56.01	41	50	2.7	
d	0422-380	4	24	42.24	-37	56	20.8	
d	0422+004	4	24	46.84	0	36	6.3	OF038, PKS 0422+004
d	0423+051	4	26	36.60	5	18	19.9	PKS 0423+051
d	0425+048	4	27	47.57	4	57	8.3	
d	0426-380	4	28	40.42	-37	56	19.6	PKS 0426-380
d	0434-188	4	37	1.48	-18	44	48.6	PKS 0434-188
d	0437-454	4	39	0.85	-45	22	22.6	
d	0438-436	4	40	17.18	-43	33	8.6	PKS 0438-436
d	0440-003	4	42	38.66	0	-17	43.4	NRAO190, PKS 0440-003

	Name		R.A.			Dec.		Alias
d	0440+345	4	43	31.63	34	41	6.7	
d	0446+112	4	49	7.67	11	21	28.6	
d	0454-810	4	50	5.44	-81	1	2.2	PKS 0454-810
d	0451-282	4	53	14.65	-28	7	37.3	PKS 0451-282
d	0454-234	4	57	3.18	-23	24	52.0	
d	0457+024	4	59	52.05	2	29	31.2	PKS 0457+024
d	0458-020	5	1	12.81	-1	59	14.3	4C-02.19
d	0458+138	5	1	45.27	13	56	7.2	
d	0459+060	5	2	15.45	6	9	7.5	
d	0500+019	5	3	21.20	2	3	4.7	
d	0502+049	5	5	23.18	4	59	42.7	
d	0506-612	5	6	43.99	-61	9	41.0	PKS 0506-612
d	0454+844	5	8	42.36	84	32	4.5	S5 0454+84
d	0506+101	5	9	27.46	10	11	44.6	
d	0507+179	5	10	2.37	18	0	41.6	PKS 0507+179
d	0511-220	5	13	49.11	-21	59	16.1	PKS 0511-220
d	0516-621	5	16	44.93	-62	7	5.4	
d	0518+165	5	21	9.89	16	38	22.0	3C138, 4C+16.12
d	0522-611	5	22	34.43	-61	7	57.1	PKS 0522-611
d	0521-365	5	22	57.98	-36	27	30.9	PKS 0521-365
d	0530-727	5	29	30.04	-72	45	28.5	PKS 0530-727
d	0528-250	5	30	7.96	-25	3	29.9	PKS 0528-250
d	0528+134	5	30	56.42	13	31	55.1	PKS 0528+134
d	0537-441	5	38	50.36	-44	5	8.9	PKS 0537-441
d	0537-158	5	39	32.01	-15	50	30.3	PKS 0537-158
d	0536+145	5	39	42.37	14	33	45.6	
d	0537-286	5	39	54.28	-28	39	56.0	PKS 0537-286
d	0539-057	5	41	38.08	-5	41	49.4	PKS 0539-057
d	0538+498	5	42	36.14	49	51	7.2	3C147, 4C+49.14
d	0544+273	5	47	34.15	27	21	56.8	
d	0552+398	5	55	30.81	39	48	49.2	B2 0552+39A
d	0556+238	5	59	32.03	23	53	53.9	
d	0600+177	6	3	9.13	17	42	16.8	
d	0605-085	6	7	59.70	-8	34	50.0	PKS 0605-085
d	0607-157	6	9	40.95	-15	42	40.7	PKS 0607-157
d	0609+607	6	14	23.87	60	46	21.8	
d	0615+820	6	26	3.00	82	2	25.6	S5 0615+82
d	0629-418	6	31	12.00	-41	54	26.9	PKS 0629-418
d	0637-752	6	35	46.51	-75	16	16.8	PKS 0637-752
d	0637-337	6	39	20.90	-33	46	0.1	PKS 0637-337

	Name		R.A.			Dec.		Alias
d	0636+680	6	42	4.26	67	58	35.6	S4 0636+68
d	0624+214	6	45	24.10	21	21	51.2	3C166, 4C+21.21
d	0642+449	6	46	32.03	44	51	16.6	B3 0642+449
d	0646-306	6	48	14.10	-30	44	19.7	PKS 0646-306
d	0650+371	6	53	58.28	37	5	40.6	S4 0650+37
d	0657+172	7	0	1.53	17	9	21.7	
d	0707+476	7	10	46.10	47	32	11.1	B3 0707+476
d	0711+356	7	14	24.82	35	34	39.8	
d	0716+714	7	21	53.45	71	20	36.4	S5 0716+71
d	0722+145	7	25	16.81	14	25	13.7	4C+14.23
d	0723-008	7	25	50.64	0	-54	56.5	PKS 0723-008
d	0718+792	7	26	11.73	79	11	31.0	
d	0727-115	7	30	19.11	-11	41	12.6	PKS 0727-115
d	0733-174	7	35	45.81	-17	35	48.5	PKS 0733-174
d	0735+178	7	38	7.39	17	42	19.0	OI158, PKS 0735+178
d	0738-674	7	38	56.50	-67	35	50.8	PKS 0738-674
d	0736+017	7	39	18.03	1	37	4.6	PKS 0736+017
d	0738+313	7	41	10.70	31	12	0.2	B2 0738+31
d	0743-673	7	43	31.61	-67	26	25.5	PKS 0743-673
d	0742+103	7	45	33.06	10	11	12.7	PKS 0742+103
d	0743-006	7	45	54.08	0	-44	17.5	4C-00.28
d	0743+259	7	46	25.87	25	49	2.1	
d	0745+241	7	48	36.11	24	0	24.1	PKS 0745+241
d	0748+126	7	50	52.05	12	31	4.8	PKS 0748+126
d	0749+540	7	53	1.38	53	52	59.6	4C+54.15
d	0754+100	7	57	6.64	9	56	34.9	PKS 0754+100
d	0805-077	8	8	15.54	-7	51	9.9	PKS 0805-077
d	0804+499	8	8	39.67	49	50	36.5	S4 0804+49
d	0805+410	8	8	56.65	40	52	44.9	B3 0805+410
d	0808+019	8	11	26.71	1	46	52.2	PKS 0808+019
d	0812+367	8	15	25.94	36	35	15.1	B2 0812+36
d	0814+425	8	18	16.00	42	22	45.4	S4 0814+42
d	0820+560	8	24	47.24	55	52	42.7	4CP56.16A
d	0821+394	8	24	55.48	39	16	41.9	4C+39.23
d	0823-500	8	25	26.87	-50	10	38.5	PKS 0823-500
d	0823+033	8	25	50.34	3	9	24.5	PKS 0823+033
d	0823-223	8	26	1.57	-22	30	27.2	PKS 0823-223
d	0826-373	8	28	4.78	-37	31	6.3	PKS 0826-373
d	0827+243	8	30	52.09	24	10	59.8	B2 0827+24
d	0829+046	8	31	48.88	4	29	39.1	PKS 0829+046

	Name		R.A.			Dec.		Alias
d	0828+493	8	32	23.22	49	13	21.0	S4 0828+49
d	0831+557	8	34	54.90	55	34	21.1	4C+55.16
d	0834-201	8	36	39.22	-20	16	59.5	PKS 0834-201
d	0833+585	8	37	22.41	58	25	1.8	S4 0833+585
d	0836+710	8	41	24.36	70	53	42.2	4C+71.07
d	0839+187	8	42	5.09	18	35	41.0	PKS 0839+187
d	0851+202	8	54	48.87	20	6	30.6	OJ287, PKS 0851+202
d	0859-140	9	2	16.83	-14	15	30.9	PKS 0859-140
d	0859+470	9	3	3.99	46	51	4.1	4C+47.29
d	0906+015	9	9	10.09	1	21	35.6	4C+01.24
d	0912+029	9	14	37.91	2	45	59.2	PKS 0912+029
d	0912+297	9	15	52.40	29	33	24.0	B2 0912+29
d	0917+449	9	20	58.46	44	41	54.0	S4 0917+44
d	0917+624	9	21	36.23	62	15	52.2	S5 0917+62
d	0920-397	9	22	46.42	-39	59	35.1	PKS 0920-397
d	0923+392	9	27	3.01	39	2	20.9	4C39.25, 4C+39.25
d	0925-203	9	27	51.82	-20	34	51.2	PKS 0925-203
d	0945+408	9	48	55.34	40	39	44.6	4C+40.24
d	0953+254	9	56	49.88	25	15	16.1	OK290, VRO 25.09.08
d	0955+476	9	58	19.67	47	25	7.8	B3 0955+476
d	0955+326	9	58	20.95	32	24	2.2	3C232, 4C+32
d	0954+658	9	58	47.24	65	33	54.8	S4 0945+65
d	1004+141	10	7	41.50	13	56	29.6	PKS 1004+141
d	1011+250	10	13	53.43	24	49	16.4	B2 1011+25
d	1012+232	10	14	47.07	23	1	16.6	4C+23.24
d	1020+400	10	23	11.57	39	48	15.4	B3 1020+400
d	1021-006	10	24	29.59	0	-52	55.5	PKS 1021-006
d	1022+194	10	24	44.81	19	12	20.4	4C+19.34
d	1030+415	10	33	3.71	41	16	6.2	VRO 10.41.03
d	1032-199	10	35	2.16	-20	11	34.4	PKS 1032-199
d	1034-293	10	37	16.08	-29	34	2.8	PKS 1034-293
d	1038+064	10	41	17.16	6	10	16.9	4C+06.41
d	1038+528	10	41	46.78	52	33	28.2	
d	1040+123	10	42	44.60	12	3	31.3	3C245, 4C+12.37
d	1039+811	10	44	23.06	80	54	39.4	S5 1039+811
d	1042+071	10	44	55.91	6	55	38.3	PKS 1042+071
d	1044+719	10	48	27.62	71	43	35.9	
d	1048-313	10	51	4.78	-31	38	14.3	PKS 1048-313
d	1049+215	10	51	48.79	21	19	52.3	4C+21.28
d	1053+704	10	56	53.62	70	11	45.9	

	Name		R.A.			Dec.		Alias
d	1053+815	10	58	11.53	81	14	32.7	
d	1055+018	10	58	29.61	1	33	58.8	4C+01.28
d	1057-797	10	58	43.31	-80	3	54.2	PKS 1057-797
d	1101-536	11	3	52.22	-53	57	0.7	PKS 1101-536
d	1104-445	11	7	8.69	-44	49	7.6	PKS 1104-445
d	1105-680	11	7	12.69	-68	20	50.7	PKS 1105-680
d	1111+149	11	13	58.69	14	42	27.0	4C-00.43
d	1116-462	11	18	26.96	-46	34	15.0	PKS 1116-462
d	1116+128	11	18	57.30	12	34	41.7	4C+12.39
d	1123+264	11	25	53.71	26	10	20.0	PKS 1123+264
d	1124-186	11	27	4.39	-18	57	17.4	PKS 1124-186
d	1127-145	11	30	7.05	-14	49	27.4	PKS 1127-145
d	1128+385	11	30	53.28	38	15	18.6	B3 1128+385
d	1130+009	11	33	20.06	0	40	52.8	PKS 1130+009
d	1143-245	11	46	8.10	-24	47	32.9	PKS 1143-245
d	1144+402	11	46	58.30	39	58	34.3	B3 1144+402
d	1144-379	11	47	1.37	-38	12	11.0	PKS 1144-379
d	1145-071	11	47	51.55	-7	24	41.1	PKS 1145-071
d	1148-001	11	50	43.87	0	-23	54.2	4C-00.47
d	1148-671	11	51	13.43	-67	28	11.1	PKS 1148-671
d	1150+812	11	53	12.50	80	58	29.2	S5 1150+812
d	1150+497	11	53	24.47	49	31	8.8	4C+49.22
d	1155+251	11	58	25.79	24	50	18.0	
d	1156-094	11	59	12.71	-9	40	52.0	PKS 1156-094
d	1156+295	11	59	31.83	29	14	43.8	4C+29.45
d	1213+350	12	15	55.60	34	48	15.2	4C+35.28
d	1215+303	12	17	52.08	30	7	0.6	B2 1215+30
d	1216+487	12	19	6.41	48	29	56.2	S4 1216+48
d	1219+285	12	21	31.69	28	13	58.5	W Com
d	1219+044	12	22	22.55	4	13	15.8	4C+04.42
d	1221+809	12	23	40.49	80	40	4.3	
d	1222+037	12	24	52.42	3	30	50.3	4C+03.23
d	1226+373	12	28	47.42	37	6	12.1	
d	1228+126	12	30	49.42	12	23	28.1	3C274, M87, Virgo A
d	1236+077	12	39	24.59	7	30	17.2	PKS 1236+077
d	1236-684	12	39	46.65	-68	45	30.9	PKS 1236-684
d	1243-072	12	46	4.23	-7	30	46.6	PKS 1243-072
d	1244-255	12	46	46.80	-25	47	49.3	PKS 1244-255
d	1252+119	12	54	38.26	11	41	5.9	PKS 1252+119

	Name			R.A.			Dec.	Alias
d	1251-713	12	54	59.92	-71	38	18.4	PKS 1251-713
d	1253-055	12	56	11.17	-5	47	21.5	3C279, 4C-05.55
d	1255-316	12	57	59.06	-31	55	16.8	PKS 1255-316
d	1257+145	13	0	20.92	14	17	18.5	PKS 1257+145
d	1302-102	13	5	33.01	-10	33	19.4	PKS 1302-102
d	1308+326	13	10	28.66	32	20	43.8	AU CVn
d	1313-333	13	16	7.99	-33	38	59.2	PKS 1313-333
d	1315+346	13	17	36.49	34	25	15.9	OP326, B2 1315+34A
d	1324+224	13	27	00.86	22	10	50.2	
d	1334-127	13	37	39.78	-12	57	24.7	PKS 1334-127
d	1338+381	13	40	22.95	37	54	43.8	
d	1342+662	13	43	45.96	66	2	25.8	
d	1342+663	13	44	8.68	66	6	11.7	
d	1347+539	13	49	34.66	53	41	17.0	1347+53, 4C+53.28
d	1349-439	13	52	56.53	-44	12	40.4	PKS 1349-439
d	1351-018	13	54	6.90	-2	6	3.2	PKS 1351-018
d	1354+195	13	57	4.44	19	19	7.4	4C+19.44
d	1354-152	13	57	11.24	-15	27	28.8	PKS 1354-152
d	1357+769	13	57	55.37	76	43	21.1	
d	1402-012	14	4	45.90	-1	30	21.9	PKS 1402-012
d	1402+044	14	5	1.12	4	15	35.8	PKS 1402+044
d	1404+286	14	7	00.39	28	27	14.7	OQ208, MRK 668
d	1406-076	14	8	56.48	-7	52	26.7	PKS 1406-076
d	1413+135	14	15	58.82	13	20	23.7	PKS 1413+135
d	1416+067	14	19	8.18	6	28	34.8	3C298
d	1418+546	14	19	46.60	54	23	14.8	S4 1418+54
d	1424-418	14	27	56.30	-42	6	19.4	PKS 1424-418
d	1430-178	14	32	57.69	-18	1	35.2	PKS 1430-178
d	1435+638	14	36	45.80	63	36	37.9	S4 1435+63
d	1435-218	14	38	9.47	-22	4	54.7	PKS 1435-218
d	1442+101	14	45	16.47	9	58	36.1	OQ172, PKS 1442+101
d	1443-162	14	45	53.38	-16	29	1.6	PKS 1443-162
d	1445-161	14	48	15.05	-16	20	24.5	PKS 1445-161
d	1448+762	14	48	28.78	76	1	11.6	
d	1451-375	14	54	27.41	-37	47	33.1	PKS 1451-375
d	1451-400	14	54	32.91	-40	12	32.5	PKS 1451-400
d	1458+718	14	59	7.58	71	40	19.9	3C309.1, 4C+71.15
d	1459+480	15	0	48.65	47	51	15.5	1459+48
d	1502+106	15	4	24.98	10	29	39.2	PKS 1502+106
d	1502+036	15	5	6.48	3	26	30.8	PKS 1502+036

	Name		R.A.			Dec.		Alias
d	1504+377	15	6	9.53	37	30	51.1	B3 1504+377
d	1504-166	15	7	4.79	-16	52	30.3	PKS 1504-166
d	1510-089	15	12	50.53	-9	5	59.8	PKS 1510-089
d	1511-100	15	13	44.89	-10	12	0.3	PKS 1511-100
d	1514+197	15	16	56.80	19	32	13.0	PKS 1514+197
d	1514-241	15	17	41.81	-24	22	19.5	AP Lib
d	1519-273	15	22	37.68	-27	30	10.8	PKS 1519-273
d	1532+016	15	34	52.45	1	31	4.2	PKS 1532+016
d	1538+149	15	40	49.49	14	47	45.9	4C+14.60
d	1547+507	15	49	17.47	50	38	5.8	
d	1546+027	15	49	29.44	2	37	1.2	PKS 1546+027
d	1548+056	15	50	35.27	5	27	10.5	4C+05.45
d	1549-790	15	56	58.87	-79	14	4.3	PKS 1549-790
d	1555+001	15	57	51.43	0	-1	50.4	PKS 1555+001
d	1600+335	16	2	7.26	33	26	53.1	
d	1604-333	16	7	34.76	-33	31	8.9	PKS 1604-333
d	1606+106	16	8	46.20	10	29	7.8	4C+10.45
d	1611+343	16	13	41.06	34	12	47.9	
d	1614+051	16	16	37.56	4	59	32.7	PKS 1614+051
d	1610-771	16	17	49.28	-77	17	18.5	PKS 1610-771
d	1616+063	16	19	3.69	6	13	2.2	PKS 1616+063
d	1619-680	16	24	18.44	-68	9	12.5	PKS 1619-680
d	1624+416	16	25	57.67	41	34	40.6	4C+41.32
d	1622-297	16	26	6.02	-29	51	27.0	PKS 1622-297
d	1633+382	16	35	15.49	38	8	4.5	
d	1637+574	16	38	13.46	57	20	24.0	S4 1637+57
d	1638+398	16	40	29.63	39	46	46.0	NRAO512
d	1642+690	16	42	7.85	68	56	39.8	4C+69.21
d	1641+399	16	42	58.81	39	48	37.0	3C345, 4C+39.48
d	1647-296	16	50	39.54	-29	43	47.0	PKS 1647-296
d	1652+398	16	53	52.22	39	45	36.6	DA426, 4C+39.49
d	1656+348	16	58	1.42	34	43	28.4	
d	1655+077	16	58	9.01	7	41	27.5	PKS 1655+077
d	1656+053	16	58	33.45	5	15	16.4	PKS 1656+053
d	1657-261	17	0	53.15	-26	10	51.7	PKS 1657-261
d	1705+456	17	7	17.75	45	36	10.6	4C+45.34
d	1705+018	17	7	34.42	1	48	45.7	PKS 1705+018
d	1706-174	17	9	34.35	-17	28	53.4	
d	1717+178	17	19	13.05	17	45	6.4	PKS 1717+178

	Name	R.A.			Dec.			Alias
d	1718-649	17	23	41.03	-65	0	36.6	NGC 6328
d	1726+455	17	27	27.65	45	30	39.7	B3 1726+455
d	1727+502	17	28	18.62	50	13	10.5	IIZW77
d	1725+044	17	28	24.95	4	27	4.9	PKS 1725+044
d	1730-130	17	33	2.71	-13	4	49.5	NRAO530, PKS 1730-132
d	1732+389	17	34	20.58	38	57	51.4	B3 1732+389
d	1738+476	17	39	57.13	47	37	58.4	S4 1738+47
d	1739+522	17	40	36.98	52	11	43.4	4C+51.37
d	1741-038	17	43	58.86	-3	50	4.6	PKS 1741-038
d	1743+173	17	45	35.21	17	20	1.4	PKS 1743+173
d	1745+624	17	46	14.03	62	26	54.7	4C+62.29
d	1749+701	17	48	32.84	70	5	50.8	S5 1749+70
d	1749+096	17	51	32.82	9	39	0.7	OT081, 4C+09.57
d	1751+441	17	53	22.65	44	9	45.7	S4 1751+44
d	1751+288	17	53	42.47	28	48	4.9	
d	1803+784	18	0	45.69	78	28	4.0	S5 1803+78
d	1800+440	18	1	32.32	44	4	21.9	B3 1800+440
d	1807+698	18	6	50.68	69	49	28.1	3C371
d	1815-553	18	19	45.40	-55	21	20.7	PKS 1815-553
d	1821+107	18	24	2.86	10	44	23.8	PKS 1821+107
d	1823+568	18	24	7.07	56	51	1.5	4C+56.27
d	1830+285	18	32	50.19	28	33	36.0	4C+28.45
d	1831-711	18	37	28.71	-71	8	43.6	PKS 1831-711
d	1845+797	18	42	8.99	79	46	17.1	3C390.3, 4C+79.18
d	1842+681	18	42	33.64	68	9	25.2	
d	1849+670	18	49	16.07	67	5	41.7	S4 1849+67
d	1856+736	18	54	57.30	73	51	19.9	
d	1901+319	19	2	55.94	31	59	41.7	3C395, 4C+31.52, 19
d	1908-201	19	11	9.65	-20	6	55.1	PKS 1908-201
d	1903-802	19	12	40.02	-80	10	5.9	PKS 1903-802
d	1920-211	19	23	32.19	-21	4	33.3	
d	1921-293	19	24	51.06	-29	14	30.1	OV236, PKS 1921-293
d	1923+210	19	25	59.61	21	6	26.2	PKS 1923+210
d	1928+738	19	27	48.50	73	58	1.6	4C+73.18
d	1925-610	19	30	6.16	-60	56	9.2	PKS 1925-610
d	1929+226	19	31	24.92	22	43	31.3	
d	1932+204	19	35	10.47	20	31	54.2	
d	1933-400	19	37	16.22	-39	58	1.6	PKS 1933-400
d	1936-155	19	39	26.66	-15	25	43.1	PKS 1936-155
d	1937-101	19	39	57.26	-10	2	41.5	PKS 1937-101

	Name		R.A.				Dec.	Alias
d	1935-692	19	40	25.53	-69	7	57.0	PKS 1935-692
d	1951+355	19	53	30.88	35	37	59.4	
d	1950-613	19	55	10.77	-61	15	19.1	PKS 1950-613
d	1954+513	19	55	42.74	51	31	48.5	PKS 1954+513
d	1954-388	19	57	59.82	-38	45	6.4	PKS 1954-388
d	1958-179	20	0	57.09	-17	48	57.7	OV198, PKS 1958-179
d	2000-330	20	3	24.12	-32	51	45.1	PKS 2000-330
d	2007+777	20	5	31.00	77	52	43.2	S5 2007+77
d	2005-489	20	9	25.39	-48	49	53.7	PKS 2005-489
d	2011-067	20	11	14.22	-6	44	3.6	OW-015
d	2008-159	20	11	15.71	-15	46	40.3	PKS 2008-159
d	2017+743	20	17	13.08	74	40	48.0	4C+74.25
d	2021+317	20	23	19.02	31	53	2.3	4C+31.56
d	2030+547	20	31	47.96	54	55	3.1	4C+54.42
d	2029+121	20	31	54.99	12	19	41.3	PKS 2029+121
d	2037+511	20	38	37.04	51	19	12.7	3C418, 4C+51.42
d	2051+745	20	51	33.74	74	41	40.5	
d	2052-474	20	56	16.36	-47	14	47.6	PKS 2052-474
d	2059+034	21	1	38.83	3	41	31.3	PKS 2059+034
d	2059-786	21	5	44.96	-78	25	34.5	PKS 2059-786
d	2106-413	21	9	33.19	-41	10	20.6	PKS 2106-413
d	2113+293	21	15	29.41	29	33	38.4	
d	2109-811	21	16	30.84	-80	53	55.2	PKS 2109-811
d	2126-158	21	29	12.18	-15	38	41.0	PKS 2126-158
d	2128-123	21	31	35.26	-12	7	4.8	PKS 2128-123
d	2131-021	21	34	10.31	-1	53	17.2	4C-02.81
d	2136+141	21	39	1.31	14	23	36.0	PKS 2136+141
d	2143-156	21	46	22.98	-15	25	43.9	PKS 2143-156
d	2144+092	21	47	10.16	9	29	46.7	PKS 2144+092
d	2142-758	21	47	12.73	-75	36	13.2	PKS 2142-758
d	2145+067	21	48	5.46	6	57	38.6	4C+06.69
d	2149+056	21	51	37.88	5	52	13.0	PKS 2149+056
d	2149-307	21	51	55.52	-30	27	53.7	PKS 2149-306
d	2146-783	21	52	3.15	-78	7	6.6	PKS 2146-783
d	2150+173	21	52	24.82	17	34	37.8	PKS 2150+173
d	2152-699	21	57	5.98	-69	41	23.7	
d	2155-152	21	58	6.28	-15	1	9.3	PKS 2155-152
d	2200+420	22	2	43.29	42	16	40.0	VR422201, BL Lac
d	2201+315	22	3	14.98	31	45	38.3	4C+31.63
d	2204-540	22	7	43.73	-53	46	33.8	PKS 2204-540

	Name	R.A.			Dec.			Alias
d	2209+236	22	12	5.97	23	55	40.5	PKS 2209+236
d	2216-038	22	18	52.04	-3	35	36.9	4C-03.79
d	2223-052	22	25	47.26	-4	57	1.4	3C446, 4C-05.92
d	2227-088	22	29	40.08	-8	32	54.4	PKS 2227-088
d	2229+695	22	30	36.47	69	46	28.1	2229+69
d	2227-399	22	30	40.28	-39	42	52.1	PKS 2227-399
d	2230+114	22	32	36.41	11	43	50.9	CTA102, 4C+11.69
d	2232-488	22	35	13.24	-48	35	58.8	PKS 2232-488
d	2234+282	22	36	22.47	28	28	57.4	B2 2234+28A
d	2233-148	22	36	34.09	-14	33	22.2	PKS 2233-148
d	2243-123	22	46	18.23	-12	6	51.3	PKS 2243-123
d	2245-328	22	48	38.69	-32	35	52.2	PKS 2245-328
d	2252-089	22	55	4.24	-8	44	4.0	PKS 2252-089
d	2253+417	22	55	36.71	42	2	52.5	B3 2253+417
d	2254+024	22	57	17.56	2	43	17.5	PKS 2254+024
d	2254+074	22	57	17.30	7	43	12.3	PKS 2254+074
d	2255-282	22	58	5.96	-27	58	21.3	PKS 2255-282
d	2311-452	23	14	9.38	-44	55	49.2	PKS 2311-452
d	2312-319	23	14	48.50	-31	38	39.5	PKS 2312-319
d	2318+049	23	20	44.86	5	13	49.9	PKS 2318+049
d	2319+272	23	21	59.86	27	32	46.4	4C+27.50
d	2320-035	23	23	31.95	-3	17	5.0	PKS 2320-035
d	2326-477	23	29	17.70	-47	30	19.1	PKS 2326-477
d	2328+107	23	30	40.85	11	0	18.7	4C+10.73
d	2329-384	23	31	59.48	-38	11	47.7	PKS 2329-384
d	2331-240	23	33	55.24	-23	43	40.7	PKS 2331-240
d	2335-027	23	37	57.34	-2	30	57.6	PKS 2335-027
d	2344+092	23	46	36.84	9	30	45.5	
d	2345-167	23	48	2.61	-16	31	12.0	PKS 2345-167
d	2351+456	23	54	21.68	45	53	4.2	4C+45.51
d	2351-154	23	54	30.19	-15	13	11.2	PKS 2351-154
d	2353-686	23	56	00.68	-68	20	3.5	PKS 2353-686
d	2355-534	23	57	53.27	-53	11	13.7	PKS 2355-534
d	2355-106	23	58	10.88	-10	20	8.6	PKS 2355-106
c	0002-478	0	4	35.66	-47	36	19.6	PKS 0002-478
c	0003+380	0	5	57.18	38	20	15.1	4C+38.02
c	0008-421	0	10	52.52	-41	53	10.8	PKS 0008-421
c	0022-423	0	24	42.99	-42	2	4.0	PKS 0022-423
c	0108+388	1	11	37.32	39	6	28.1	
c	0116+319	1	19	35.00	32	10	50.1	4C31.04

	Name		R.A.			Dec.		Alias
c	0118-272	1	20	31.66	-27	1	24.7	PKS 0118-272
c	0138-097	1	41	25.83	-9	28	43.7	PKS 0138-097
c	0153-410	1	55	37.06	-40	48	42.4	
c	0202-765	2	2	13.69	-76	20	3.1	PKS 0202-765
c	0237-027	2	39	45.47	-2	34	40.9	
c	0241+622	2	44	57.70	62	28	6.5	
c	0252-712	2	52	46.16	-71	4	35.3	
c	0317+188	3	19	51.26	19	1	31.3	
c	0334-546	3	35	53.92	-54	30	25.1	PKS 0334-546
c	0334+014	3	37	17.11	1	37	22.8	
c	0355-483	3	57	21.92	-48	12	15.2	PKS 0355-483
c	0355+508	3	59	29.75	50	57	50.2	NRAO150, 4C+50.11
c	0400-319	4	2	21.27	-31	47	25.9	
c	0403-132	4	5	34.00	-13	8	13.7	PKS 0403-132
c	0405-385	4	6	59.04	-38	26	28.0	PKS 0405-385
c	0405-123	4	7	48.43	-12	11	36.7	
c	0407-658	4	8	20.38	-65	45	9.1	PKS 0407-658
c	0431-512	4	32	21.18	-51	9	25.2	PKS 0431-512
c	0503-608	5	4	1.70	-60	49	52.5	PKS 0503-608
c	0517-726	5	16	37.72	-72	37	7.5	
c	0529+075	5	32	39.00	7	32	43.3	
c	0611+131	6	13	57.69	13	6	45.4	
c	0614-349	6	16	35.98	-34	56	16.6	PKS 0614-349
c	0615-365	6	17	32.32	-36	34	14.8	PKS 0615-365
c	0622-441	6	23	31.79	-44	13	2.5	PKS 0622-441
c	0647-475	6	48	48.45	-47	34	27.2	PKS 0647-475
c	0648-165	6	50	24.58	-16	37	39.7	PKS 0648-165
c	0700-465	7	1	34.55	-46	34	36.6	PKS 0700-465
c	0736-332	7	38	16.95	-33	22	12.8	PKS 0736-332
c	0809-493	8	11	8.80	-49	29	43.5	PKS 0809-493
c	0818-128	8	20	57.45	-12	58	59.2	PKS 0818-128
c	0842-754	8	41	27.04	-75	40	27.9	PKS 0842-754
c	0850+581	8	54	42.00	57	57	29.9	4C+58.17
c	0936-853	9	30	32.57	-85	33	59.7	PKS 0936-853
c	0952+179	9	54	56.82	17	43	31.2	0952+172, PKS 0952+179
c	0959-443	10	1	59.91	-44	38	0.6	PKS 0959-443
c	1038+529	10	41	48.90	52	33	55.6	
c	1045-188	10	48	6.62	-19	9	35.7	PKS 1045-188
c	1101-325	11	3	31.53	-32	51	16.7	PKS 1101-325
c	1117+146	11	20	27.81	14	20	55.0	4C+14.41

	Name	R.A.			Dec.			Alias
c	1128-047	11	31	30.52	-5	0	19.7	PKS 1128-047
c	1147+245	11	50	19.21	24	17	53.8	B2 1147+24
c	1206-399	12	9	35.24	-40	16	13.1	PKS 1206-399
c	1213-172	12	15	46.75	-17	31	45.4	PKS 1213-172
c	1215-457	12	18	6.25	-46	0	29.0	PKS 1215-457
c	1221-829	12	24	54.38	-83	13	10.1	PKS 1221-829
c	1234-504	12	37	15.24	-50	46	23.2	
c	1307+121	13	9	33.93	11	54	24.6	4C+12.46
c	1320-446	13	23	4.25	-44	52	33.8	PKS 1320-446
c	1328+307	13	31	8.29	30	30	33.0	3C286,4C+30.26
c	1334-649	13	37	52.44	-65	9	24.9	PKS 1334-649
c	1409+218	14	11	54.86	21	34	23.4	
c	1417+273	14	19	59.30	27	6	25.6	4C+27.28
c	1420+326	14	22	30.38	32	23	10.4	B2 1420+32
c	1424+240	14	27	0.39	23	48	0.0	PKS 1424+240
c	1432+200	14	34	39.79	19	52	0.7	PKS 1432+200
c	1433+304	14	35	35.40	30	12	24.5	
c	1540-828	15	50	59.14	-82	58	6.8	PKS 1540-828
c	1555-140	15	58	21.95	-14	9	59.1	
c	1656+477	16	58	2.78	47	37	49.2	S4 1656+47
c	1733-565	17	37	35.77	-56	34	3.2	PKS 1733-565
c	1740-517	17	44	25.45	-51	44	43.8	PKS 1740-517
c	1748-253	17	51	51.26	-25	24	0.1	
c	1758-651	18	3	23.50	-65	7	36.8	PKS 1758-651
c	1814-637	18	19	35.00	-63	45	48.2	PKS 1814-637
c	1817-254	18	20	57.85	-25	28	12.6	
c	1829-718	18	35	37.20	-71	49	58.2	PKS 1827-718
c	1936-623	19	41	21.77	-62	11	21.1	PKS 1936-623
c	1943+228	19	46	6.25	23	0	4.4	
c	1955+335	19	57	40.55	33	38	27.9	
c	2005+403	20	7	44.95	40	29	48.6	
c	2023+336	20	25	10.84	33	43	0.2	
c	2037-253	20	40	8.77	-25	7	46.7	PKS 2037-253
c	2048+312	20	50	51.13	31	27	27.4	CL4
c	2054-377	20	57	41.60	-37	34	3.0	PKS 2054-377
c	2058-425	21	1	59.11	-42	19	16.2	PKS 2058-425
c	2115-305	21	18	10.60	-30	19	11.6	PKS 2115-305
c	2155-304	21	58	52.06	-30	13	32.1	PKS 2155-304
c	2210-257	22	13	2.50	-25	29	30.1	PKS 2210-257

	Name		R.A.				Dec.	Alias
c	2211-388	22	14	38.57	-38	35	45.0	PKS 2211-388
c	2259-374	23	2	23.89	-37	18	6.8	PKS 2259-374
c	2300-307	23	3	5.82	-30	30	11.5	PKS 2300-307
c	2320+506	23	22	25.98	50	57	52.0	
c	2325-150	23	27	47.96	-14	47	55.8	PKS 2325-150
c	2329-162	23	31	38.65	-15	56	57.0	PKS 2329-162
c	2333-528	23	36	12.14	-52	36	22.0	PKS 233-528
o	0019+000	0	22	25.43	0	14	56.1	4C+00.02
o	0024+348	0	26	41.73	35	8	42.3	OB338
o	0036-216	0	38	29.90	-21	20	5.0	PKS 0036-216
o	0218+357	2	21	5.47	35	56	13.7	
o	0218+35A	2	21	5.47	35	56	13.7	
o	0218+35B	2	21	5.47	35	56	14.1	
o	0237-233	2	40	8.17	-23	9	15.7	
o	0250+178	2	53	34.88	18	5	42.5	
o	0316+413	3	19	48.16	41	30	42.1	3C84,PerA,NGC1275
o	0335-122	3	37	55.56	-12	4	12.5	
o	0336-017	3	39	0.80	-1	33	7.0	
o	0411+054	4	14	37.59	5	34	46.2	
o	0420-625	4	20	56.13	-62	23	39.7	
o	0428+205	4	31	3.76	20	37	34.3	
o	0430+052	4	33	11.10	5	21	15.6	3C120, BW Tau
o	0434+299	4	38	4.91	30	4	32.4	
o	0454-463	4	55	51.27	-46	15	58.1	
o	0515-674	5	15	37.54	-67	21	27.8	
o	0537-692	5	36	57.06	-69	13	24.7	
o	0558-504	5	59	46.82	-50	26	52.6	PKS 0558-504
o	0629+104	6	32	15.33	10	22	2.2	4C+10.20
o	0710+439	7	13	38.16	43	49	17.2	S4 0710+43
o	0727-365	7	29	5.39	-36	39	45.1	
o	0902+343	9	5	30.11	34	7	57.2	B2 0902+34
o	0919-260	9	21	29.35	-26	18	43.4	PKS 0919-260
o	0941-080	9	43	36.95	-8	19	30.9	PKS 0941-080
o	0954+556	9	57	38.17	55	22	58.0	4C+55.17
o	1031+567	10	35	7.04	56	28	46.8	S4 1031+56
o	1226+023	12	29	6.70	2	3	8.6	3C273B,4C+02.32
o	1245-197	12	48	23.90	-19	59	18.7	PKS 1245-197
o	1323+321	13	26	16.51	31	54	9.5	
o	1328+254	13	30	37.69	25	9	11.0	4C+25.43
o	1329-665	13	32	37.55	-66	46	50.1	

	Name	R.A.			Dec.			Alias
o	1345+125	13	47	33.36	12	17	24.2	4C+12.50
o	1352-104	13	52	6.84	-10	26	21.3	PKS 1352-104
o	1355-416	13	59	0.18	-41	52	52.6	PKS 1355-416
o	1421-490	14	24	32.30	-49	13	49.0	PKS 1421-178
o	1511+238	15	13	40.19	23	38	35.2	4C+23.41
o	1607+268	16	9	13.32	26	41	29.0	CTD93, PKS 1607+268
o	1622-253	16	25	46.89	-25	27	38.3	PKS 1622-253
o	1634+628	16	34	33.80	62	45	35.9	3C343, 4C+62.26
o	1637+626	16	38	28.20	62	34	44.3	3C343.1, 4C+63.27
o	1709-342	17	13	9.91	-34	18	28.9	
o	1710-269	17	13	31.25	-26	58	52.3	
o	1710-323	17	13	50.79	-32	26	12.0	
o	1714-336	17	17	36.00	-33	42	8.2	
o	1741-312	17	44	23.58	-31	16	36.0	
o	1756-663	18	1	18.08	-66	23	1.0	PKS 1756-663
o	1813-241	18	16	49.60	-24	5	59.2	
o	1826+796	18	23	14.11	79	38	49.0	
o	1827-360	18	30	58.88	-36	2	30.2	PKS 1827-360
o	1829-106	18	32	20.84	-10	35	11.3	
o	1830-211	18	33	39.90	-21	3	40.0	PKS 1830-210
o	1830-21A	18	33	39.89	-21	3	40.7	
o	1830-21B	18	33	39.94	-21	3	40.0	
o	1848+333	18	50	4.79	33	21	45.8	
o	1855+031	18	58	2.34	3	13	16.4	
o	1934+207	19	36	48.02	20	51	36.8	
o	1934-638	19	39	25.03	-63	42	45.6	PKS 1934-638
o	1947+079	19	50	5.54	8	7	14.0	PKS 1947+079
o	2021+614	20	22	6.68	61	36	58.8	S4 2021+61
o	2027+383	20	28	54.11	38	32	47.7	
o	2044-168	20	47	19.66	-16	39	5.8	PKS 2044-168
o	2100+468	21	2	17.04	47	2	16.2	
o	2121+053	21	23	44.52	5	35	22.1	OX036,PKS 2121+053
o	2128+048	21	30	32.88	5	2	17.5	PKS 2128+048
o	2134+004	21	36	38.59	0	41	54.2	
o	2251+158	22	53	57.75	16	8	53.6	3C454.3, 4C+15.76
o	2310-417	23	12	55.61	-41	26	56.1	PKS 2310-417
o	2314+038	23	16	35.09	4	5	19.8	2314+03, 4C+03.57
o	2322-411	23	25	3.42	-40	51	30.1	PKS 2322-411
o	2337+264	23	40	29.03	26	41	56.8	
o	2352+495	23	55	9.46	49	50	8.3	S4 2352+49

Author Index

Index